Nuclear Strategy, Arms Control, and the Future

Also of Interest

†Available in hardcover and paperback.

About the Book and Editors

Balanced and comprehensive in approach, this text assembles classic statements on nuclear strategy and arms control made by Soviet and U.S. policymakers, military thinkers, and opinion leaders during the last forty years. Major Soviet statements, rarely appearing in translation, reflect the disagreement over whether "victory" or "parity" is the goal of Soviet nuclear strategy and forces. Taken as a whole, the selections record the concerns and hopes of government leaders who bear responsibility for protecting their nation's security in the nuclear age.

The general introduction is structured and written in a straightforward, succinct fashion that helps the student master the seemingly inchoate mass of ideas surrounding the arms race. The development of Soviet and U.S. policies and postures since the Cold War is recounted. The ramifications of such concepts as counterforce strategy, massive retaliation, assured destruction, deterrence, stability, and the strategic defense initiative are logically and thoroughly explained so that students with no background can easily grasp the discussions that follow. The introduction also explores the intricacies of arms control negotiations, as well as the pitfalls that have been and will be encountered by the superpowers. With their analysis of the arms race, the editors bridge the gap between antinuclear activists and those who are legally charged with the defense of their country.

Intended for use as a basic text in courses on national security, arms control, and peace studies, this collection of primary sources encourages students to reassess their own perceptions of the arms race. Each chapter contains its own extensive introduction, and each part has a selected bibliography that can be used for further study. The glossary at the end of the text provides a comprehensive dictionary of arms control and disarmament terms. The historic nature of the selections makes this book a valuable resource for scholars, researchers, and libraries as well.

P. Edward Haley is director of the Keck Center for International Strategic Studies and chairman of the International Relations Committee, Claremont McKenna College. He has worked on the staffs of members of the U.S. Senate and House of Representatives and was an international affairs fellow of the Council on Foreign Relations. *David M. Keithly* is administrative assistant, the Keck Center for International Strategic Studies. He holds a doctorate in Government from Claremont Graduate School. *Jack Merritt* is professor emeritus of physics, Claremont McKenna College. Prior to coming to Claremont, he served as an administrative officer at the U.S. Atomic Energy Commission and the National Science Policy Organization of the Bureau of Budget.

Nuclear Strategy, Arms Control, and the Future

edited by P. Edward Haley, David M. Keithly, and Jack Merritt

Routledge
Taylor & Francis Group

LONDON AND NEW YORK

First published 1985 by Westview Press, Inc.

Published 2018 by Routledge
52 Vanderbilt Avenue, New York, NY 10017
2 Park Square, Milton Park, Abingdon, Oxon OX14 4RN

Routledge is an imprint of the Taylor & Francis Group, an informa business

Library of Congress Cataloging in Publication Data
Main entry under title:
Nuclear strategy, arms control, and the future.
  Bibliography: p.
  1. Nuclear warfare—Addresses, essays, lectures.
2. Nuclear disarmament—Addresses, essays, lectures.
3. Military policy—Addresses essays, lectures.
I. Haley, P. Edward. II. Keithly, David M.
III. Merritt, Jack.
U263.N765 1985   358.39   85-8903
ISBN 13: 978-0-367-00603-7 (hbk)
ISBN 13: 978-0-367-15590-2 (pbk)

—to—
Laura and Catherine,
Wayne and Wes,
future students

Contents

PART TWO
THE TWO FACES OF
SOVIET NUCLEAR STRATEGY

PART THREE
INTERMEDIATE-RANGE NUCLEAR
FORCE CONTROVERSIES

PART FOUR
ARMS CONTROL

PART FIVE
MORALITY AND NUCLEAR WEAPONS

PART SIX
STRATEGY AND ARMS CONTROL
IN THE FUTURE

Chapter 11
Epilogue: The Soviet Union and the United States Resume Arms Control Negotiations 354

Credits

The selections in this book are taken from the following sources. Permissions to reprint are gratefully acknowledged.

The discussion of nuclear weapons in Europe in the editors' introduction ("The Fundamentals of Nuclear Strategy and Arms Control") is based on a paper by P. Edward Haley, "Nuclear Weapons, Democratic Statesmanship, and the Security of Europe." It was prepared for presentation at a conference in honor of the fortieth anniversary of VE Day that was sponsored by the Washington Institute for Values in Public Policy, Washington, D.C., May 7–9, 1985.

1. Reprinted from *Foreign Relations of the United States: 1948*, 1 (pt. 2), pp. 624–628. NSC-30 was dated September 10, 1948.
2. Records of the Joint Chiefs of Staff on deposit in the Modern Military Records Branch, National Archives, Washington, D.C. JCS 1952/1 was dated December 21, 1948.
3. Reprinted from *Foreign Relations of the United States: 1950*, 1, pp. 237–292. NSC-68 was dated April 14, 1950.
4. Records of the Organizational Research and Policy Division of the Office of the Chief of Naval Operations (Op-23), Naval Historical Center, Washington, D.C. The selection was dated May 11, 1949.
5. Reprinted from *Foreign Relations of the United States: 1950*, 1, Department of State Atomic Energy Files, January 20, 1950, Memorandum by the Counselor (Kennan).
6. Reprinted from *Department of State Bulletin*, March 29, 1954, vol. 30, pp. 459–464.
7. Excerpted from Henry A. Kissinger, *Nuclear Weapons and Foreign Policy* (New York: Council on Foreign Relations, 1975), by permission of the Council on Foreign Relations, Inc. A Council on Foreign Relations book, published by Westview Press, 1984.
8. Reprinted from *Public Papers of the Presidents of the United States*, John F. Kennedy, 1961, document no. 99.

9. Reprinted from *Department of State Bulletin*, July 9, 1962, vol. 47, pp. 66–69.

10. Reprinted from *Department of State Bulletin*, October 9, 1967, vol. 57, pp. 443–449.

11. Reprinted from *Authorization for Military Procurement, Research, and Development, 1969*, Hearings Before the Committee on Armed Services, United States Senate, S3293.

12. Press Conference of U.S. Secretary of Defense, January 10, 1974, at the National Press Club, Washington, D.C.

13. Statement Before the Senate Foreign Relations Committee on "Nuclear War Strategy, PD-59," September 16, 1980.

14. Reprinted from Desmond Ball, "Can Nuclear War Be Controlled?" Adelphi Paper no. 169 (London: International Institute for Strategic Studies, 1981).

15. Reprinted from Colin S. Gray, "Nuclear Strategy: The Case for a Theory of Victory," *International Security* 4 (Summer 1979):63–87. By permission of The MIT Press, Cambridge, Massachusetts. Copyright 1980 by the President and Fellows of Harvard College and of the Massachusetts Institute of Technology.

16. Reprinted by permission from Jeffrey Richelson, "PD-59, NSDD-13, and the Reagan Strategic Modernization Program," *Journal of Strategic Studies* 6, no. 2 (June 1983):125–146, published by Frank Cass & Co. Ltd., 11 Gainsborough Road, London E11, England. Copyright Frank Cass & Co. Ltd.

17. Excerpted from Marshal V. D. Sokolovsky, "The Nature of Modern War," *Soviet Military Strategy*, Moscow, 1962, by permission of the Rand Corporation. Copyright 1963 the Rand Corporation.

18. Reprinted from Colonel A. A. Sidorenko, *The Offensive*, Moscow, 1970. Published under the auspices of the U.S. Air Force, 1970.

19. Reprinted from Col. General N. A. Lomov, *The Revolution in Military Affairs*, Moscow, 1973. Published under the auspices of the U.S. Air Force, 1973.

20. Excerpted from Marshal N. V. Ogarkov, "Military Strategy," *Soviet Military Encyclopedia*, vol. 7 (Moscow: Voyenizdat, 1979), pp. 555–565, in Leon Gouré and Michael J. Deane, "The Soviet Strategic View," *Strategic Review* 8, no. 3 (Summer 1980):92–95, by permission of *Strategic Review*, 1980. Copyright *Strategic Review*, 1980.

21. Excerpted from Major General S. K. Il'in, *Morale Factors in Modern Wars* (Moscow: Voyenizdat, 1979), in Leon Gouré and Michael J. Deane, "The Soviet Strategic View," *Strategic Review* 8, no. 3 (Summer 1980):95–96, by permission of *Strategic Review*, 1980. Copyright *Strategic Review*, 1980.

22. Reprinted from Fritz W. Ermath, "Contrasts in American and Soviet Strategic Thought," *International Security* 3 (Fall 1978):138–155, by permission of The MIT Press, Cambridge, Massachusetts. Copyright 1979 by the President and Fellows of Harvard College and of the Massachusetts Institute of Technology.

23. Reprinted from *The Current Digest of the Soviet Press* 29, no. 3 (February

40. United States Statutes at Large, 92d Cong., 2d sess. (1972), vol. 86.

41. Excerpted from Paul H. Nitze, "Assuring Strategic Stability in an Era of Détente," *Foreign Affairs* 54 (January 1976):207–232, by permission of *Foreign Affairs*, January 1976. Copyright 1976 by the Council on Foreign Relations, Inc.

42. Reprinted from *Origins* 13 (May 19, 1983). Excerpts from "The Challenge of Peace: God's Promise and Our Response," copyright © 1983 by the United States Catholic Conference. All rights reserved. A complete copy of the pastoral letter may be ordered from the Office of Publishing and Promotions Services, U.S.C.C., 1312 Massachusetts Ave. N.W., Washington, D.C., 20005.

43. Reprinted from Albert Wohlstetter, "Bishops, Statesmen, and Other Strategists on the Bombing of Innocents," *Commentary* 75 (June 1983):15–35, by permission; all rights reserved.

44. Reprinted from *Gagner la Paix*, Déclaration de l'Assemblée plénièr de l'épiscopat français (November 1983), by permission of the Conference of French Catholic Bishops. A complete copy of the text, which was translated by the French Bishops Conference, may be obtained from the General Secretary of the Conference, 106, rue du Bac, 75341 Paris Cedex 07, France.

45. Reprinted from *New York Times*, April 15, 1983. Copyright © 1983 by the New York Times Company. Reprinted by permission.

46. Reprinted with permission from Andrei Sakharov, "The Danger of Thermonuclear War," *Foreign Affairs* 61 (Summer 1983):1001–1016. Copyright Andrei Sakharov, 1983.

47. Report of the President's Commission on Strategic Forces (the Scowcroft Commission), April 1983.

48. Address to the Nation, *Weekly Compilation of Presidential Documents*, March 28, 1983, pp. 442–448.

49. Reprinted from Sidney D. Drell and Wolfgang K. H. Panofsky, "The Case Against Strategic Defense: Technical and Strategic Realities," *Issues in Science and Technology* 1 (Fall 1984):45–65. Excerpted by permission from *Issues in Science and Technology* 1, no. 1 (Fall 1984):45–65. Copyright © 1984 National Academy of Sciences.

50. Reprinted from *The Current Digest of the Soviet Press* 33, no. 8 (March 25, 1981). Translation copyright 1981 by *The Current Digest of the Soviet Press*, published weekly at the Ohio State University; reprinted by permission.

51. Reprinted from McGeorge Bundy, George F. Kennan, Robert S. McNamara, and Gerard Smith, "Nuclear Weapons and the Atlantic Alliance," *Foreign Affairs* 60 (Summer 1982):754–768. Condensed and reprinted by permission of *Foreign Affairs*, Spring 1982. Copyright 1982 by the Council on Foreign Relations, Inc.

52. Excerpted from Karl Kaiser, Georg Leber, Alois Mertes, and Franz-Josef Schulze, "Nuclear Weapons and the Preservation of Peace," *Foreign Affairs* 60 (Summer 1982):1157–1170, by permission of *Foreign Affairs*,

Preface

The purpose of this book is to make readily available to a wide audience interested in nuclear questions a group of "classic" statements on nuclear strategy and arms control. Our working definition of "classic" is: everything changed because the statement was made, or the statement embodied the beliefs of many people in and out of government.

Wherever possible, statements by civilian and military leaders have been used. It is they who, on our behalf, daily grapple with and bear the moral weight of decisions about nuclear war and arms control. As the book reveals, they are conscious of the great responsibilities they carry and have brought a seriousness and dedication to their work that sometimes are lost in the heat of controversy about the subject of their labors.

The opinions of academics and other analysts play a vital part in the development and correction of the views and actions of Western policymakers. A number of classic academic statements are included in the book. More will be found in the sources included in the bibliography at the end of each chapter.

A large place has been reserved in the book for statements by Soviet political and military leaders. Too often nuclear questions are discussed and the most far-reaching conclusions reached as if the entire issue were an internal U.S., or at most, Western concern and the Soviet government had no firm and consequential views on the subject. Not only is this not the case, but from the beginning of our nuclear era Soviet leaders have approached nuclear strategy in ways fundamentally different from those followed by U.S. leaders. That the Soviet approach is so different makes it all the more important to consider what *Soviet* leaders have thought and written about nuclear war and arms control, rather than to work from secondary sources.

The idea for the book came from a class—Nuclear Weapons and Arms Control—started three years ago and still being taught by P. Edward Haley, a political scientist, and Jack Merritt, a physicist, with the assistance of David M. Keithly. The three of us soon discovered the interest of our students in these kinds of statements and the useful role they could play in stimulating discussion and understanding. We also quickly realized how difficult and time-consuming it was to assemble them all and make them available. Taken together these two observations made clear that we had a book.

But that was not all. As our editing chores progressed, it became clear to us that the book might serve an even more important purpose outside the classroom. An alarming misunderstanding has developed between the general public, including members of the antinuclear movement, and the civilians and soldiers who bear official and personal responsibility for national defense. The public, particularly those actively concerned with nuclear weapons, tend to relegate nuclear policymakers to one or another demeaning category: foolish, criminal, or insane. In response, the policymakers sometimes view the activists as ignorant busybodies meddling with vital and dangerous matters that they do not understand.

If both groups keep this view of each other, political and military paralysis is the likely result. There is common ground: Both groups desperately wish to avoid nuclear war. Yet how many nuclear activists know or would believe that Secretary of Defense Robert McNamara is convinced that he persuaded Presidents Kennedy and Johnson in the early 1960s never to initiate nuclear war? On the other side, how many policymakers will accept that nuclear strategy is not a priestly calling, that although complex, its intricacies can be grasped by bright and determined lay people—and their ranks include Nobel laureates—who in turn might have something of value to contribute to U.S. and allied strategy? It thus became clear that our second major purpose was to provide a book that could bridge the gap in outlook and understanding between activists and officials, between—in Freeman Dyson's words—warriors and victims.

The warm response of Fred Praeger and Miriam Gilbert to the book shortened the time to publication. However, it is obvious that a book of this kind is never really finished. The worldwide concern with its subject matter and the technological, political, and military factors driving the arms race continue to produce major changes in every area covered and in many we were obliged to omit or cover only in the bibliography. We realize that we have made a beginning only, and we cordially invite our readers to send us their comments and recommendations regarding additional or alternate "classics."

We thank Laura Anderson, Libby Barstow, Joan Edgar, Anita Elsbee, Alice Levine, Gwendolyn Lohman, Ruth Palmer, and the students in Political Science 138 for their assistance in the preparation of this book.

P. Edward Haley
David M. Keithly
Jack Merritt

Nuclear Strategy, Arms Control, and the Future

The Fundamentals of Nuclear Strategy and Arms Control

An analysis of the nuclear arms race between the United States and the Soviet Union must be capable of bridging the gap in outlook between antinuclear activists and those who are charged legally and morally with the defense of their country. It is not an easy bridge to build, for it must span the void between two worlds: the world of the victims and the world of the warriors. "There is," as nuclear physicist Freeman Dyson observed, "prejudice and antipathy on both sides. The military establishment looks on the peace movement as a collection of ignorant people meddling in a business they do not understand, while the peace movement looks on the military establishment as a collection of misguided people protected by bureaucratic formality from all contact with human realities."[1] The warriors usually ignore the victims, and the victims end up talking only to themselves because they fear the warriors and their nuclear strategy.

Moved by their fears, victims are inevitably led to say, "Better no defense at all than a nuclear defense." Warriors regard this as folly or worse and reply, "Nuclear defense is at least defense, and it is better than surrender." In their mutual distrust and incomprehension, warriors and victims reject each others' proposals and, in this way, produce a political stalemate that undermines the protection of the values and security of the United States and its allies and slows the development of military alternatives that would reduce the risk and costs of nuclear war.

In an attempt to break the stalemate, this book provides a discussion of the strategies adopted by the United States and the Soviet Union and the strategic choices that the two nations have made. The purpose is to provide both warriors and victims with answers to two crucially important questions: Why has the arms race reached its current inflated level? And why is it so difficult to bring about significant reductions in the arsenals of both sides? By coming to a common understanding of these two issues, victims and warriors—and all the rest of us—may find ways to agree on a better and safer defense of the values and security of this country and its allies.

THE STRATEGIC OUTLOOK AND STRATEGIC CHOICES

At one level, the United States and the Soviet Union approach international politics in a similar fashion. Both nations have set their nuclear strategies in accordance with what might be called the "strategic outlook." Although each would interpret the origins of conflict differently, both the U.S. and Soviet governments believe that the international system lacks the reliable institutions and procedures that are used to settle disputes within a given society. As a result, both governments expect one another and all other nations to resort to war to achieve their objectives and to defend themselves against aggression. Proponents of the strategic outlook in both nations find proof of its validity in the past and the present—in the Peloponnesian Wars between the ancient Greeks and in the repeated battles of today's Arab-Israeli conflict. The strategic outlook, which is strongly supported by history and the national experience of both the Soviet Union and the United States, has been an important factor in the nuclear buildup.

The strategic choices of the two sides have also contributed to the nuclear arms race. On the U.S. side there were two crucial decisions: the first, made in the 1950s and 1960s, was to maintain nuclear superiority; and the second, made in the 1970s, was to resort to limited nuclear strike options. On the Soviet side there was one crucial decision: the development of a nuclear counterforce, or war-fighting, strategy, the goal of which is the destruction of the opposition's military capabilities and will to resist. Based on the strategic outlook, these choices have resulted in the accumulation of tens of thousands of nuclear warheads. Former U.S. Secretary of Defense Harold Brown (who held office from 1977 to 1981) estimated that in 1983 the United States and the Soviet Union together possessed more than twenty-one thousand strategic nuclear weapons capable of being delivered across intercontinental distances against each other's homeland.

In our nuclear era it is impossible not to focus on the deadly aspect of force. What good are all the missiles, submarines, and bombers if we all die in a nuclear war? The question is central to any discussion of nuclear strategy and arms control. But to pose it in this manner is to misunderstand the positive element in the use of force and to distort the purpose of foreign policy.

The Nature of Force and Its Role in Foreign Policy

Force has a dual character. It can be used to defend values as well as destroy them. Nowhere is this more evident than in the invention during the past century of enormously powerful chemical and nuclear explosives. Great crimes have been committed with modern weapons; but the same kinds of weapons have been employed to defeat aggression and to destroy evil regimes, such as Hitler's Nazi state in World War II. The same science that produces bombs and artillery shells also provides the explosives to build highways, railroads, airfields, and skyscrapers. Thus, force has a positive as well as negative potential; but it is policy—not force—that determines the goals and conditions a nation seeks abroad.

Throughout history, the aim of foreign policy has always been more than the survival of the members of a political and cultural unit. The preservation of a nation's values, sovereignty, and territory has always merged with physical survival to form the fundamental goal of foreign policy.

The reason that men and women sacrifice their lives to defeat an enemy can be expressed crudely in this way: "We would rather die than live under the control of the enemy." The unselfishness of this reasoning—and it is fundamental to a nation's ability to defend itself—hides a discrepancy between the supreme sacrifice of some of the members of the community and the survival of the rest. In addition, there has always been a discrepancy between the sacrifice of individual lives and the quality of the solution obtained by the sacrifice. Never was the gap between promise and reality greater than at the end of World War I. Millions died in the carnage; millions more were maimed or starved. Entire societies, Western culture itself, had been turned upside down and deeply harmed, seemingly without purpose. None of the war aims of any of the combatants was worth a fraction of the losses.

Remembering the carnage, millions of citizens in Western Europe and the United States developed such a revulsion against war during the 1920s and 1930s that the democracies were unable to prevent the rise and conquests of Adolf Hitler's Nazi regime in Europe and an expansionist military Japanese state in Asia. Eventually, a grand alliance of Britain, the United States, and the Soviet Union was able to defeat the German and Japanese conquerors in 1945. In order to shock the Japanese into surrender and save U.S. and Allied lives, the United States dropped nuclear bombs on the Japanese cities of Hiroshima and Nagasaki in August 1945. The Japanese surrendered almost immediately. Since those fateful days, the world has lived in the shadow of nuclear war.

Despite the toll in lives and destruction, World War II left a different impression than did World War I on victors and vanquished. In the second great war, the outcome, particularly the destruction of Hitler and the Nazi regime, seemed worth the terrible costs. Unlike the first world war, the second indicated that the use of force to protect the innocent and defenseless could be just and worthwhile.

This consideration restored a completeness to the view of military strategy that had been lost in the senseless slaughter of World War I. However, the creation and use of nuclear weapons threatened this hard-won lesson. Because of their enormous destructiveness, nuclear weapons can cause losses out of all proportion to the ends that can be achieved by their use. They return the world to the battlefields of World War I—but there is a difference.

Nuclear weapons threaten not just to repeat the experience of World War I, but to multiply it many times in the space of minutes and hours. Some scientists take seriously the possibility—and the U.S. Defense Department has agreed—that the explosion of a large number of nuclear weapons over cities in the Northern Hemisphere could produce a "nuclear winter." The smoke and ash produced by city fires would literally blot out the sun and cause freezing temperatures over much of the globe, including most of the world's best food-producing areas. What conceivable national values and objectives could be served by such horror? At the same time one asks that question, one must recognize that nuclear weapons have not altered the *goals* of foreign policy—the protection of national values and sovereignty as well as the preservation of physical existence. For the United States it is not enough to avoid nuclear disaster if freedom is sacrificed in the bargain.

The Tools of Foreign Policy and the Nature of Deterrence

The instruments of foreign policy are two: persuasion and coercion—or, diplomacy and strategy, to use the terms of the French scholar of international and nuclear issues, Raymond Aron.[2] Persuasion is limited to asking that something be done and offering incentives for its accomplishment. When one government begins to threaten another for noncompliance, it has crossed the line separating persuasion from coercion.

Given this limited set of choices, leaders naturally make elaborate plans by which to achieve their objectives. When plans involve coercion—the actual or threatened use of force—those plans become strategy. Strategy is thus the use or threatened use of military force to achieve national objectives. (It is important to understand the word "use" in its fullest sense. When a cannon or rocket is fired at an enemy position, force is "used." And force is also used when an enemy is told that a certain military capability will be employed.)

The threat to use force is of vital importance in the nuclear age, for it is the essence of deterrence. Precisely because the detonation of nuclear weapons would cause such heavy damage, the threat to explode them as a means of deterring aggression has become vastly more important than the threat to employ conventional weapons. The two critical factors of deterrence are capability and intention. A government protecting its interests by a threat must be seen by the enemy to have the forces needed (or capability) to make good the threat and be perceived to have the intention to use those forces. A failure—for whatever reason (even if it is only perceived)— in either capability or intention necessarily reduces deterrence. The overall goal of deterrence is to raise the costs of aggression higher than the value of the benefits that would be gained if the aggression were to succeed.

The cruel logic of deterrence is a source of weapons proliferation, for as both the Soviet Union and the United States acquired the capability to devastate each other's society, their willingness to use nuclear weapons declined. In order to prevent the collapse of deterrence, both governments sought to strengthen the other's perception of their willingness to use nuclear weapons as a last resort by taking a variety of steps, including increasing the numbers of warheads in their forces, reducing warhead size, and increasing warhead accuracy. (A warhead is the explosive part of a missile or bomb.) The two governments have also strengthened their capabilities by attempting to make their nuclear forces invulnerable. Missiles have been put in reinforced concrete silos in the ground (the technical name is intercontinental ballistic missile or ICBM) and in submarines capable of escaping detection and launching their missiles from underwater at sea (the technical name is submarine-launched ballistic missile or SLBM). Bombers are kept in the air to be safe against surprise attack. In the early 1960s and again in the early 1970s, both sides considered and came close to developing an extensive system of radars and interceptor missiles to shoot down attacking missiles in flight (the technical name is antiballistic missile defense or ABM). In 1972, the Soviet Union and the United States signed a treaty limiting ABM systems.

A severe misapprehension governs most discussions of strategy and of military operations as a whole. The most successful strategists are not those who win battles with great loss of life, but those who achieve the goals of war without fighting at all. The secondary goal of strategy is to put one's own nation in a position so advantageous that should war occur it will achieve its objectives with the lowest possible losses.

In summary, the Soviet and U.S. governments believe that nations seek their foreign policy objectives through a combination of persuasion and coercion. Their national goals transcend survival and include the preservation of national values and independence. In this quest, they must rely on their own resources—leadership, popular support, and military and industrial strength—and those of whatever allies they can find. They apply those resources according to the strategic outlook. However peaceful the scene, Soviet and U.S. leaders recall many instances in which the strong sought to impose their will on the weak.

The Legacy of World War II:
The Iron Curtain and the Cold War

As World War II ended, there was a widespread and sincere hope shared by millions of ordinary citizens and some government officials that it would be possible for the United States and the Soviet Union, together with Britain and France, to establish a constructive pattern of relations and to settle all major international disputes through the United Nations. This hope faltered, then vanished, as Stalin challenged U.S. interests in Europe, Asia, and the Mediterranean and gradually fastened Soviet domination on the nations of Central Europe that had been occupied by the Red Army at the close of the war. After 1945, in Churchill's memorable phrase, an "Iron Curtain" of Soviet tyranny had descended across Europe. West of the curtain, there was freedom of expression, political organization, and religion. To the east, there were the increasingly familiar and horrible signs of Soviet control: a single monolithic party, loyal to the Soviet Union; control of virtually all segments of public, social, and family life by the secret police; the denial of freedom of speech and of political and religious organization.

The division of Europe into Soviet and Western spheres had its origins first in the location of the various Allied armies when the Nazi authorities surrendered in May 1945. That division was perpetuated by the Soviet Union's decision to impose its control over all the territories occupied by the Red Army. Only eastern Austria would escape; the Soviet Union chose to leave Austria in 1955 in exchange for the permanent neutralization of the country.

The eastern third of Germany remained behind the Iron Curtain. However, there was an anomaly in the Soviet domination of eastern Germany. According to Soviet-British-U.S. agreements made during the war, Britain, the United States, and France received zones of occupation in the former German capital of Berlin, although it lay well within the sphere of Germany occupied by the Red Army and was controlled by Soviet secret police and a German regime loyal to Moscow. The eastern and western zones of occupation would soon become the separate states of West Germany (the Federal Republic)—allied to the United States, Britain, and France; and East Germany (the Democratic Republic)—allied to the Soviet Union and its other client states in Central Europe.

The division of the city of Berlin into eastern and western zones also grew out of the wartime agreements and the position of the Red and Allied armies at the end of hostilities. By chance—and it was to prove to be a momentous oversight—the Western Allies failed to reach clear agreements with the Soviet Union before the war ended on unimpeded access on the ground to their zones of occupation in West Berlin.

The United States soon realized that Europe and Japan could not recover from their wartime losses without massive U.S. assistance. In addition, the U.S. government feared that the Soviet Union would play on the economic distress of the Europeans and Japanese with threats and subversion and, in this way, bring these nations within the Soviet orbit as well. The United States sought a pluralistic world open to trade and investment according to the practices of constitutional, limited government and free market economics.

Initially, the U.S. countermeasures took the form of the Marshall Plan for Western Europe and an end to the austere and punitive occupation regimes for defeated West Germany and Japan. Under the Marshall Plan, the United States supplied the foreign exchange in dollars that the Europeans needed to purchase the machinery and other goods they required to rebuild their economies and national life. By ending the occupation regimes in West Germany and Japan, the United States could harness to global economic recovery the productive labor and markets of these two industrial giants of Europe and Asia. The overall result was calculated to create a "free world" of such vitality and attractiveness that it would isolate the Soviet Union and compel it to abandon its expansionist plans.

The determination of the United States was matched by Stalin's conviction that the world was divided into two camps—capitalist and communist—and that a showdown was imminent. On this ground, Stalin justified the harsh measures he enforced in the Soviet Union and Eastern Europe. There was to be no postwar relaxation, no opening of Soviet life to the West. Stalin clamped down on intellectual, political, social, and economic life inside the Soviet orbit and ordered the propaganda and military organizations of the Soviet state to proclaim defiance of the West. The peace that had been the hope and expectation after the war became instead a Cold War.

THE ESSENCE OF NUCLEAR STRATEGY

A curious asymmetry developed between Soviet and U.S. military forces at the end of World War II. Until 1949 the United States alone possessed nuclear weapons. For nearly a decade afterward, only the United States had a long-range bomber fleet with which to deliver those weapons. After the war, the United States returned virtually all its men and women in uniform to civilian life. From 1945 to 1950, the armed forces of the United States declined from more than 12 million to less than 1.5 million. U.S. Army personnel alone fell from 8.3 million to 593,000. In Europe, the number of U.S. forces dropped from 3.1 million to 391,000, and few of these were combat ready.

The Soviet Union, on the other hand, lagged nearly five years behind the United States in acquiring nuclear weapons and nearly fifteen years behind in acquiring the ability to deliver those weapons against the United States. At the same time, the Soviets kept millions of men in uniform and deployed many of its troops well forward in Central Europe. Publicly, Soviet leaders denigrated the importance of nuclear weapons. In secret, they made enormous sacrifices to develop their own nuclear arsenal as quickly as possible. In measuring the degree of effort the Soviet Union has made to overtake and in some areas surpass the United States in nuclear weapons, one must remember two factors: that its gross national product is less than half that of the United States; and that it simultaneously had to recover from the enormous losses suffered by its people, industry, and agriculture during the war against the Nazis from 1941 to 1945.

As relations between the Soviet Union and the United States deteriorated in the late 1940s, a dangerous flash point was reached during the Berlin Blockade of 1948-1949. The Western allies were determined to unify the zones of Germany they controlled and to create a new state in western Germany, which they would then integrate into their economic and military arrangements in Western Europe. In an effort to block this development, the Soviet dictator, Joseph Stalin (who was in power from about 1929 to 1953), cut all land communications from the West with the city of Berlin, leaving it isolated deep within the Soviet occupation zone. Stalin apparently hoped to halt the progress toward West German unity and to force a settlement of German affairs more favorable to the Soviet Union. The United States and its allies regarded Soviet goals and actions as unacceptable. Rather than resort to war, President Harry S. Truman (who held office from April 1945 to 1953) undertook to supply food, fuel, medicine, and other necessities to Berlin by a round-the-clock airlift.

The success of the airlift and the growing opposition to Soviet policy throughout Europe persuaded Stalin to lift the Berlin blockade without having achieved any of his objectives. However, the crisis drove U.S. strategists to seriously consider the possibility that war might break out in Europe and to formulate their plans accordingly. Given the military asymmetry between the United States and the Soviet Union, it is not surprising that the new U.S. war plans called for the delivery of nuclear strikes against major industrial centers in the Soviet Union. The development in the early 1950s of the much more powerful and relatively less expensive fusion, or hydrogen, bombs appeared to confirm the correctness of this strategy, for it promised to make possible the substitution of firepower for manpower. (The West now lacked the conventional military forces to defeat a Soviet invasion and rebuilding the forces would have required the rearmament of Germany—which European nations were still reluctant to do.) The plan also took into account the Soviet Union's limited nuclear capabilities to strike at the United States.

An Early U.S. Strategy: Massive Retaliation

Thus, the basis was laid for a U.S. strategy of massive retaliation. As long as the United States possessed a monopoly of nuclear weapons and the means to deliver them, it would incorporate the threat to use these weapons effectively into its strategy. To do so offered two great advantages. First, the Soviets could not respond in kind. Soviet political and military leaders had no choice but to take U.S. threats seriously, for they could not answer them. Second, if the U.S. nuclear monopoly could be used to deter military adventures by the Soviet Union, China, and their allies, the United States would be spared the stupendous costs in money and social discipline that would be required to maintain several million men under arms and to keep them ready to fight in distant theaters such as Europe and Asia.

The Communist invasion of Korea in 1950 and, later in the decade, the belligerence of Soviet Premier Nikita Khrushchev (who held power from about 1955 to 1964) about Soviet missile strength provoked strong responses from the United States. Under the shadow of Korea and Khrushchev's bluster, the United States developed the Strategic Air Command (SAC; long-range bombers carrying nuclear weapons) and later deployed hundreds of intercontinental bombers, ICBMs, and SLBMs. Although the Soviet Union responded by steadily improving its nuclear capabilities, concentrating first on intermediate-range rockets targeted on Western Europe, the United States possessed a significant strategic nuclear advantage over the Soviet Union until well into the 1960s. At the time of the Cuban missile crisis in 1962, for example, the United States possessed more than one thousand intercontinental bombers, land-based missiles, and submarine-launched missiles, while the Soviet Union had fewer than one hundred missiles, and those were not protected by underground silos.[3]

After the Cuban missile crisis, the nuclear balance began to change rapidly. Within five years, the Soviet Union had acquired the capability to devastate the United States, regardless of the weight of attack the United States sent against it. By 1972, the Soviet Union had deployed half again as many ICBMs as the United States and would surpass it in SLBMs in three years (see Table 1).

The Formulation of Soviet Strategy: War-fighting

In closing the gap, the Soviet Union had chosen to build different kinds of weapons and to deploy them in different ways than the United States. It favored very large missiles, for example, and placed far more of them on land than did the United States. It soon became clear that the two countries had fundamentally different views of nuclear strategy and had decided to develop their missile forces along very different lines. The Soviet government chose to follow a war-fighting counterforce strategy, which aimed to destroy U.S. nuclear and conventional forces, leadership, and will to resist. Beginning in the 1960s, the United States chose to rely increasingly on the strategy of deterrence of war by threatening, in the last resort, to attack Soviet cities and industry. The difference in strategies added a difficult problem of mutual understanding to the already great threat that both countries felt because of the immense nuclear arsenal of the other. What appeared to be a sensible and safe measure according to the strategy of one side was regarded as a dangerous and destabilizing initiative by the other. The different strategies and strategic choices of the two sides had a deeply unsettling effect on their overall political relationship and gravely complicated the willingness of the two to accept significant reductions in the arms control negotiations they entered beginning in 1969.

Rethinking U.S. Strategy

The anticipated Soviet ability to devastate the United States provoked major changes in U.S. strategy and strategic choices. The administrations of Presidents John F. Kennedy (1961-1963) and Lyndon B. Johnson (1963-1969) had to initiate the changes. A desire to reduce the nation's reliance on nuclear weapons prompted a change in U.S. strategy from massive retaliation to "flexible response." The goal of "flexible response" was to escape the dilemma—paralysis or catastrophe—caused by the growth, anticipated and real, in Soviet ability to devastate the United States. The immediate need was to raise the nuclear threshold in two ways: to make U.S. weapons invulnerable by putting them in silos and submarines and to increase U.S. and NATO conventional forces in order to make self-defense possible without having to resort immediately to nuclear weapons in case of war.

The logic of flexible response could have led the United States and its European allies to reduce their reliance on battlefield and short-range nuclear

TABLE 1
U.S. and USSR Launcher Strength, 1962-1984

	1962	1963	1964	1965	1966	1967	1968	1969	1970	1971	1972	1973
United States												
ICBMs	294	424	834	854	904	1,054	1,054	1,054	1,054	1,054	1,054	1,054
SLBMs	144	224	416	496	592	656	656	656	656	656	656	656
Long-range bombers	600	630	630	630	630	600	545	560	550	360	390	397
USSR												
ICBMs	75	100	200	270	300	460	800	1,050	1,300	1,527	1,527	1,575
SLBMs	some	100	120	120	125	130	130	160	280	448	500	628
Long-range bombers	190	190	190	190	200	210	150	150	150	140	140	140

	1974	1975	1976	1977	1978	1979	1980	1981	1982	1983	1984
United States											
ICBMs	1,054	1,054	1,054	1,054	1,054	1,054	1,054	1,052	1,052	1,045	1,037
SLBMs	656	656	656	656	656	656	656	576	520	568	592
Long-range bombers	397	397	387	373	366	365	338	316	316	272	241
USSR											
ICBMs	1,618	1,527	1,477	1,350	1,400	1,398	1,398	1,398	1,398	1,398	1,398
SLBMs	720	784	845	909	1,028	1,028	1,028	989	989	980	981
Long-range bombers	140	135	135	135	135	156	156	150	150	143	143

Source: The Military Balance 1975-1976, 1983-1984, 1984-1985 (London: International Institute for Strategic Studies, annual).

forces (which had been seen as the trigger or "trip-wire" to unleash massive retaliation). In fact, during the 1960s the number of nuclear weapons in Europe tripled to 7,200 warheads (later to be reduced by 1,000), with much of the increase coming precisely in the most vulnerable battlefield weapons. This was a strange development, because nuclear war games in the United States and Europe as early as the mid-1950s revealed that enormous collateral damage to NATO countries, especially West Germany, would result from the use of a fraction of the "theater" nuclear weapons then in Europe.

NATO's response to the loss of superiority was to embrace flexible response in 1967 and, through a prolonged process of interallied consultation, study, and planning, to change the character of its nuclear strategy. Under the new guidelines, developed mostly in the 1970s, NATO would treat nuclear weapons initially as an instrument to force the aggressor to stop his invasion and to withdraw by raising the prospect of the resort to general nuclear war. Presumably, a few targets in Poland, East Germany, or Czechoslovakia would be attacked with nuclear weapons. The point was not to explode a weapon in this manner but to deter Soviet aggression by maintaining the capability and will to act in this way.

Initially, in implementing flexible response, President Kennedy chose simply to maintain U.S. superiority by greatly multiplying the numbers of modern powerful missiles in the U.S. arsenal and shortening the time for deployment. The Kennedy and Johnson administrations essentially doubled the number of ICBMs and SLBMs planned by the Eisenhower administration (1953–1961) and caused them to be built and put in place within five years, by 1966. However, the effect of this huge and rapid buildup was only to slow rather than to prevent the achievement of rough nuclear equality by the Soviet Union. As the realization of this dawned on defense planners in the Kennedy and Johnson administrations, they recognized that merely making further additions to the U.S. nuclear arsenal would neither reduce the damage the United States would suffer in a nuclear exchange with the Soviet Union nor significantly increase the damage the United States could inflict on that country. They believed that the further augmentation of U.S. offensive nuclear missiles would yield diminishing returns for the assets committed to the buildup. Merely to build more and more offensive launchers would not only be intolerably expensive, but it offered no answer to the question "How much is enough?" In addition, it would ruin any hope of successful arms control negotiations. A new concept, Assured Destruction, appeared to offer a solution to all these problems.

THE CONCEPT OF ASSURED DESTRUCTION

Assured Destruction was the declaratory nuclear strategy of the United States (what the U.S. government said it would do in case of war) for about a decade. According to Secretary of Defense Robert McNamara (1961–1967) and his advisers, however damaging a Soviet attack might be, the United States would make certain that it always possessed the weapons needed to destroy the Soviet Union as a viable society, or, as it was termed, to inflict "unacceptable damage" on that country. It was assumed that the destruction of approximately one-third of the population and two-thirds of the industry of the Soviet Union would cause the collapse of the Soviet Union as a modern industrial country. McNamara and his advisers believed that this capacity, dubbed Assured Destruction, was sufficient to deter a direct Soviet attack against the United States and its most important allies, Western Europe and Japan, and to prevent the Soviet Union from deriving political, psychological, and military advantage from its nuclear weapons.

Even assuming the need to build extra missiles, planes, and submarines in order to provide insurance against destruction of some weapons by Soviet attack and losses due to malfunction, the total U.S. arsenal could be kept relatively small because the number of essential Soviet targets was not great. In practice, the number of U.S. warheads was never limited to what was needed to inflict "unacceptable" loss on Soviet population and industry. Much larger numbers of nuclear warheads were built and aimed at "military" targets, including Soviet nuclear weapons. But first *priority* went to assuring the ability to inflict "unacceptable" damage.

A relatively small total of missiles and bombers was attractive to U.S. strategists for two reasons. (1) It put a reasonable ceiling on defense expenditures. Beyond a certain point additional strategic weapons would only make the rubble bounce. (2) More important, McNamara and the presidents he served purposely refrained from pushing U.S. strategic nuclear deployments as fast as the technical and industrial advantages of the United States would have allowed in hopes they could persuade the Soviet Union to apply the same Assured Destruction standard to their own strategic forces, thus turning Assured Destruction into Mutual Assured Destruction, or MAD as critics termed it. This the Soviets doggedly refused to do.

There appear to be at least three reasons for the Soviet refusal to follow the lead of U.S. strategists and adopt Assured Destruction. In the most fundamental sense, Assured Destruction leaves the survival of the Soviet regime in the hands of the U.S. government. While a war-fighting strategy cannot prevent the United States from inflicting irreparable damage on Soviet society in an all-out nuclear exchange, in all other situations the war-fighting approach holds out the promise, even if illusory, of victory. Second, Soviet military and political thought has developed along a more traditional military path than that of the United States, which owes far more to the innovations and political clout of civilian thinkers and policy-makers. Soviet strategists are frank about their indebtedness to the nineteenth-century German strategist, Karl von Clausewitz, and their entire approach to war is marked by traditional military thought. Last, even if only superficially important, the Soviet leaders' preference for victory as expressed in their war-fighting strategy is ideologically consistent with Marxism-Leninism, which is predicated on the superiority of communism and the historical inevitability of its triumph. To adopt another approach to warfare, particularly one predicated on making nuclear war unwinnable, would be to contradict one of the fundamental tenets of Soviet ideology. In this ideological sense, the victory promised by the Soviet strategy of nuclear war-fighting is not only that gained on the battlefield, but the mastery of an inferior social system by a superior one.

The Stability Factor and ABM Defense

By the time the Nixon administration took office in 1969, U.S. strategic modernization (the replacement of older missiles, bombers, and submarines with more modern versions) was slowed as a result of the growing concern of policymakers and strategists with nuclear "stability." Stability exists when there is no incentive for one side to resort to a preemptive nuclear strike. Obviously, stability would be most important and desirable during a crisis between the United States and the Soviet Union. If stability did not exist and the Soviet leaders should conclude, argued supporters of Assured Destruction, that the United States was seeking to destroy virtually all Soviet nuclear missiles, in a crisis they might decide to launch the weapons before the United States had destroyed them.

The watchwords of stability are "safe weapons, unsafe cities." Weapons are deemed safe if they are relatively inaccurate and are as well protected as possible. Cities are unsafe if they are undefended and must be left so: Their vulnerability serves to dampen the incentive to resort to war in a crisis—escalation would endanger the defenseless cities—and allows both sides to be certain they can inflict unacceptable damage on one another.

The proponents of stability favor inaccurate weapons, because if either side has the ability to destroy the other's strategic forces, then it might conclude there was an advantage to striking first. In conformance with the logic of stability, U.S. policymakers consciously refrained from building and deploying the large numbers of extremely accurate missiles and warheads necessary to attack Soviet missile silos all through the 1970s, even though the country was capable of deploying such weapons well before the Soviet Union (see Table 2). Again, the signal to the Soviets was clear: "Join us in avoiding such destabilizing weapons." The Soviets refused and stuck to a war-fighting counterforce strategy (attacking U.S. missiles, bombers, and submarines) that required large missiles with many highly accurate warheads.

Strategic weapons, the theorists argued, must be as well protected as possible. Initially, this realization led both sides to downgrade the importance of intercontinental bombers (which could be caught on the runway) and to move their missiles into hardened (reinforced) silos underground. At the time (from the late 1950s throughout the early 1970s), both governments recognized that eventually the silos would be vulnerable to attack by highly accurate warheads, but concluded that the gain in safety against surprise attack outweighed both the large monetary cost and eventual vulnerability of the silos.

The search for an invulnerable base for strategic weapons naturally led to submarines. Submarines are extremely difficult for an enemy to find and destroy. Putting strategic nuclear missiles in submarines created the least

TABLE 2
Strategic Nuclear-Capable Delivery Vehicles
of the United States and the Soviet Union

Category and Type	First Year Deployed	Range (km)	Throw-weight (000 lb)	CEP (m)	Launcher Total 7/84	Warhead Details and Comments
United States						
Land-based						
ICBM						
LGM-25C *Titan II*	1962	15,000	8.3	1,300	37	1 x 9MT W-53/Mk 6. Being phased out.
LGM-30F *Minuteman II*	1966	11,300	1.6	370	450	1 x 1.2MT W-56/Mk 11C. (442 launchers if 8 are comms vehs.)
LGM-30G *Minuteman III*	1970	14,800	1.5	280	250	3 x 170KT W-62/Mk 12 MIRV
	1980	12,900	2.4	220	300	3 x 335 KT W-78/Mk 12A MIRV
Sea-based						
SLBM						
UGM-73A *Poseidon C-3*	1971	4,600	3.3	450	304	10 x 40-50KT W-68. Max 14 MIRV.
Trident C-4	1980	7,400	3.0+	450	288	8 x 100KT W-76 MIRV
Air						
Long-range bombers						
B-52G	1959	12,000	0.95	45	151	12 ALCM, 4 SRAM, 4 x B-28/-43/-57/-61 bombs
B-52H	1962	16,000	0.95	45	90	4 SRAM, 4 x B-28/-43/-57/-61 bombs
Soviet Union						
Land-based						
ICBM						
SS-11 *Sego* mods 1/2	1966/73	10/13,000	2	1,400	520	1 x 1MT
mod 3	1975	8,800	2.5	1,100		3 x 100-300KT
SS-13 *Savage*	1968	10,000	1	2,000	60	1 x 750KT
SS-17 (RS-16) mod 1	1975	10,000	6	450	150	4 x 750KT MIRV
mod 2 (cold launch)	1977	11,000	3.6	450		1 x 6MT
mod 3	1982	10,000	–	–		4 x 20KT MIRV
SS-18 (RS-20) mod 1	1975	12,000	16.5	450	308	1 x 20MT
mod 2 (cold launch)	1977	11,000	16.7	450		8 x 900KT MIRV
mod 3	1979	10,500	16	350		1 x 20MT
mod 4	1982	11,000	16.7	300		10 x 500KT MIRV
(mod 5)	(1985)	(9,000)	(16)	(250)		(?10 x 750KT)MIRV
SS-19 (RS-18) mod 2 (hot launch)	1979	10,000	7.5	300	360	1 x 5MT
mod 3	1982	10,000	8	300		6 x 550KT MIRV

TABLE 2 (continued)

Category and Type	First Year Deployed	Range (km)	Throw-weight (000 lb)	CEP (m)	Launcher Total 7/84	Warhead Details and Comments
Soviet Union						
Sea-based						
SLBM						
SS-N-5 *Serb*	1964	1,400	n.a.	2,800	45	1 x 1MT
SS-N-6 mod 1	1968	2,400	1.5	900	368	1 x 1MT. Liquid-fuel.
Sawfly mod 2	1973	3,000	n.a.	900		1 x 1MT. Liquid-fuel.
mod 3	1974	3,000	1.5	1,400		2 warheads. Liquid-fuel.
SS-N-8 mod 1	1972	7,800	1.5	1,300	292	1 x 1MT
mod 2	n.a.	9,100	8	900		1 x 800MT
SS-N-17	1977	3,900	2.5	1,500	12	1 x 1MT. Solid-fuel.
SS-N-18 mod 1	n.a.	6,500	5	1,400	224	3 x ?200KT. Solid-fuel.
mod 2	1978	8,000	n.a.	600		1 x 450KT
mod 3	n.a.	6,500	n.a.	600		7 x 200KT MIRV
SS-N-20	1981	8,300	n.a.	n.a.	40	9 x ?200KT MIRV
Air			Max speed (Mach)	Weapon Load (000 lb)		
Long-range bombers						
Tu-95 *Bear B/C*	1956	12,800	0.78	40	100	1-2 AS-3/-4 ALCM, 2-3 bombs
Mya-4 *Bison*	1956	11,200	0.87	20	43	4 bombs
Medium-range bombers						
Tu-16 *Badger*	1955	4,800	0.8	20	410	1-2 AS-2/-3/-6 ALCM, 1 bomb. (*Badger G* carries (?6) AS-5.) Air Force (220), Navy (190).
Tu-22 *Blinder*	1962	4,000	1.5	12	160	1 AS-4 ALCM, 1 bomb. Air Force (125), Navy (35).
Tu-26 *Backfire*	1974	8,000	2.5	17.5	235	1 or 2 AS-4 ALCM, 2 bombs. Air Force (130), Navy (105).

Source: The Military Balance 1984–1985 (London: International Institute for Strategic Studies, 1984), p. xxxviii.

vulnerable—and therefore most stable—weapons system possible. There are some drawbacks, however. When they are submerged, submarines are hard to reach by radio. When they near the surface to transmit or receive messages or to determine their location—essential to missile accuracy—they risk detection. Moreover, the radio transmitters on land that are used to communicate with the submarines are vulnerable to attack. Nevertheless, on balance, the gain in stability outweighed the drawbacks.

The desire to maintain stability became the most telling argument against deploying any kind of antiballistic missile defense because it could be extended to protect cities as well as missiles. In the United States, the logic of stability was also used to justify a refusal to provide any systematic and effective civil defense against nuclear attack. The proponents of stability argued that in a crisis the possessor of an effective city defense could perceive an advantage in striking first. The attacking country, it was argued, would send its city dwellers into fall-out shelters and then try to disarm the enemy to destroy much of its offensive capability. Such a first strike would leave the victim only a remnant of its forces, which the attacker's ABM could defeat. In this scenario, the disarmed nation would be deterred from retaliating against the attacker's cities because retaliation would bring a response against its own cities.

The Soviet-U.S. ABM treaty of 1972 codified this view of stability. Severe limits were placed on missile defense. Each side was allowed no more than one hundred defensive missiles around each of two sites—a missile field and the national capital. Later the two governments decided to restrict their defense deployments to one site—either a missile field or the national capital. The United States declined to defend either. The Soviet Union exercised its option and built a missile defense around Moscow within the limits set by the ABM treaty. The Soviet government has also devoted far greater resources than the United States to civil defense measures against nuclear attack. These measures cannot deny the United States the capability to inflict unacceptable damage on the Soviet Union. However, in the event of less than an all-out attack against cities, the civil defense measures would provide significant protection to civilians. The keen Soviet concern with civil defense has been regarded as unsettling if not destabilizing by U.S. policymakers.

Missiles for Assured Destruction

Because U.S. policymakers preferred Assured Destruction, they made no attempt to match the total renovation of nuclear missiles that was carried out by Soviet leaders in the 1970s. Not content with superiority in delivery vehicles, the Soviet government then replaced its older missiles with three new and more powerful land-based missiles—SS-17, SS-18, and SS-19—armed with more accurate and independently targetable warheads on each missile, and all capable of destroying U.S. ICBMs. These warheads, known as MIRVs (multiple, independently targetable reentry vehicles), can be aimed at separate targets. MIRVs appeal to defense policymakers because the capability to strike separate targets increases the destructiveness of each missile firing.

The ICBM the United States chose to deploy during the 1970s, the Minuteman III missile with a Mark 12 warhead, was insufficient in numbers, accuracy, and yield to destroy a significant portion of Soviet missile silos. Thousands of warheads were added to the U.S. arsenal with the deployment of the submarine-launched Poseidon C3 MIRV, during the 1970s. However, the C3 also lacked the accuracy and destructive power needed to threaten Soviet silos (see Table 2). Finally, in the early 1980s, the United States began to deploy a more accurate and potent warhead for the Minuteman, the Mark 12A. Even with this addition, the United States could not pose a threat to Soviet ICBMs comparable to the threat Soviet missiles posed to those of the United States.

Assured Destruction: Pro and Con

There is much that is attractive about Assured Destruction. It limits strategic nuclear arms, although the limit is high. It is not an offensive strategy either in theory or in practice. It is devoted to the prevention of nuclear war rather than its conduct. The missiles and bombers that were developed by the United States to implement Assured Destruction were too small and too inaccurate to significantly reduce the damage the Soviet Union could inflict on the United States, let alone disarm the Soviet military in a first strike.

However, there are two major problems with the approach. The first problem is that according to Assured Destruction, in order to prevent nuclear war or any major aggression, the United States must target industrial and population centers. If war should occur, and neither conventional nor limited nuclear responses should end it favorably, the president would be forced to decide whether to make good the threat by attacking Soviet industrial and population centers. He would be most reluctant to "push the button," regardless of the strategic balance between the Soviet Union and the United States, but without superiority (the ability to devastate the Soviet Union without suffering a catastrophic reprisal), it would be most unlikely that he would launch the weapons. Even in the early 1960s, when U.S. superiority over the Soviet Union was greatest, Secretary McNamara believed that he had persuaded Presidents Kennedy and Johnson never to initiate the use of nuclear weapons. As McNamara put it: "At that time, in long private conversations with successive presidents—Kennedy and Johnson—I recommended, without qualification, that they never initiate, under any circumstances, the use of nuclear weapons. I believe they accepted my recommendation."[4]

The second problem with Assured Destruction is that to a degree unrealized by its authors, it depends for success on U.S. superiority, which existed for fifteen to twenty years after World War II. The loss of U.S. superiority undermined Assured Destruction in two ways: by exposing a morally dubious reliance on attacks against cities, which began to be taken seriously by large numbers of people only when the Soviets could do it too; and by showing that the growth of Soviet nuclear capabilities as a whole, and in particular Soviet counterforce weapons, increased *Soviet* deterrence of the United States and could deny the United States the value of a counter-city threat.

Nevertheless, the U.S. government clung at least in public to Assured Destruction despite these problems. As Ben Lambeth, a strategic analyst at the Rand Corporation, observed, "Although commonly called a 'strategy,' 'assured destruction' was by itself an antithesis of strategy. . . . It ceased to be useful precisely where military strategy is supposed to come into effect: at the edge of war."[5]

Loss of U.S. Superiority Weakens Assured Destruction

When the strategy of Assured Destruction was developed, the United States enjoyed a commanding superiority over the Soviet Union in every category used to measure nuclear military power: numbers of strategic launchers, numbers of warheads, deliverable megatonnage, throw-weight, and accuracy of warheads. By the mid-1970s, the Soviets had surpassed the United States in numbers of launchers and throw-weight and was challenging the United States in megatonnage and accuracy. It thus became theoretically possible that by using the large numbers of extremely accurate land-based missiles against U.S. nuclear weapons, Soviet leaders might prevent the United States from exercising its Assured Destruction capability.

U.S. policymakers began publicly to express their concern about this situation during the first Nixon administration (1969–1973) when the Soviet Union began to deploy the very large SS-9 missile. The U.S. secretary of defense in the early 1970s, Melvin Laird, stressed that the Soviet Union could choose to improve the warheads of the SS-9 and obtain a destabilizing counterforce weapon. As it turned out, what the Soviet government did made Laird appear to be an optimist. Instead of improving the SS-9, the Soviet Union decided to deploy more than three hundred of an even more deadly ICBM, the SS-18, the largest nuclear missile in the world. This move caused even greater alarm among U.S. strategists and policymakers. In addition, the Soviets began building SS-17s and SS-19s, ICBMs that also have counterforce capability.

The SS-18's throw-weight of 16,500 pounds is twice as great as the largest U.S. missile, the MX. Its warheads range from one of 20 megatons to eight MIRV of 900 kilotons or ten of 500 kilotons. Much of the great lift (throw-weight) of the missile is presently unused, which suggests that the Soviet government could choose suddenly to add as many as twenty more warheads to each missile. The SS-18 is a "cold-launch missile," which means that the rocket engine is not ignited until after the missile has been "popped" from its silo using gas pressure. This might allow the Soviets to reuse or "reload" the silo with another missile stored nearby. The possibility of reloading silos encourages the United States to treat Soviet proposals for reductions in its own forces with great caution.

THE UNITED STATES RESPONDS
TO NEW SOVIET CAPABILITIES

In the early 1970s, the U.S. government decided it must respond to the dramatic growth in Soviet capabilities to attack U.S. missile silos. Faithful to the logic of stability—the need to avoid incentives for a first strike—the United States chose to alter its targeting doctrine rather than to develop a U.S. counterforce strategy and capability. As then-Secretary of Defense James Schlesinger announced publicly in 1974, the United States began to change the Single Integrated Operation Plan (SIOP) to allow commanders to make limited nuclear attacks in the belief that this would be the best way to conclude a war before all-out nuclear war occurred. The change was supported by the Carter administration (1977–1981) and continued by the Reagan administration (1981–).

Limited Strike Options

Although it was not formally stated, the development of Limited Strike Options (LSO) marked a significant addition to and in a sense a departure from Assured Destruction. The United States has plainly shifted away from threats to attack Soviet population and industry as a means of deterring a nuclear war and preventing the Soviets from trying to exploit a perception that they hold a nuclear advantage over the United States. With the shift in declaratory policy came a shift in the restraints that had bound weapons procurement up until this time. In effect, it was now permitted for the United States to acquire more and more accurate missiles, for only these would be useful in limiting unintended damage to civilian targets in the event of a nuclear exchange. The aspect of this change that alarmed proponents of Assured Destruction and stability was that such weapons were indistinguishable from those that would be suitable for a first strike against Soviet hard targets (missile silos and shelters for political and military leaders). Planning for limited strikes carried with it the possibility that a nuclear war might occur over a prolonged period. Accordingly, great care and resources were devoted to the protection of communications and intelligence gathering during nuclear attack.

Missiles for Limited Strikes

In accordance with U.S. options, the United States began to produce a new, more powerful ICBM—the MX (Missile Experimental); to deploy hundreds of extremely accurate air-launched cruise missiles (ALCMs); and to speed up the deployment of a much larger submarine, the Trident, armed with the C4 missile. In the 1990s, the C4 missile will be replaced by the more advanced D5. These systems provide the only military response to the destabilizing capabilities the Soviet government developed during the 1970s and 1980s that does not conform to the U.S. policy of Assured Destruction and stability.

The new U.S. weapons offer significant military advantages over existing strategic weapons for attacking Soviet hard targets. The MX carries ten

300-kiloton warheads instead of the three of the Minuteman. Its throw-weight or useful payload—warheads, decoys, or other penetration aids—is three times greater than the Minuteman's, and its warheads have a circular error probable (CEP) of 120 meters as compared to the 220 to 280 meters of the Minuteman. (CEP is a statistical measure of the accuracy of warheads. It is the radius of the circle that includes 50 percent of the impact points of incoming warheads.)

The Reagan administration plans to deploy one hundred MX missiles. Congress has voted funds to build twenty-one MX missiles and decided in the spring of 1985 to fund the next twenty-one. At first, following the lead of House Armed Services chairman, Congressman Les Aspin, and a number of senators including Sam Nunn of Georgia and William Cohen of Maine, Congress supported the construction of additional MX missiles when the administration could report "good faith" efforts in the new round of arms control negotiations with the Soviet Union. However, opposition to the MX has continued to develop in the Senate, and it is far from certain that the Congress will permit the deployment of more than forty MX missiles.

The air-launched cruise missile is a pilotless drone aircraft that is launched from a bomber. Its range with a 200-kiloton nuclear warhead is about 1,500 miles. Its accuracy, due to terminal guidance, is a phenomenal 10 to 30 meters, giving it a greater capability against Soviet silos than even the MX. According to the logic of crisis stability, the cruise missile is preferable to the MX, for the cruise takes a long time to travel to its target and the bomber carrying it can be recalled. Over the coming decade, the United States intends to acquire a very large number of ALCMs. More than 1,000 have been deployed and the estimates vary as to the total planned. A low estimate, provided by former Secretary of Defense Harold Brown, is 2,400; a higher estimate, mentioned in Reagan administration budget requests, is between 4,000 and 5,000 ALCMs.

The Trident Ohio-class submarine is much quieter and larger than its predecessor, the Poseidon. It has twenty-four missile tubes; the Poseidon has sixteen. The Trident was built to carry the C4 and D5 missiles. These missiles have such a long range that the submarines carrying them do not require overseas bases. The missiles' range increases the areas of ocean the Trident submarines may use by factors of ten to twenty. The D5 will carry 9 to 10 warheads of 150 to 600 kilotons with a degree of accuracy that would enable them to destroy any Soviet hard targets. Four Trident submarines are now in service with 96 C4 missiles. Twenty are to be deployed by 1998, with the D5 missile to be fitted on the ninth submarine; the previous eight with the C4 missile would then be rebuilt to carry the newer SLBM. Each submarine has 24 missile tubes, which, with eight warheads each, would put 192 warheads in each Trident. As impressive as they were, neither the change to LSO nor the new missiles provided a militarily and politically convincing answer to the growing Soviet capability to attack U.S. strategic nuclear forces.

The Debate Continues

It was on this shortcoming of Assured Destruction—the growing Soviet counterforce capability—that critics concentrated. Their criticisms provided the intellectual basis for opposition to the SALT II treaty and for both the Reagan administration's major arms buildup and its deep interest in antimissile defense. The critics objected to Assured Destruction and to the notion that arms control negotiations were restraining Soviet nuclear programs and thereby enhancing U.S. security. To the critics, the Soviets were using the negotiations to prevent the United States from responding to their counterforce buildup. In addition, they argued, the logic of stability prohibited the United States from building its own counterforce capability on the grounds that such an initiative would be destabilizing. Tied up in negotiations, handcuffed by an inadequate strategy, the United States, in the critics' eyes, was self-deterred and increasingly vulnerable to Soviet coercion, nuclear blackmail, and even—in their worst fears—to a disarming nuclear strike. In 1980, Ronald Reagan called this possibility a "window of vulnerability" for Soviet aggression, and his claims helped elect him. The political and intellectual debate in the West continues to turn on this shortcoming as well as on the claims of supporters of Assured Destruction that it is not a shortcoming at all.

In response, the supporters of stability and Assured Destruction argued that the critics had made a point that was at best only hypothetically valid. It is extremely unlikely, they maintained, that the Soviet Union would ever attack U.S. land-based missiles. The uncertainties are enormous. The Soviet missiles would have to perform at maximum reliability and accuracy. The United States would have to permit destruction of all its missiles before firing any. As the Scowcroft Commission (appointed by President Reagan to report on the MX) observed, timing problems prevent the Soviets from disarming the United States. If Soviet missiles are fired at the U.S. missile silos, in the half hour before they arrive a substantial number of U.S. bombers armed with ALCMs could be flown out of danger. If the bombers, in turn, were attacked by missiles from Soviet submarines close to U.S. shores, these missiles would impact well before the Soviet countersilo missiles. Alerted by the impact, the United States would have fifteen minutes in which to launch its own missiles against Soviet targets. Above all, Soviet leaders would have to make the attacks against U.S. strategic weapons without knowing whether the U.S. would respond with attacks on their cities and industry. Even limited attacks against U.S. missile fields and submarine bases would cause millions of American deaths. Wouldn't an American president respond ferociously to such losses? The supporters of Assured Destruction said, "Certainly!" The critics answered, "Probably not, and we don't blame him, for such a response would bring ruin to U.S. cities as well."

The supporters of Assured Destruction and stability then advanced three points: No changes in strategy would protect the United States from catastrophic damage; the United States must therefore continue to emphasize the prevention of war; the best path to this goal is to look to its capacity for stable second-strike retaliation.

Reagan's Strategic Defense Initiative

To this, Assured Destruction critics and the Reagan administration have replied that the United States should build a strategic defense of its own and allied territory. In President Reagan's words, as he described the Strategic Defense Initiative, "Would it not be better to save lives than to avenge them?" Immediately dubbed Star Wars by its opponents, the administration's plan was ambitious and ambiguous. In his March 1983 speech announcing the program, the president certainly raised the possibility of constructing a missile-proof defense of U.S. cities:

Up until now, we have increasingly based our strategy of deterrence upon the threat of retaliation. But what if free people could live secure in the knowledge that their security did not rest on the threat of instant U.S. retaliation to deter a Soviet attack; that we could intercept and destroy strategic ballistic missiles before they reached our soil or that of our allies? I know that this is a formidable task . . . but is it not worth every investment necessary to free the world from the threat of nuclear war?

The president's plan drew sharp criticism from nuclear scientists, such as Sidney Drell and Wolfgang Panofsky, and from former policymakers, such as Secretary of Defense Harold Brown and National Security Adviser McGeorge Bundy. In their immediate objection to the Strategic Defense Initiative, they repeated the arguments that had been used two decades earlier against the first proposals to build an antiballistic missile defense: It will not work, it will trigger an arms race, and it will be destabilizing. It cannot, they repeated, prevent the destruction of the United States in an all-out war. The emphasis, therefore, must remain on prevention of war rather than protection from missiles. The surest prevention is to maintain the Assured Destruction capability.

Under attack, the president's advisers tried to scale down the whole idea. Of course, they said, the research and development would take years. It will be possible to protect missiles long before we can make cities safe against nuclear attack. This is because a missile defense that destroyed 50 percent of incoming warheads could be valuable, but a city defense would have to be able to destroy virtually every incoming warhead. The senior arms control adviser in the Reagan administration, Paul Nitze, introduced two additional and very demanding criteria that any defensive system would have to meet before it was deployed. Such a system would have to be able to withstand Soviet attack and it would have to be cheaper to deploy than the offensive countermeasures needed to defeat it. Nitze also suggested that during the transition from total reliance on offensive deterrence to at the least a mixed defensive-offensive deterrent, the United States would negotiate with the Soviet Union in order to assure that the change would occur by mutual agreement and would not upset the strategic relationship between the two countries. In the interim, a period of potentially indefinite duration, as Nitze said on ABC television in March 1985, the United States "would have to rely on the threat of devastating nuclear retaliation, that is the ultimate threat which has ensured the peace. It's all we can rely upon for the indefinite future, because we don't know how to do anything else."

The effect of the qualifications of Nitze and the other presidential advisers was to postpone the deployment of a defensive system for many years, perhaps until well into the twenty-first century; before then only a system to defend missile silos could be deployed. But a larger effect passed almost unnoticed. By unequivocally committing the United States to strategic defense, the Reagan administration had established its good faith and commitment to avoiding nuclear war. In the bargain, it appeared to have made continued reliance on offensive nuclear deterrence politically acceptable, as the price that would have to be paid until the United States and its allies could arrive at a feasible strategic defensive system and negotiate its deployment with the Soviet Union. For fiscal year 1986, the administration asked Congress to increase the number of MX missiles to be constructed from twenty-one to forty-eight (cost: $1.5 billion) and to triple spending on strategic defense, raising the total funds for defense to nearly $4 billion annually.

The Star Wars defense actually consists of many different systems—lasers, rockets, radars, and power sources—linked by computers to shoot down attacking missiles and warheads. The defense would operate against the missiles at different phases in their flight toward the United States. Immediately after launch, in the three- to five-minute boost phase, while the missile's rockets were still firing, and during the warheads' twenty-minute flight in space, the demands on the defense are greatest and, accordingly, the necessary technologies are most exotic: lasers based in space or on the ground, "rail-guns" that use electromagnetism to accelerate projectiles to fantastic speeds, and streams of subatomic particles generated in nuclear accelerators. In general, these technologies are still in the early phases of development and they pose enormous engineering problems that must be solved before effective weapons can be produced. This difficulty will considerably lengthen the time to deployment of a nationwide defense.

In the two-minute terminal phase, when the warheads reenter the atmosphere, the defense would employ existing and less sophisticated technologies to shoot down the warheads with a missile. In mid-1984, a nonnuclear U.S. Army missile intercepted a warhead. Although the intercept was achieved in a test rather than under simulated combat conditions, the intercept shows that the terminal phase technology has already passed an important milestone. Some experts believe an effective terminal-phase defense for protecting missile sites could be deployed within ten years.[6]

Among engineers and scientists in the defense industry there is considerable optimism. Rockwell International's vice president for advanced systems development and former chief scientist for the U.S. Air Force, Michael I. Yarymovych, told the *New York Times*, "As the idea gets dropped on the table, everybody empties out their file cabinets and conjures up ideas—'Star Wars'—and it's so easy for the critics to shoot it down, it's totally immature. But as time goes on, you'll see the critics becoming more and more confused and quiet. Because as times goes on people will start coming in with new thoughts that are totally different from the old ones."[7]

The proponents of stability remain adamantly and vocally opposed to strategic defense. As physicist and former Secretary of Defense Harold Brown said, referring to President Reagan's vision of protecting cities, "The combinations of limitations—scientific, technological, systems engineering costs—and especially the potential countermeasures make the prospect of a perfect or near-perfect defense negligibly low."[8]

With informed opinion so sharply divided, the debate has stalled, and policymakers, strategists, and scientists have begun choosing sides. Part of the reason for the intensity of the disagreements is that all parties sense that the administration has proposed, in Congressman Les Aspin's words, "a revolutionary change" in U.S. strategic policy by moving to diminish the country's total reliance on offensive nuclear weapons.[9] In the meantime, the administration has refined its proposals somewhat and begun to stress the long-term character of its strategic defense program.

In a sense, the strategic arguments will be judged and decided in the political arena, for the U.S. Congress will determine whether and at what level to fund research and deployment of strategic defense. The Reagan administration is confident that it will obtain the funds to continue the research programs. When it entered arms control negotiations with the Soviet Union in January 1985, the administration seemed committed to preserving its research effort on strategic defense.

At this stage, one may make several observations. The critics have identified at least three serious strategic deficiencies in Assured Destruction. First, the prevention of war by holding populations hostage is repugnant and profoundly unsettling to all who think seriously about nuclear questions. Second, in their preoccupation with deterring Soviet attack, the supporters of Assured Destruction neglected a significant part of the psychological dimension of *Soviet* deterrence of the United States. The extraordinary accumulation of highly accurate, land-based Soviet missiles has deeply alarmed the U.S. government for more than a decade. U.S. leaders are determined to achieve a satisfactory response to that threat not because they are misguided or evil but because they are legally and personally bound to defend the country. At a time of Soviet-U.S. parity in some areas and Soviet advantage in others, Assured Destruction no longer reassures many U.S. policymakers, whatever its effect on the Soviet leadership.

Third, the current orientation of U.S. strategy toward counterforce and defense against strategic missile attack is in part a delayed reaction to the persistent adherence of the Soviet Union to a war-fighting, counterforce strategy of its own. McNamara's hope has proved to be an illusion. The Soviet government has never adopted Assured Destruction. It continues to base its defense on the possibility of nuclear war and the necessity of being prepared to wage such a war successfully. The most daunting element of this repeatedly expressed belief is the willingness of Soviet leaders to contemplate launching their nuclear weapons when *they* have become convinced that an attack is imminent. John Erickson, a British analyst of the Soviet military, has argued that the Soviets would probably not use their counterforce weapons in a preemptive strike against the United States unless they were convinced that an attack on the Soviet Union was likely. In that event, the Soviets could resort to a "counterpreparation" (*kontrpodgotovka*) nuclear attack, in order to limit damage to their own forces and territory by destroying U.S. weapons before they could be used. The Soviet conviction that it can and will prevail in nuclear war has also been given concrete expression at every level of the Soviet military forces, from the construction of a large land-based missile force with its highly accurate warheads to the great attention in Soviet armored divisions to decontamination and protection from fall-out.

In some ways, as Freeman Dyson observed, the Soviet preference for counterforce is a more attractive strategic approach than Assured Destruction, although Dyson rejected both as unstable in a crisis. To Dyson, the problem with Soviet strategy is that Soviet leaders might wrongly conclude an attack was imminent and launch missiles that cannot be recalled. The difficulty with U.S. strategy is that U.S. leaders might resort to first use of nuclear weapons in order to avoid a conventional defeat, in Europe, for example, and in this way begin nuclear escalation. The clash between Soviet and U.S. strategies is well captured by Dyson: "The Soviet concept of counterforce says: 'Whatever else happens, if you drive us to war, we shall survive.' The U.S. concept of Assured Destruction says: 'Whatever else happens, if you drive us to war, you shall not survive.' "[10]

The clash of strategic concepts has led U.S. policymakers to question and to begin to modify Assured Destruction, initially by allowing for limited and prolonged nuclear strikes. The difference in strategies has also enormously complicated arms control negotiations and has reduced the results from them to a kind of codification of existing arsenals. At the same time, one must assume that the adoption of a counterforce strategy by the Soviet Union does not make it impossible to reach mutually advantageous agreements with them. The ABM treaty limiting ballistic missile defense—and only the Soviet Union has deployed one around Moscow—contradicts counterforce, and yet in accordance with its provisions the Soviet Union—like the United States—has left all its cities (except Moscow) undefended. *In itself*, the Soviet adoption of counterforce does not prove aggressive intention. It does guide Soviet behavior in peace and war, and it is precisely this that U.S. strategists have found most difficult to deal with.

THE AIMS AND DISAPPOINTMENTS OF ARMS CONTROL NEGOTIATIONS

Disarmament is the *elimination* or reduction of weapons. Arms control is the *limitation* of weapons with specific objectives in mind. From the start of serious arms control negotiations for the Partial Nuclear Test Ban Treaty of 1963 to the conclusion of SALT II in 1979, U.S. policymakers believed that the main purpose of arms control is to enhance strategic stability.

According to this view, the basis of stability is the existence of survivable nuclear retaliatory forces and vulnerable societies on both sides. Actions by either side that change either of these elements—by endangering the other's nuclear retaliatory force or by significantly limiting the damage to cities—cause instability. Arms control can improve stability, as strategist Jerome Kahan observed, by "the elimination of vulnerable systems on both sides and the prohibition of weapons that might threaten the survivability of deterrent forces."[11]

Other objectives of arms control are to reduce the damage of nuclear war, to build mutual confidence, and to bring down the costs of strategic programs.[12]

In addition, during the 1970s the United States attempted to use arms control negotiations to solve the strategic problems caused by Soviet attainment of nuclear parity and its adoption of a war-fighting strategy. In practice, this meant that the United States attempted to persuade the Soviet Union to give up its large land-based missiles that threatened U.S. ICBMs in exchange for a commitment by the United States to forgo building such systems as the MX missile. The Soviets rejected this effort, calculating that domestic opposition to nuclear modernization would prevent the United States from building an effective equivalent to their own ICBM force. Their calculation proved partially accurate, although as the 1970s ended, the United States began to deploy highly accurate cruise missiles, a new, more accurate warhead (the Mark 12A) for its Minuteman missiles, and the Trident submarine and to develop the D5 missile.

In general, the Soviet government regards arms control negotiations with a good deal more caution and realism than they are regarded in the West. The negotiations obviously have worked to complicate and delay Western and U.S. efforts to respond to major increases in Soviet nuclear forces, and one must assume that the Soviets have taken advantage of this effect in their approach to arms control. But in the ABM treaty the Soviet government accepted limitations of a kind that are inconsistent with its nuclear war-fighting strategy. All its cities (except Moscow) remain vulnerable to U.S. attack, and this vulnerability surely operates against a decision to begin a major conventional, let alone nuclear, attack on vital Western interests. It is slender but real evidence of the possibility that the two sides may conclude agreements that are mutually advantageous.

From its first days in office the Reagan administration took a much more skeptical view of arms control than had its predecessors. While the administration clearly did not wish to act in ways that would encourage a Soviet first strike, it was less inclined to believe that arms control could enhance stability or even that meaningful agreements could be concluded with the Soviet Union.[13] That the Reagan administration entered strategic and intermediate-range nuclear arms control talks with the Soviet Union in 1982 (broken off by the Soviets in 1983) and returned to the bargaining table in Geneva in 1985 may not be evidence that its views of arms control have changed. Instead, the administration appears to have entered both sets of negotiations with much larger and vital concerns in mind: the need to win and hold public opinion within the NATO countries and the recognition that important additions to U.S. and allied military forces could only be won from national legislatures if the requests for them were accompanied by a significant arms control initiative.

The goals of arms control promise much. Little has been achieved. In twenty years the United States and the Soviet Union have agreed to hold equal numbers of strategic launchers, have put a ceiling on the number of launchers that may carry MIRVs, and have agreed to limit their deployment of ABM systems to one on each side. A number of other agreements have been concluded—prohibiting nuclear testing in the atmosphere and committing the two nations to oppose nuclear proliferation, for example. However, these agreements have more of a political than a military impact and, while not unimportant, are not significant in relation to general arms control.

Obstacles on the Road to Arms Control

There appear to be four main reasons for the absence of genuinely significant arms control accomplishments. The most important reason is that neither Soviet nor U.S. leaders have been willing to concede anything in arms control negotiations that would significantly reduce the extremely generous margins they have made for security through nuclear strength. Both governments have subjected every arms control measure to this stringent test and have approved only those that did not interfere in any important way with their unilateral capacity to defend themselves. This is the primary reason that the number of nuclear weapons on both sides has reached such high levels by mutual agreement.

This situation in turn becomes a political problem in Western Europe and the United States because critics on the left and right can and do attack arms control as meaningless: Enormous efforts expended over several years end in agreements that allow tens of thousands of warheads on each side and that have such generous provisions for strategic modernization that the two nations are free to develop their forces exactly as they intended before signing the agreement.

A second impediment to arms control lies in the fundamentally different strategic views of the two sides. Measures that are natural and appropriate according to Soviet counterforce strategy are deeply alarming by U.S. standards of second-strike deterrence and stability, and vice versa. Describing the contrast, the Stanford Arms Control Group observed:

the essence of U.S. doctrine . . . is to deter nuclear war through the credible threat of inflicting unacceptable damage on the enemy, should deterrence fail. The objective is to enhance the credibility and stability of deterrence so that strategic nuclear weapons will not have to be used. . . .

Above all else, the Soviets seek to prevent nuclear war. Yet they believe that this objective requires, rather than rules out, effective war-fighting capabilities and defensive measures. Soviet leaders support this logic even while recognizing that it could never do enough to save the Soviet homeland from the untold devastation in the event of a nuclear attack. They do so perhaps because they consider other options, such as relying on the threat of mutual destruction with the hope of encouraging American restraint, to be unsound and patently unreliable.[14]

Given this discrepancy in fundamental strategic outlook, it is extremely unlikely that formal arms control negotiations can do other than codify decisions that the two powers have concluded work to their unilateral advantage.

The international political scene is a third impediment to arms control. The process and the agreements are obviously vulnerable to swings in U.S.-Soviet relations. The Soviet invasion of Afghanistan, for example, ended the chance for formal approval of SALT II by the U.S. government. (That both sides have continued to observe at least the main provisions of that agreement suggests the possibility of relying on informal and even unilateral actions rather than formal agreements to achieve arms control objectives.)

The fourth obstacle to arms control—with an even greater impact than Soviet-U.S. relations—is the effect of factional politics in the United States. The openness of the political system allows numerous interest groups to influence negotiations and debate on agreements. Contenders for the presidential nomination—and the campaign season is at least two years long in the United States—cannot resist politicizing the arms control debate. Moreover, one must not assume that arms control can escape becoming a pawn in internal political maneuvering in the Soviet Union.

In light of all these circumstances, one is not encouraged to expect too much from arms control negotiations. Nevertheless, there are great advantages to negotiations. They keep the two super-nuclear powers talking, which is clearly desirable. While negotiations are under way, both sides have shown a willingness to moderate their rhetoric and, to a very limited extent, their actions. In addition, the agreements justify significant savings. The ABM treaty allowed both sides to forgo the deployment of systems that, in retrospect, would have soon been obsolete because of progress in information-processing and optical technologies. Even if both sides should decide to deploy a missile defense, savings were facilitated by the ABM treaty.

Unilateral Arms Control: A Dubious Prospect

The lack of progress in negotiated arms control raises the question: Can meaningful reductions in nuclear weapons be achieved by unilateral measures?

There are at least four varieties of unilateral reductions. The first is reductions or eliminations that are dictated by an antinuclear movement that becomes politically irresistible in some or many Western democracies. A second type is reductions as example; that is, inviting the Soviet Union to follow suit by demonstrating peaceful intention and good faith. The reductions might be reversed if the Soviets refused to follow suit. Third are reductions as part of a tacit understanding with the Soviet Union about the nature of unilateral nuclear strategies and weapons development programs—the build-down concept is an intriguing example. Reductions that occur because of unilateral improvements in nuclear and other defense-related technologies are the fourth form.

The first two of these measures—reductions imposed by domestic political opposition and reductions as an example—violate the strategic outlook to such an extent that they could only be achieved at the potential cost of disrupting the social contract within the Western democracies and endangering their independence and way of life. Should the antinuclear movement succeed in forcing Western governments to engage in unilateral nuclear disarmament—as advocated, for example, by the British Labour Party—the advocates of national defense would immediately intensify a countermovement of their own. Even if the clash of the two groups remained peaceful and the constitutional order remained intact, the West's capacity for individual and collective resistance to Soviet expansionism would be substantially weakened. To those who prize resistance to nuclear weapons above all else, this would not necessarily be regarded as unfortunate. But this strategy amounts to the politics of making things worse in order for them to become better (*politique de pire*). The trouble with such an approach to political life is that one cannot easily predict what the pieces will look like after the social contract has been broken and, even more importantly, what kinds of political and military groups will have grown powerful during the collapse and, therefore, come to dominate the reconstruction of a new order.

If the Western democracies were to embark unilaterally on arms reductions dictated by antinuclear sentiment rather than strategic calculation or technological innovation, the Soviet Union would have absolutely no incentive to follow suit. Political opposition groups of any kind—let alone antinuclear movements—are simply not allowed to exist in the Soviet Union. There would be no counterpart in the Soviet Union to the political coercion felt by Western leaders. Were Soviet leaders then to refrain from exploiting the resulting strategic advantage that had been handed to them, they would become the first leaders in history to behave in this way. It is safe to say that nothing in the behavior, background, or political outlook of Soviet leaders suggests that they would refrain from exploiting such an advantage.

Once before in recent history, during the 1920s and 1930s, the Western democracies actually took this path and refused either to arm themselves adequately or to pursue an activist and interventionist foreign policy. They were tragically unready—materially, but even more important intellectually and spiritually—to answer Hitler's demonic aggression with defiance and resistance and superior force applied early with restraint and telling effect. The Nazi-Soviet pact to divide and destroy Poland in 1939 is one of the most chilling examples of the consequences that followed the democracies' abandonment of the strategic outlook. At a stroke, and because the democracies could not stop it, a proud independent nation lost its independence and most of Europe's Jews fell into Hitler's hands.

One need not equate the Soviet leadership with that of Hitler's Germany or militarist Japan to see the validity of the lesson that was learned by many during and after the ghastly outcome of World War II in Europe and Asia. If the West strays from the strategic outlook, it surrenders the stage to atavistic destructive elements in the world and must then allow and suffer horrible losses before it can restore a just and decent order to international affairs.

The third approach to reduction—by tacit agreement—has been given an intriguing twist by U.S. strategists and has received some attention from the Reagan administration as well. Some members of the Reagan administration have argued publicly that it is probably not possible for the United States to achieve agreement on mutually advantageous arms control measures with the Soviet Union. Instead, Kenneth Adelman, director of the Arms Control and Disarmament Agency in the administration, has suggested: "To me, the most promising of innovative thoughts is arms control through individual but (where possible) parallel policies, i.e., arms control without agreements (treaties, in particular). In simple terms, each side would take measures which enhance strategic stability and reduce nuclear weapons in consulation with each other—but not necessarily in a formalized signed agreement."[15]

The trouble with this approach is that it sounds like a bargaining ploy. It is simply too casual to accomplish anything but to demonstrate to the Soviets that there will be no going back to the previous U.S. attitude of "better a flawed agreement that gives the Soviets an advantage than no agreement at all."

Of these four, only the last—reductions made possible by technological breakthroughs—has actually resulted in significant alterations in the U.S. and Soviet nuclear arsenals. Over the past three decades, because of improvements in accuracy, yield, and delivery systems, the United States has made significant unilateral reductions in numbers of strategic launchers, megatonnage, and tactical nuclear weapons in Europe. The deployment of ballistic missiles with accurate warheads, for example, allowed the United States to reduce the number of bombers in SAC. Because the bombers carried very large-yield weapons, their retirement caused a significant decline in the overall megatonnage in the U.S. nuclear arsenal.

A strategist at the Council on Foreign Relations, Alton Frye, has made a stimulating attempt to link technological modernization with stabilizing reductions. In an approach called "strategic build-down" (see "Strategic Build-Down: A Context for Restraint" in this volume), Frye argued in favor of developing a formula for the replacement of older weapons with fewer modern ones. The formula is weighted to encourage the deployment of stabilizing instead of destabilizing systems. The Reagan administration has officially accepted the "build-down" concept as part of its arms control approach.

A similar positive development may follow from the extraordinary progress recently achieved in what are called precision-guided munitions (PGMs).

The easiest way to enhance the destructiveness of a weapon is to improve the accuracy with which it strikes its target. Because of their capabilities (such as following maps of the ground over which they fly), PGMs can literally fly within a few feet of the intended target. They therefore may lead to the replacement of nuclear warheads with conventional ones, particularly for use at short and intermediate ranges for tactical purposes.

That technological innovation has actually led to significant nuclear reductions by both sides—the Soviets have also moved away from very large-yield weapons and for the same reasons as the United States—runs counter to the view of many that the world is caught in a military-industrial death trap. It is a profoundly encouraging development, for it reconfirms the idea that science and engineering have positive as well as negative sides, that their creations can be allies as well as enemies of humanity.

THE FUTURE OF STRATEGY AND ARMS CONTROL

It appears increasingly likely that greater changes will occur in U.S. strategy than in Soviet strategy in the next decade. The unwillingness of the Soviet Union to adopt Assured Destruction has had its effect. The United States is being drawn closer to the Soviet strategy of counterforce and war-fighting, with its necessary corollary of providing a defense of population. U.S. policymakers continue to adhere to the logic of stability or, stated another way, have adopted a strategy of modified Assured Destruction, one that allows limited nuclear strikes and deemphasizes the targeting of civilians. Moreover, most of the new strategic weapons of the United States—the ALCM, the MX ICBM, the Trident II D5 SLBM—point toward a U.S. war-fighting strategy. The increasing interest of U.S. policymakers in strategic defense also shows a greater interest in counterforce strategy.

What one makes of this change in policy, should it continue, is another matter. The advocates of stability and modified Assured Destruction, and there are many with very impressive scientific and political credentials, are deeply concerned. They fear the change will result in a burst of defense spending as the Soviet Union counters the U.S. innovations, and they worry that the countermeasures will be destabilizing. An attempt to defend U.S. cities, for example, would be doubly harmful in their view because it would provoke Soviet offensive measures to counter the defense and, in a crisis, could influence Soviet leaders to doubt their ability to deter a U.S. attack, therefore encouraging a "counterpreparation" nuclear strike.

In response, one might make two observations. If the United States continues to adopt a strategy similar to that of the Soviet Union, it might actually increase the chance that the two sides could agree on significant arms control measures. At present, the two are attempting to equate apples and oranges, intellectually as well as practically. More important, the emphasis in U.S. strategy on the destruction of innocent bystanders has become increasingly difficult to sustain because of both greater Soviet capabilities and the possibility of "nuclear winter." It is clearly desirable to make great reductions in the numbers of nuclear weapons on both sides. But as an interim measure, would it not be an improvement to move away, to the greatest extent possible, from conscious reliance on the threat to destroy human beings as the basis of U.S. strategy? Such a change in strategy could be facilitated by recent technical improvements, particularly in the accuracy with which weapons may be delivered against targets over very long distances. Because of the dramatic gains in warhead accuracy, it is not impossible to imagine the use of conventional explosives for purposes that now require nuclear munitions. Even the increasing interest in strategic defense may prove beneficial rather than harmful to arms control. It seems unlikely, for example, that either the Soviet or U.S. government would permit truly massive reductions in nuclear weapons unless they could be assured that the relatively few that remained would be safe against preemptive attack. City defense could become feasible if the defense had to deal with only a small number of attacking missiles and warheads.

In these small ways, at least, strategy and technology can become the allies of arms control. That was always the hope of the advocates of Assured Destruction. We must not abandon that hope. If it is difficult to imagine a future without nuclear weapons, it is necessary to continue to hope and believe in a much safer future, one in which the United States and the Soviet Union discover strategies that will allow the defense of their vital interests without the threat or possibility of destroying millions of people.

NOTES

1. Freeman Dyson, *Weapons and Hope* (New York: Harper & Row, 1984), p. 7.

2. Raymond Aron, *Peace and War: A Theory of International Relations*, an abridged version (Garden City, N.Y.: Doubleday, 1966), pp. 19–43.

3. On Soviet strategic deception in the late 1950s, see Arnold Horelick and Myron Rush, *Strategic Power and Soviet Foreign Policy* (Chicago: University of Chicago Press, 1965). Regarding the Cuban missile crisis: In October 1962, the Kennedy administration discovered that the Soviet Union had brought approximately forty ballistic missiles into Cuba with ranges from 1,000 to 2,200 miles. The missiles lacked protection against attack, and therefore had to be considered first-strike weapons. After careful deliberation, the United States decided to impose a naval blockade around Cuba and to demand the complete removal of the Soviet missiles. Meanwhile, President Kennedy ostentatiously prepared the United States to go to war against Cuba to destroy the missiles and against the Soviet Union, in case Premier Khrushchev defended his new ally. With the blockade in place and Soviet ships approaching the picket line of U.S. destroyers, Khrushchev relented and agreed to withdraw the missiles. In exchange, Kennedy privately promised not to overthrow the Cuban regime and to withdraw U.S. missiles from Turkey. The latter promise was made easier by the growth in U.S. SLBMs, which could replace the capability lost in the withdrawal of the vulnerable ground-based missiles. Although Khrushchev had won a promise from Kennedy to leave Castro alone, the affair was a public and vivid humiliation of the Soviet government and its military power. Within two years Khrushchev was overthrown, and the Soviet Union had embarked on a massive buildup of nuclear missiles of its own. For more information, see the large body of literature on the Cuban missile crisis, in particular, Herbert S. Dinerstein, *The*

Making of a Missile Crisis, October 1962 (Baltimore: Johns Hopkins University Press, 1976); and Graham T. Allison, *Essence of Decision: Explaining the Cuban Missile Crisis* (Boston: Little, Brown and Co., 1971).

4. Robert S. McNamara, "The Military Role of Nuclear Weapons," *Foreign Affairs* 62, no. 1 (Fall 1983):79.

5. Benjamin S. Lambeth, *Selective Nuclear Options in American and Soviet Strategic Policy* (Santa Monica, Calif.: Rand Corporation, 1976), p. 14; quoted in Lawrence Freedman, *The Evolution of Nuclear Strategy* (New York: St. Martin's Press, 1981), p. 260.

6. See the six-part series on strategic defense, "Weapons in Space: The Controversy over 'Star Wars,'" by Wayne Biddle, Philip M. Boffey, William J. Broad, Leslie H. Gelb, and Charles Mohr, March 3–8, 1985, in the *New York Times*.

7. "Weapons in Space," *New York Times*, March 5, 1985.

8. "Weapons in Space," *New York Times*, March 3, 1985.

9. Unpublished address delivered by Congressman Les Aspin to Face-to-Face Luncheon, January 16, 1985.

10. Dyson, *Weapons and Hope*, pp. 231–232.

11. Jerome Kahan, *Security in the Nuclear Age: Developing U.S. Strategic Arms Policy* (Washington, D.C.: Brookings, 1975), p. 277.

12. Ibid.

13. Kenneth L. Adelman, "Arms Control with and Without Agreements," *Foreign Affairs* 63, no. 2 (Winter 1984/85):259.

14. Stanford Arms Control Group, Coit Blacker, and Gloria Duffy (eds.), *International Arms Control: Issues and Agreements*, 2d ed. (Stanford, Calif.: Stanford University Press, 1984), pp. 211, 215.

15. Adelman, "Arms Control with and Without Agreements," p. 259.

PART ONE

THE DEVELOPMENT OF U.S. NUCLEAR STRATEGY

CHAPTER 1

The Early Years: 1946–1953

The selections covering the early years reveal much about U.S. nuclear strategy during a period when the United States possessed a monopoly of nuclear weapons and the continental United States was immune to nuclear attack. The selections also anticipate in an uncanny manner the problems U.S. policymakers would encounter and the reservations many would develop as they struggled to derive benefits from the application of nuclear weapons to the country's foreign and military policies.

A proposal by the U.S. Air Force was approved by the Joint Chiefs of Staff (JCS) as JCS 1952 and became part of an Emergency War Plan for the period of the Berlin Blockade of 1948–1949. JCS 1952 (selection 2) shows that if the Red Army had invaded Western Europe, the U.S. military planned to launch an immediate nuclear attack against the seventy most important urban-industrial centers in the Soviet Union. The U.S. government continued to hold this view, while increasing the arsenal of nuclear weapons available for initial and follow-on attacks, despite the disadvantages revealed by the Harmon Committee's study submitted in May 1949.

The drawbacks identified by Lt. General H. R. Harmon of the Harmon Committee were both psychological and military. Wholesale nuclear attacks on Soviet cities would confirm the Communist regime's propaganda and strengthen the will of the Soviet people to resist, without destroying either communism or the power of the Soviet government to dominate the people of the country. The use of nuclear weapons by the United States would assure that the Soviet Union would retaliate in kind whenever it acquired nuclear weapons. The Soviet armed forces would not be prevented from making large-scale rapid military advances into Western Europe, although they would encounter serious supply problems, particularly shortages of high-test aviation gasoline and other refined petroleum products. Even so, the committee concluded that the use of nuclear weapons against Soviet cities was the only means available to the Western alliance to shock and damage the Soviet Union's war-making capabilities in the event of a Soviet invasion of Western Europe. The Harmon Committee recommended that the U.S. store of weapons should be increased as rapidly as possible and, in the event of war, should be used to the maximum extent possible.

George Kennan's January 1950 memorandum for Secretary of State Dean Acheson, "International Control of Nuclear Weapons," was doubtless composed in the shadow of the kinds of discussions and recommendations contained in JCS 1952 and the Harmon Committee report. Characteristically, Kennan went to the heart of the problem. What is the U.S. attitude toward nuclear weapons? Kennan asked. Everything depended on the answer to that question. Are we to plan to use them in battle as a kind of superpowerful long-range artillery? Or are they to be kept in relatively small numbers to prevent the use of such weapons against the United States and its allies?

Kennan not only presented an eloquent brief in favor of restricting the weapons to a "deterrent/retaliatory" role—in effect an early version of Mutual Assured Destruction—but made an equally impressive argument in favor of adopting a policy of "no first use." Four decades later, he and three colleagues would make the same appeal. Kennan believed that atomic weapons served no useful political purpose and could only prevent a nuclear attack on the United States and its allies. Planning for their military use—developing them according to criteria of military efficiency—could only lead to a profound confusion of Western policy and, Kennan feared, to the eventual use of the weapons in a horrible human cataclysm of death and loss. Even though he favored "no first use," Kennan distrusted the ability of U.S. policymakers to remain true to this self-denial. Even the uncertainties of an international agreement on control of nuclear weapons were preferable to reliance on a unilateral commitment to no first use. Kennan's advice was not heeded, and U.S. strategy took another path—the one outlined by National Security Council memo NSC-68, adopted in the spring of 1950.

The systematic analysis and recommendations of NSC-68 have become familiar to students of U.S. strategy and diplomacy. Its purpose was certainly substantive. With Paul Nitze among its principal authors, NSC-68 recorded the agreement of senior U.S. foreign-policy makers about the nature and magnitude of the Soviet threat to U.S. interests and the steps that should be taken to meet that threat. However, another important purpose of the document was bureaucratic. As Dean Acheson observed, NSC-68 was meant "to . . . bludgeon the mass mind of 'top government'" into a condition that would allow the president to conduct an effective foreign policy.[1]

The significance of NSC-68 is that it shows U.S. strategy moving toward symmetry with that of the Soviet Union. As John Lewis Gaddis has written, there is an oscillation in U.S. strategy between symmetry and asymmetry.[2] At times the U.S. feels able to play its strengths against the weaknesses of the Soviet Union. In this manner, the United States can avoid mirroring the Soviet military establishment. Above all it can escape the tremendous expense and onerous social regimentation necessary to field an extremely large standing army fully prepared for the conduct of major conventional war. The early U.S. advantage in nuclear weapons was the greatest possible asymmetry. In effect, NSC-68 argued that it was not that simple. More, much more, would have to be spent on the armed forces, and the United States would have to develop much larger and more potent conventional units.

NOTES

1. Dean Acheson, *Present at the Creation: My Years in the State Department* (New York: W. W. Norton, 1969), p. 374.
2. John Lewis Gaddis, *Strategies of Containment: A Critical Appraisal of Postwar American National Security Policy* (New York: Oxford University Press, 1982).

1. U.S. Policy on Atomic Weapons (NSC-30)

National Security Council

THE PROBLEM

1. To determine the advisability of formulating, at this time, policies regarding the use of atomic weapons.

ANALYSIS

2. The decision to employ atomic weapons is a decision of highest policy. The circumstances prevailing when war is joined cannot be wholly forecast with any greater certainty than can the arrival of war. It appears imprudent either to prescribe or to prohibit beforehand the use of any particular weapons when the character of future conflict is subject only to imperfect prediction. In this circumstance, a prescription preceding diagnosis could invite disaster.

3. If war itself cannot be prevented, it appears futile to hope or to suggest that the imposition of limitations on the use of certain military weapons can prevent their use in war.

4. The United States has nothing presently to gain, commensurable with the risk of raising the question, in either a well-defined or an equivocal decision that atomic weapons would be used in the event of war. An advance decision that atomic weapons will be used, if necessary, would presumably be of some use to the military planners. Such a decision does not appear essential, however, since the military can and will, in its absence, plan to exploit every capability in the form of men, materials, resources and science this country has to offer.

5. In this matter, public opinion must be recognized as a factor of considerable importance. Deliberation or decision on a subject of this significance, even if clearly affirmative, might have the effect of placing

before the American people a moral question of vital security significance at a time when the full security impact of the question had not become apparent. If this decision is to be made by the American people, it should be made in the circumstances of an actual emergency when the principal factors involved are in the forefront of public consideration.

6. Foreign opinion likewise demands consideration. Official discussion respecting the use of atomic weapons would reach the Soviets, who should in fact never be given the slightest reason to believe that the U.S. would even consider not to use atomic weapons against them if necessary. It might take no more than a suggestion of such consideration, perhaps magnified into a doubt, were it planted in the minds of responsible Soviet officials, to provide exactly that Soviet aggression which it is fundamentally U.S. policy to avert. . . .

10. The United States has offered, along with all other nations, to eliminate atomic weapons from national armaments if and when a fully effective, enforceable system of international control is put into effect. In the meantime United States policy should ensure that no commitment be made in the absence of an established and acceptable system of international control of atomic energy which would deny this country the right to employ such weapons in the event of actual hostilities. The actual decision to employ weapons should be made by the Chief Executive and in the light of prevailing circumstances.

11. The time and circumstances under which atomic weapons might be employed are incapable of accurate determination prior to the evident imminence of hostilities. The type and character of targets against which atomic weapons might be used is primarily a function of military selection in the preparation and planning of grand strategy. In this case, however, there is the additional requirement for blending a political with a military responsibility in order to assure that the conduct of war, to the maximum extent practicable, advances the fundamental and lasting aims of U.S. policy.

CONCLUSIONS

12. It is recognized that, in the event of hostilities, the National Military Establishment must be ready to utilize promptly and effectively all appropriate means available, including atomic weapons, in the interest of national security and must therefore plan accordingly.

13. The decision as to the employment of atomic weapons in the event of war is to be made by the Chief Executive when he considers such decision to be required.

14. In the light of the foregoing, no action should be taken at the present time:

a. To obtain a decision either to use or not to use atomic weapons in any possible future conflict;

b. To obtain a decision as to the time and circumstances under which atomic weapons might or might not be employed.

2. Evaluation of Current Strategic Air Offensive Plans (JCS 1952/1)

Joint Chiefs of Staff

THE PROBLEM

1. To evaluate the chances of success in delivering a powerful strategic air offensive against vital elements of the Soviet war-making capacity as contemplated in current war plans; to consider the risks involved in the planned strategic air operations; and to appraise any adverse effect on this offensive of the continuation of the Berlin air lift at its contemplated level until war occurs.

ASSUMPTIONS

2. War will occur prior to 1 April 1949.

3. Atomic bombs will be used to the extent determined to be practicable and desirable.

4. The Berlin air lift will be continued at the contemplated level until the outbreak of hostilities.

5. The strategic air offensive will be implemented on a first-priority basis.

EVALUATION

32. The factors affecting our chances of success in delivering a powerful strategic air offensive against the vital elements of the Soviet warmaking capacity as affected by the Berlin air lift operations and as compared with the risk involved in such an air offensive can be summarized as follows:

a. After due consideration of the number of atomic bombs available, the radii of action of Allied air forces, the estimated bombing accuracy, the available weight of attack and the time required for realization of effects, the highest priority target system is that system constituted by the major Soviet urban-industrial concentrations. Destruction of this system should so cripple the Soviet industrial and control centers as to reduce drastically the offensive and defensive power of their armed forces.

b. Target folders and navigation charts will be available by 1 February 1949 for operations against the first seventy cities. Currently available aeronautical charts (scale 1:1,000,000) are sufficiently accurate to permit aerial navigation to any desired point in the USSR.

c. An effective program of attack against petroleum refining targets in the USSR and Soviet occupied areas would, after depletion of inventories, practically destroy the offensive capabilities of the USSR and seriously cripple its defensive capabilities.

d. Major attacks against the Soviet hydro-electric system can be undertaken at about D + 8 months and completed by approximately D + 14 months.

e. Attacks on the inland transportation system as a major objective can be undertaken by approximately D + 14 to D + 16 months with the effect of disrupting the entire system within two months. The transportation offensive would seriously disrupt the entire Soviet economy.

f. The Soviet early warning system is composed primarily of World War II lend-lease or captured equipment. It is doubtful that the Soviets have overcome the previous lag in electronic development which was detrimental to their antiaircraft artillery and fighter control system.

g. Soviet antiaircraft cannot maintain effective fire at 30,000 feet, the operational altitude of attacking bombers. The proximity fuze, even if available to the Soviets, would be relatively ineffective due to Soviet radar deficiency and allied jamming operation. Consideration of these facts leads to the conclusion that the total effectiveness of Soviet ground defenses will be below that encountered and overcome by the Allies in World War II.

h. It is not believed that the Soviets have developed a fighter suitable for night and all-weather operations, or that they possess an effective system of fighter control. The large areas to be defended as compared to the number of aircraft operationally available and the inherent difficulties, time-wise, of effectively intercepting high-altitude, high-speed bombers operating at night or in bad weather reduce appreciably the hazards to the delivery of the air offensive.

i. Soviet jamming of U.S. Air Force radar bombing sets, if attempted, would be ineffective.

j. Soviet counter air operations against Allied base areas will be limited initially by the range of bombing aircraft. It can be stated, in addition, that all available intelligence and operational data make it difficult to foresee development of a Russian capability for making United Kingdom bases untenable before D + 45 to 60 days at the earliest, and of making Cairo-Suez bases untenable before D + 4 to 6 months. These estimates are based on intelligence data regarding Soviet capabilities only and do not reflect the effects of Allied opposition. Losses due to the Soviet air offensive against Allied bases are not expected to be of such severe nature as to preclude conducting the strategic air offensive.

k. The missions as planned to implement the atomic offensive and the tactics and techniques to be used give maximum chances of successful penetration, successful attack on selected targets, and minimum risk to Allied aircraft. Where necessary or operationally desirable, daylight attacks can and would be conducted. However, night and weather operations, jamming capabilities, speed and altitude of the bombers would all be used in reducing the risks involved.

l. For the initial atomic attacks a possible 25 percent attrition loss has been accepted for planning purposes, which still leaves ample capability for delivery of the entire stockpile of atomic bombs. As the effects of the atomic offensive are reflected in the Soviet air defense system, aircraft losses should be reduced. . . .

33. The following conclusions may be drawn from an evaluation of the foregoing:

a. A powerful strategic air offensive against vital elements of the Soviet war-making capacity could be delivered as planned.
b. Certain risks, such as tenability of base areas and attrition of personnel and materiel during operations, do exist, but these risks are not unreasonable and can be taken without unduly jeopardizing the successful execution of the strategic air offensive.
c. The risks existing during the early phases of the campaign should decrease appreciably within a relatively short time due to the cumulative effects of the strategic air offensive.
d. Preliminary studies of the strategic air lift requirements to implement the short-range emergency war plans and the effects of continuation of the Berlin air lift on U.S. Air Force capabilities indicate that the strategic air offensive, if accorded first priority, could be delivered as planned even though the Berlin air lift is continued at its contemplated level.

3. U.S. Objectives and Programs for National Security (NSC-68)

National Security Council

III. FUNDAMENTAL DESIGN OF THE KREMLIN

The fundamental design of those who control the Soviet Union and the international communist movement is to retain and solidify their absolute power, first in the Soviet Union and second in the areas now under their control. In the mind of the Soviet leaders, however, achievement of this design requires the dynamic extension of their authority and the ultimate elimination of any effective opposition to their authority.

The design, therefore, calls for the complete subversion or forcible destruction of the machinery of government and structure of society in the countries of the non-Soviet world and their replacement by an apparatus and structure subservient to and controlled from the Kremlin. To that end Soviet efforts are now directed toward the domination of the Eurasian land mass. The United States, as the principal center of power in the non-Soviet world and the bulwark of opposition to Soviet expansion, is the principal enemy whose integrity and vitality must be subverted or destroyed by one means or another if the Kremlin is to achieve its fundamental design. . . .

With particular reference to the United States, the Kremlin's strategic and tactical policy is affected by its estimate that we are not only the greatest immediate obstacle which stands between it and world domination, we are also the only power which could release forces in the free and Soviet worlds which could destroy it. The Kremlin's policy toward us is consequently animated by a peculiarly virulent blend of hatred and fear. Its strategy has been one of attempting to undermine the complex of forces, in this country and in the rest of the free world, on which our power is based. In this it has both adhered to doctrine and followed the sound principle of seeking maximum results with minimum risks and commitments. The present application of this strategy is a new form of expression for traditional Russian caution. However, there is no justification in Soviet theory or practice for predicting that, should the Kremlin become convinced that it could cause our downfall by one conclusive blow, it would not seek that solution.

In considering the capabilities of the Soviet world, it is of prime importance to remember that, in contrast to ours, they are being drawn upon close to the maximum possible extent. Also in contrast to us, the Soviet world can do more with less—it has a lower standard of living, its economy requires less to keep it functioning and its military machine operates effectively with less elaborate equipment and organization. . . .

The greatest vulnerability of the Kremlin lies in the basic nature of its relations with the Soviet people.

That relationship is characterized by universal suspicion, fear and de-nunciation. It is a relationship in which the Kremlin relies, not only for its power but its very survival, on intricately devised mechanisms of coercion. The Soviet monolith is held together by the iron curtain around it and the iron bars within it, not by any force of natural cohesion. These artificial mechanisms of unity have never been intelligently challenged by a strong outside force. The full measure of their vulnerability is therefore not yet evident.

The Kremlin's relations with its satellites and their peoples is likewise a vulnerability. Nationalism still remains the most potent emotional-political force. The well-known ills of colonialism are compounded, however, by the excessive demands of the Kremlin that its satellites accept not only the imperial authority of Moscow but that they believe in and proclaim the ideological primacy and infallibility of the Kremlin. These excessive re-quirements can be made good only through extreme coercion. The result is that if a satellite feels able to effect its independence of the Kremlin, as Tito was able to do, it is likely to break away.

In short, Soviet ideas and practices run counter to the best and potentially the strongest instincts of men, and deny their most fundamental aspirations. Against an adversary which effectively affirmed the constructive and hopeful instincts of men and was capable of fulfilling their fundamental aspirations, the Soviet system might prove to be fatally weak.

The problem of succession to Stalin is also a Kremlin vulnerability. In a system where supreme power is acquired and held through violence and intimidation, the transfer of that power may well produce a period of instability.

In a very real sense, the Kremlin is a victim of its own dynamism. This dynamism can become a weakness if it is frustrated, if in its forward thrusts it encounters a superior force which halts the expansion and exerts a superior counterpressure. Yet the Kremlin cannot relax the condition of crisis and mobilization, for to do so would be to lose its dynamism, whereas the seeds of decay within the Soviet system would begin to flourish and fructify.

The Kremlin is, of course, aware of these weaknesses. It must know that in the present world situation they are of secondary significance. So long as the Kremlin retains the initiative, so long as it can keep on the offensive unchallenged by clearly superior counter-force—spiritual as well as material—its vulnerabilities are largely inoperative and even concealed by its successes. The Kremlin has not yet been given real reason to fear and be diverted by the rot within its system. . . .

The two fundamental requirements which must be met by forces in being or readily available are support of foreign policy and protection against disaster. To meet the second requirement, the forces in being or readily available must be able, at a minimum, to perform certain basic tasks:

a. To defend the Western Hemisphere and essential allied areas in order that their war-making capabilities can be developed;
b. To provide and protect a mobilization base while the offensive forces required for victory are being built up;
c. To conduct offensive operations to destroy vital elements of the Soviet war-making capacity, and to keep the enemy off balance until the full offensive strength of the United States and its allies can be brought to bear;
d. To defend and maintain the lines of communication and base areas necessary to the execution of the above tasks; and
e. To provide such aid to allies as is essential to the execution of their role in the above tasks.

In the broadest terms, the ability to perform these tasks requires a build-up of military strength by the United States and its allies to a point at which the combined strength will be superior for at least these tasks, both initially and throughout a war, to the forces that can be brought to bear by the Soviet Union and its satellites. In specific terms, it is not essential to match item for item with the Soviet Union, but to provide an adequate defense against air attack on the United States and Canada and an adequate defense against air and surface attack on the United Kingdom and Western Europe, Alaska, the Western Pacific, Africa, and the Near and Middle East, and on the long lines of communication to these areas. Furthermore, it is mandatory that in building up our strength, we enlarge upon our technical superiority by an accelerated exploitation of the scientific potential of the United States and our allies.

Forces of this size and character are necessary not only for protection against disaster but also to support our foreign policy. In fact, it can be argued that larger forces in being and readily available are necessary to inhibit a would-be aggressor than to provide the nucleus of strength and the mobilization base on which the tremendous forces required for victory can be built. For example, in both World Wars I and II the ultimate victors had the strength, in the end, to win though they had not had the strength in being or readily available to prevent the outbreak of war. In part, at least, this was because they had not had the military strength on which to base a strong foreign policy. At any rate, it is clear that a substantial and rapid building up of strength in the free world is necessary to support a firm policy intended to check and to roll back the Kremlin's drive for world domination. . . .

CONCLUSIONS

The foregoing analysis indicates that the probable fission bomb capability and possible thermonuclear bomb capability of the Soviet Union have greatly intensified the Soviet threat to the security of the United States. This threat is of the same character as that described in NSC 20/4 (approved by the President on November 24, 1948) but is more immediate than had previously been estimated. In particular, the United States now faces the contingency that within the next four or five years the Soviet Union will possess the military capability of delivering a surprise atomic attack of such weight that the United States must have substantially increased general air, ground, and sea strength, atomic capabilities, and air and civilian defenses to deter war and to provide reasonable assurance, in the event of war, that it could survive the initial blow and go on to the eventual attainment of its objectives. In turn, this contingency requires the intensification of our efforts in the fields of intelligence and research and development.

Allowing for the immediacy of the danger, the following statement of Soviet threats, contained in NSC 20/4, remains valid:

14. The gravest threat to the security of the United States within the foreseeable future stems from the hostile designs and formidable power of the U.S.S.R., and from the nature of the Soviet system.

15. The political, economic, and psychological warfare which the U.S.S.R. is now waging has dangerous potentialities for weakening the relative world position of the United States and disrupting its traditional institutions by means short of war, unless sufficient resistance is encountered in the policies of this and other non-communist countries.

16. The risk of war with the U.S.S.R. is sufficient to warrant, in common prudence, timely and adequate preparation by the United States.

    a. Even though present estimates indicate that the Soviet leaders probably do not intend deliberate armed action involving the United States at this time, the possibility of such deliberate resort to war cannot be ruled out.

b. Now and for the foreseeable future there is a continuing danger that war will arise either through Soviet miscalculation of the determination of the United States to use all the means at its command to safeguard its security, through Soviet misinterpretation of our intentions, or through U.S. miscalculation of Soviet reactions to measures which we might take.

17. Soviet domination of the potential power of Eurasia, whether achieved by armed aggression or by political and subversive means, would be strategically and politically unacceptable to the United States.

18. The capability of the United States either in peace or in the event of war to cope with threats to its security or to gain its objectives would be severely weakened by internal developments, important among which are:

a. Serious espionage, subversion and sabotage, particularly by concerted and well-directed communist activity.

b. Prolonged or exaggerated economic instability.

c. Internal political and social disunity.

d. Inadequate or excessive armament or foreign aid expenditures.

e. An excessive or wasteful usage of our resources in time of peace.

f. Lessening of U.S. prestige and influence through vacillation or appeasement or lack of skill and imagination in the conduct of its foreign policy or by shirking world responsibilities.

g. Development of a false sense of security through a deceptive change in Soviet tactics. . . .

The Analysis also confirms that our objectives with respect to the Soviet Union, in time of peace as well as in time of war, as stated in NSC 20/4 (para. 19), are still valid, as are the aims and measures stated therein (paras. 20 and 21). Our current security programs and strategic plans are based upon these objectives, aims, and measures:

19.

a. To reduce the power and influence of the U.S.S.R. to limits which no longer constitute a threat to the peace, national independence and stability of the world family of nations.

b. To bring about a basic change in the conduct of international relations by the government in power in Russia, to conform with the purposes and principles set forth in the U.N. Charter.

In pursuing these objectives, due care must be taken to avoid permanently impairing our economy and the fundamental values and institutions inherent in our way of life.

20. We should endeavor to achieve our general objectives by methods short of war through the pursuit of the following aims:

a. To encourage and promote the gradual retraction of undue Russian power and influence from the present perimeter areas around traditional Russian boundaries and the emergence of the satellite countries as entities independent of the U.S.S.R.

b. To encourage the development among the Russian peoples of attitudes which may help to modify current Soviet behavior and permit a revival of the national life of groups evidencing the ability and determination to achieve and maintain national independence.

c. To eradicate the myth by which people remote from Soviet military influence are held in a position of subservience to Moscow and to cause the world at large to see and understand the true nature of the U.S.S.R. and the Soviet-directed world communist party, and to adopt a logical and realistic attitude toward them.

d. To create situations which will compel the Soviet Government to recognize the practical undesirability of acting on the basis of its present concepts and the necessity of behaving in accordance with precepts of international conduct, as set forth in the purposes and principles of the U.N. Charter.

21. Attainment of these aims requires that the United States:

a. Develop a level of military readiness which can be maintained as long as necessary as a deterrent to Soviet aggression, as indispensable support to our political attitude toward the U.S.S.R., as a source of encouragement to nations resisting Soviet political aggression, and as an adequate basis for immediate military commitments and for rapid mobilization should war prove unavoidable.

b. Assure the internal security of the United States against dangers of sabotage, subversion, and espionage.

c. Maximize our economic potential, including the strengthening of our peacetime economy and the establishment of essential reserves readily available in the event of war.

d. Strengthen the orientation toward the United States of the non-Soviet nations; and help such of those nations as are able and willing to make an important contribution to U.S. security, to increase their economic and political stability and their military capability.

e. Place the maximum strain on the Soviet structure of power and particularly on the relationships between Moscow and the satellite countries.

f. Keep the U.S. public fully informed and cognizant of the threats to our national security so that it will be prepared to support the measures which we must accordingly adopt.

In the light of present and prospective Soviet atomic capabilities, the action which can be taken under present programs and plans, however, becomes dangerously inadequate, in both timing and scope, to accomplish the rapid progress toward the attainment of the United States political, economic, and military objectives which is now imperative.

A continuation of present trends would result in a serious decline in the strength of the free world relative to the Soviet Union and its satellites. This unfavorable trend arises from the inadequacy of current programs and plans rather than from any error in our objectives and aims. These trends lead in the direction of isolation not by deliberate decision but by lack of the necessary basis for a vigorous initiative in the conflict with the Soviet Union.

Our position as the center of power in the free world places a heavy responsibility upon the United States for leadership. We must organize and enlist the energies and resources of the free world in a positive program for peace which will frustrate the Kremlin design for world domination by creating a situation in the free world to which the Kremlin will be compelled to adjust. Without such a cooperative effort, led by the United States, we will have to make gradual withdrawals under pressure until we discover one day that we have sacrificed positions of vital interest. . . .

In summary, we must, by means of a rapid and sustained build-up of the political, economic, and military strength of the free world, and by means of an affirmative program intended to wrest the initiative from the Soviet Union, confront it with convincing evidence of the determination and ability of the free world to frustrate the Kremlin design of a world dominated by its will. Such evidence is the only means short of war which eventually may force the Kremlin to abandon its present course of action and to negotiate acceptable agreements on issues of major importance.

The whole success of the proposed program hangs ultimately on recognition by this Government, the American people, and all free peoples, that the cold war is in fact a real war in which the survival of the free world is at stake. Essential prerequisites to success are consultations with Congressional leaders designed to make the program the object of non-partisan legislative support, and a presentation to the public of a full explanation of the facts and implications of the present international situation. The prosecution of the program will require of us all the ingenuity, sacrifice, and unity demanded by the vital importance of the issue and the tenacity to persevere until our national objectives have been attained.

4. Evaluation of Effect on Soviet War Effort Resulting from the Strategic Air Offensive

Lt. General H. R. Harmon

THE PROBLEM

1. To evaluate the effect on the war effort of the U.S.S.R. of the Strategic Air Offensive contemplated in current war plans, including an appraisal of the psychological effects of atomic bombing on the Soviet will to wage war. . . .

FACTS BEARING ON THE PROBLEM

3. The plan for the strategic air offensive . . . contemplates two distinct phases:

a. An initial phase, consisting of a series of attacks primarily with atomic bombs on 70 target areas (presently planned by the Strategic Air Command to be accomplished in approximately 30 days).

b. A second phase, consisting of a continuation of the initial attacks with both atomic and conventional weapons. . . .

CONCLUSIONS

8. It is concluded that complete and successful execution of the initial atomic offensive against the U.S.S.R., as planned, would probably affect the war effort, and produce psychological effects upon the Soviet will to wage war as set forth below. . . .

EFFECT ON INDUSTRIAL CAPACITY

9. Physical damage to installations, personnel casualties concentrated in industrial communities, and other direct or indirect cumulative effects would result in a 30 to 40 percent reduction of Soviet industrial capacity. This loss would not be permanent and could either be alleviated by Soviet recuperative action or augmented depending upon the weight and effectiveness of follow-up attacks.

10. Of outstanding importance is the prospect that the petroleum industry in the U.S.S.R. would suffer severe damage especially in refining capacity. The supply of high-test aviation gasoline would become rapidly critical.

PERSONNEL CASUALTIES

11. The initial atomic offensive could produce as many as 2,700,000 mortalities, and 4,000,000 additional casualties, depending upon the effectiveness of Soviet passive defense measures. A large number of homes would be destroyed and the problems of living for the remainder of the 28,000,000 people in the 70 target cities would be vastly complicated.

PSYCHOLOGICAL EFFECTS

12. The atomic offensive would not, per se, bring about capitulation, destroy the roots of Communism or critically weaken the power of Soviet leadership to dominate the people.

13. For the majority of Soviet people, atomic bombing would validate Soviet propaganda against foreign powers, stimulate resentment against the United States, unify these people and increase their will to fight. Among an indeterminate minority, atomic bombing might stimulate dissidence and the hope of relief from oppression. Unless and until vastly more favorable opportunities develop for them, the influence of these elements will not appreciably affect the Soviet war effort.

14. A psychological crisis will be created within the U.S.S.R. which could be turned to advantage by the Allies through early and effective exploitation by armed forces and psychological warfare. Failing prompt and effective exploitation, the opportunity would be lost and subsequent Soviet

psychological reactions would adversely affect the accomplishment of Allied objectives.

EFFECTS ON THE SOVIET ARMED FORCES

15. The capability of Soviet armed forces to advance rapidly into selected areas of Western Europe, the Middle East and Far East would not be seriously impaired, but capabilities thereafter would progressively diminish due to the following factors:

a. The supply of petroleum products of all types will rapidly become critical to all branches of Soviet armed forces, resulting in:
   (1) Greatly reducing the mobility of the Army.
   (2) Reducing the scale of operations by the Soviet Navy and merchant shipping, although submarine warfare would probably be unaffected.
   (3) Seriously reducing air operations involving training, transport, support of ground and naval forces, and independent offensive action, although proper allocation of fuel would allow continued operations by air defense forces. *Note:* The point at which capabilities of Soviet armed forces would diminish to a critical degree would depend upon many variable or unpredictable factors, most important of which is the level of stockpiles prevailing at the initiation of hostilities.
b. After consumption of initial stocks of basic equipment and consumable supplies, progressive shortages of a wide variety of items, particularly aircraft, would handicap operations and affect morale of the armed forces. . . .

16. The Soviet High Command would be forced quickly to re-estimate their strategic position and make important decisions regarding operational plans under difficult circumstances. They would probably limit, postpone, or abandon certain campaigns, but it is impossible to predict what specific decisions would be made.

17. Atomic bombing would open the field and set the pattern for all adversaries to use any weapons of mass destruction and result in maximum retaliatory measures within Soviet capabilities.

GENERAL

18. Atomic bombing will produce certain psychological and retaliatory reactions detrimental to the achievement of Allied war objectives and its destructive effects will complicate post-hostilities problems. However, the atomic bomb would be a major element of Allied military strength in any war with the U.S.S.R., and would constitute the only means of rapidly inflicting shock and serious damage to vital elements of the Soviet war-making capacity. In particular, an early atomic offensive will facilitate greatly the application of other Allied military power with prospect of greatly

lowered casualties. Full exploitation of the advantages to be obtained is dependent upon the adequacy and promptness of associated military and psychological operations. From the standpoint of our national security, the advantages of its early use would be transcending. Every reasonable effort should be devoted to providing the means to be prepared for prompt and effective delivery of the maximum numbers of atomic bombs to appropriate target systems.

5. International Control
of Atomic Energy

George F. Kennan

The Policy Planning Staff has been asked to re-examine the present position of the United States with respect to the international control of atomic energy, and to assess the adequacy of this position in the light of present circumstances, particularly the demonstrated Soviet atomic capability. The following paper is intended to contribute to this re-examination. . . .

The real problem at issue, in determining what we should do at this juncture with respect to international control, is the problem of our attitude toward weapons of mass destruction in general, and the role which we allot to these weapons in our own military planning. Here, the crucial question is: Are we to rely upon weapons of mass destruction as an integral and vitally important component of our military strength, which we would expect to employ deliberately, immediately, and unhesitatingly in the event that we become involved in a military conflict with the Soviet Union? Or are we to retain such weapons in our national arsenal only as a deterrent to the use of similar weapons against ourselves or our allies and as a possible means of retaliation in case they are used? According to the way this question is answered, a whole series of decisions are influenced, of which the decision as to what to do about the international control of atomic energy and the prohibition of the weapon is only one.

We must note, by way of clarification of this question, that barring some system of international control and prohibition of atomic weapons, it is not questioned that *some* weapons of mass destruction must be retained in the national arsenal for purposes of deterrence and retaliation. The problem is: for what purpose, and against the background of what subjective attitude, are we to develop such weapons and to train our forces in their use?

We may regard them as something vital to our conduct of a future war—as something without which our war plans would be emasculated and ineffective—as something which we have resolved, in the face of all the moral and other factors concerned, to employ forthwith and unhesitatingly at the outset of any great military conflict. In this case, we should take the consequences of that decision now, and we should obviously keep away from any program of international dealings which would bring us closer to the possibility of agreement on international control and prohibition of the atomic weapon.

Or we may regard them as something superfluous to our basic military posture—as something which we are compelled to hold against the possibility that they might be used by our opponents. In this case, of course, we take care not to build up a reliance upon them in our military planning. Since they then represent only a burdensome expenditure of funds and effort, we hold only the minimum required for the deterrent-retaliatory purpose. And we are at liberty, if we so desire, to make it our objective to divest ourselves of this minimum at the earliest moment by achieving a scheme of international control.

We should remember that more depends on this basic decision than simply our stance toward the problems of international control. It must also have an important effect on our domestic atomic energy program, and particularly on what we do about the superbomb. If we decide to hold weapons of mass destruction only for deterrent-retaliatory purposes, then the limit on the number and power of the weapons we should hold is governed by our estimate as to what it would take to make attack on this country or its allies by weapons of mass destruction a risky, probably unprofitable, and therefore irrational undertaking for any adversary. In these circumstances, the problem of whether to develop the superbomb and other weapons of mass destruction becomes only a question of the extent to which they would be needed to achieve this purpose. It might be, for example, that the present and prospective stockpile of conventional bombs, combined with present and prospective possibilities for delivery, would be found adequate to this purpose and that anything further in the way of mass destruction weapons would be redundant, or would fall into an area of diminishing returns.

If, on the other hand, we are resolved to use weapons of mass destruction deliberately and prior to their use against us or our allies, in a future war, then our purpose is presumably to inflict maximum destruction on the forces, population and territory of the enemy, with the least expenditure of effort, in full acceptance of the attendant risk of retaliation against us, and in the face of all moral and political considerations. In this case, the only limitations on the number and power of mass destruction weapons which we would wish to develop would presumably be those of ordinary military economy, such as cost, efficiency, and ease of delivery. . . .

It flows from the above discussion that if, as I understand to be the case at the present moment, we are not prepared to reorient our military planning and to envisage the renunciation, either now or with time, of our reliance on "first use" of weapons of mass destruction in a future war, then we should not move closer than we are today to international control. To do so would be doubly invidious; for not only would we be moving toward a situation which we had already found unacceptable, but we would meanwhile be making that situation even more unacceptable by increasing our reliance on plans incompatible with it.

If our military plans are to remain unchanged in this respect, then it is probably best for us to rest on the present U.N. majority proposals, not pressing them with any particular vigor, but taking care not to undermine them by any statements which would suggest a lack of readiness on our part to accept them should they find acceptance in the Soviet camp. It is true that this position is somewhat disingenuous, since if the Russians should accept what we are ostensibly urging them to accept, we might be acutely embarrassed. But the danger of their accepting it is not serious. And in the present circumstances any new departure, involving even the suggestion of a withdrawal from the U.N. proposals or of a willingness to consider other ones, would result in much confusion, as between ourselves and our friends, which would be both difficult to dispel and unnecessary.

Unless, therefore, we are prepared to alter our military concepts as indicated above, thereby placing ourselves in a position where we could afford to take these weapons or leave them as the fortunes of international negotiation might determine, I urge that we consider the question of the desirability of some new international approach to have been studied and answered in the negative, and that we bury the subject of international control as best we can for the present.

The remaining discussion in this paper accordingly relates only to what we might do if we *had* reviewed our military concepts, if we *had* come to the conclusion that we would no longer rely on mass destruction weapons in our planning for a future war, and if we *had* resolved to work ourselves out of our present dependence on those weapons as rapidly as possible. . . .

It is entirely possible that war may be waged against us again, as it has been waged against us and other nations within our time, under these concepts and by these weapons. If so, we shall doubtless have to reply in kind, for that may be the price of survival. I still think it vital to our own understanding of what it is we are about that we not fall into the error of initiating, or planning to initiate, the employment of these weapons and concepts, thus hypnotizing ourselves into the belief that they may ultimately serve some positive national purpose. I doubt our ability to hold the respective weapons in our national arsenal, to fit them into our military and political plans, to agree with our allies on the circumstances of their use, and to entertain the prospect of their continued cultivation by our adversaries, without backsliding repeatedly into this dangerous, and possibly mortal, error. In other words, even if we were to conclude today that "first use" would not be advantageous, I would not trust the steadfastness of this outlook in a situation where the shadow of uncontrolled mass destruction weapons continues to lie across the peoples of the world. . . .

CHAPTER 2

The Years of Plenty: 1954–1973

Although determined to seize the psychological and foreign policy initiative, the Eisenhower administration disagreed with symmetry—that is, the development of large conventional and nuclear forces mirroring the Soviet military establishment. In its view, such an approach was based on a misunderstanding of the strategic strengths and weaknesses of friend and foe and promised to bankrupt the United States. The threat was long term, John Foster Dulles argued in a series of forceful presentations in spring 1954. The Soviet Union and China possessed significant advantages in geography and population that, put to use by their repressive regimes, allowed them to field large armies and to conduct a sustained campaign of subversion against the existing political order around the world. To meet a threat so "centralized and vast," the United States must work together with other nations. But this community of nations must not try to match the Communists "man for man and tank for tank," must never allow the aggressor to set the conditions of battle and to exhaust and bankrupt the United States and its allies.

The proper response, according to Dulles, was to capitalize on the special advantages of the United States, particularly its air and naval power and atomic weapons. These assets would be used not to wage war, although that might be necessary, but to deter aggression. If the Soviet Union knew in advance that its aggression would be answered with punishment far more costly than any benefits it could hope to achieve, it would not resort to war. "The main reliance," as Dulles said in "Foreign Policy for Security and Peace," "must be on the power of the free community to retaliate with great force by mobile means at places of its own choice." Dulles left no doubt that the heart of this deterrent strategy was the threat that the United States might attack the industrial centers of China and the Soviet Union. The extreme form of asymmetry Dulles described affected the strength and size of U.S. armed forces. The ground units of the army and the funds needed to sustain them were cut deeply, while the air force and—

to an extent—the navy were strengthened. They would conduct strategic nuclear war.

It was a bold plan and not an ineffective one. But its assets were wasting. By the late 1950s the public was learning of the growing nuclear power of the Soviet Union and beginning to hear of the dilemmas of nuclear stalemate, which had concerned specialists for several years. Henry Kissinger captured the public and governmental concern in his book *Nuclear Weapons and Foreign Policy.* In Kissinger's terms, the growth of Soviet nuclear capabilities meant that "the only outcome of an all-out war will be that *both* contenders must lose." In words reminiscent of Kennan's, although he would have shunned the association, Kissinger observed that all-out war no longer served any useful political purpose. But in that case, what was to be the relation between force and diplomacy? And how would the United States avoid a paralysis of will engendered by contemplation of the horrors of nuclear holocaust?

Kissinger answered that the United States must develop strategy, doctrine, and capability to wage limited war. Only in this way could it restore the disproportion between costs and gains necessary to deter aggression. However, the new U.S. strategy must apply in conflicts with the Soviet Union, another nuclear power. Kissinger doubted that a war between the United States and the Soviet Union would remain conventional. This should not detract from the willingness to engage in limited war. In Kissinger's view, a willingness to fight limited nuclear war was not only necessary to avoid political and military paralysis but was advantageous for the United States and its allies. Such a willingness emphasized their advantages in industrial potential and flexible decisionmaking and deprived the Soviet Union and China of their manpower advantage.

"Flexible response," the name assigned to the Kennedy administration's overall military strategy, enhanced the capacity of the United States for limited war. Unlike its predecessor, however, the Kennedy administration once again found itself drawn toward the pole of symmetry. The United States was to have much larger and more potent armed forces of all kinds and a significant capability to intervene simultaneously in several different conflicts, including counterguerrilla war.

In the realm of nuclear strategy the Kennedy administration initially sought to prolong U.S. superiority over the Soviet Union. The president and his advisers recognized that the Soviet Union could inflict major damage on the United States by nuclear attack, but they concluded that nuclear superiority had important foreign policy and military uses. President Kennedy's first budget message, delivered in March 1961, and Secretary of Defense Robert S. McNamara's early posture statements show this development clearly.

Two practical decisions were taken immediately: to accelerate the shift to land-based and submarine-based intercontinental nuclear missiles and to continue to maintain a large strategic bomber force. When the Kennedy administration took office the United States had deployed a force of approximately 160 land- and sea-based missiles and 600 long-range bombers. Within a year the number of missiles had grown to 340; within two years to 650. By 1967 the missile force stabilized at 1,710 launchers. The bomber force was to remain essentially unchanged. During the Kennedy and Johnson years the United States never possessed fewer than twice as many missiles as the Soviet Union. For most of the two administrations the overall margin was far greater than this, because the Soviets declined to deploy much of a bomber force at any stage. U.S. superiority was so great, and the Soviet force so rudimentary and vulnerable, that in its first years the Kennedy administration contemplated the use of part of the "excess" U.S. strategic forces in "damage limitation," to attack the nuclear forces of the Soviet Union. Secretary McNamara outlined this concept in a speech at the University of Michigan, "Defense Arrangements of the North Atlantic Community," in April 1962.

The Soviets responded to the U.S. buildup and to their embarrassment in Cuba with increases in Soviet missile strength and deployment of missiles in submarines and hardened silos. The Soviet buildup foreshadowed a time when, regardless of the resources the United States committed to strategic offensive missiles and damage limitation, it would be unable to reduce the capacity of the Soviet Union to damage civil society in the United States. "It will become increasingly difficult," Secretary McNamara told the Congress in his posture statement for 1963, "to destroy a sufficiently large proportion of the Soviets' strategic nuclear forces to preclude major damage to the United States, regardless of how large or what kind of strategic forces we build." At the same time, the president and his advisers sought a principle by which to limit the augmentation of U.S. strategic offensive forces. Otherwise, they feared, the missile and bomber forces would claim an ever-larger share of the defense budget without increasing security, while at the same time denying the country the symmetrical development of the armed forces they had decided to seek.

Both needs were met by the method chosen: the development of an assured second-strike force, a capability to devastate the Soviet Union and its allies, regardless of the weight of attack they brought against the United States. A second-strike posture negated any Soviet inclination to strike first. It also allowed the Johnson administration to stop deploying nuclear missile launchers in 1967 and, in effect, to invite the Soviet Union to follow suit (see Part Two). The new strategy, named "Assured Destruction," appeared to give an unchallengeable response to the age-old question of how much is enough. The answer: enough to inflict "unacceptable damage" on an attacker. What was unacceptable depended on tolerance of suffering, but it was presumably the loss of the enemy's major urban-industrial centers. The strategic nuclear forces necessary to destroy a large proportion of those centers, with appropriate additions for reliability and enemy counteraction, were an Assured Destruction capability.

In his final posture statement before leaving the Johnson administration in 1968, Secretary McNamara summarized the offense-dominated strategy he had helped to design:

Having wrestled with this problem for the last seven years, I am convinced that our forces must be sufficiently large to possess an "Assured Destruction" capability. By this I mean an ability to inflict at all times and under all foreseeable conditions an unacceptable degree of damage upon any single aggressor or combination of aggressors—even after absorbing a surprise attack. One can add many refinements to this basic concept, but the fundamental principle involved is simply this. It is the clear and present ability to destroy the attacker as a viable 20th century nation and an unwavering will to use these forces in retaliation to a nuclear attack upon ourselves or our allies that provides the deterrent, and not the ability partially to limit damage to ourselves.

The same considerations that led the Kennedy and Johnson administrations to adopt Assured Destruction caused them to increase their interest in both antiballistic missile (ABM) defense and in nuclear arms control. It should be remembered that the Johnson administration decided to build a light ABM system and to initiate sustained arms control negotiations with the Soviet Union. As Secretary McNamara observed, the factors that make it impossible for the United States to prevent the Soviet Union from destroying U.S. civil society also make it more desirable than ever before to curb the arms race and reduce the numbers of strategic nuclear offensive weapons.

In his remarks, "The Dynamics of Nuclear Strategy," McNamara explained that ABM was not inconsistent with maintenance of Assured Destruction capabilities, for the defensive missiles would protect land-based offensive missiles. Arguing indirectly against ABM, he pointed out that the United States in the early 1960s had built a greater force than it needed because of uncertainty about Soviet intentions. If the United States were to construct an ABM, he said, it could cause policymakers in the United States and elsewhere to exaggerate the security the system provided, and it could spark a new and explosive arms race with the Soviet Union in defensive and offensive countermeasures. McNamara's full statement, and his other presentations on ABM, were plainly intended to make the case against deployment, even if—bowing to fierce congressional pressure—his texts seemed to endorse the construction of a limited missile defense and the U.S. government's decision to deploy such a defense.

6. Foreign Policy for Security and Peace

John Foster Dulles

THE NATURE OF THE THREAT

The threat we face is not one that can be adequately dealt with on an emergency basis. It is a threat that may long persist. Our policies must be adapted to this basic fact.

The Soviet menace does not reflect the ambitions of a single ruler and cannot be measured by his life expectancy. There is no evidence that basic Soviet policies have been changed with the passing of Stalin. Indeed, the Berlin conference of last February gave positive evidence to the contrary. The Soviet Communists have always professed that they are planning for what they call "an entire historical era."

The assets behind this threat are vast. The Soviet bloc of Communist-controlled countries—a new form of imperialist colonialism—represents a vast central land mass with a population of 800 million. About 10 million men are regularly under arms, with many more trained millions in reserve. This land force occupies a central position which permits of striking at any one of about 20 countries along a perimeter of some 20 thousand miles. It is supplemented by increasing air power, equipped with atomic weapons, able to strike through northern Arctic routes which bring our industrial areas in range of quick attack.

The threat is not merely military. The Soviet rulers dispose throughout the world of the apparatus of international communism. It operates with trained agitators and a powerful propaganda organization. It exploits every area of discontent, whether it be political discontent against "colonialism" or social discontent against economic conditions. It seeks to harass the existing order and pave the way for political coups which will install Communist-controlled regimes.

By the use of many types of maneuvers and threats, military and political, the Soviet rulers seek gradually to divide and weaken the free nations and to make their policies appear as bankrupt by overextending them in efforts which, as Lenin put it, are "beyond their strength." Then, said Lenin, "our victory is assured." Then, said Stalin, will be the "moment for the decisive blow."

It is not easy to devise policies which will counter a danger so centralized and so vast, so varied and so sustained. It is no answer to substitute the glitter of steel for the torch of freedom.

An answer can be found by drawing on those basic concepts which have come to be regularly practiced within our civic communities. There we have almost wholly given up the idea of relying primarily on house-by-house defense. Instead, primarily reliance is placed upon the combining of two concepts, namely, the creation of power on a community basis and the use of that power so as to deter aggression by making it costly to an aggressor. The free nations must apply these same principles in the international sphere.

The cornerstone of security for the free nations must be a collective system of defense. They clearly cannot achieve security separately. No single nation can develop for itself defensive power of adequate scope and flexibility. In seeking to do so, each would become a garrison state and none would achieve security.

This is true of the United States. Without the cooperation of allies, we would not even be in a position to retaliate massively against the war industries of an attacking nation. That requires international facilities. Without them, our air striking power loses much of its deterrent power. With them, strategic air power becomes what Sir Winston Churchill called the "supreme deterrent." He credited to it the safety of Europe during recent years. But such power, while now a dominant factor, may not have the same significance forever. Furthermore, massive atomic and thermonuclear retaliation is not the kind of power which could most usefully be evoked under all circumstances.

Security for the free world depends, therefore, upon the development of collective security and community power rather than upon purely national potentials. Each nation which shares this security should contribute in accordance with its capabilities and facilities. . . .

STRATEGY TO DETER AGGRESSION

The question remains: How should collective defense be organized by the free world for maximum protection at minimum cost? The heart of the problem is how to deter attack. This, we believe, requires that a potential aggressor be left in no doubt that he would be certain to suffer damage outweighing any possible gains from aggression.

This result would not be assured, even by collective measures, if the free world sought to match the potential Communist forces, man for man and tank for tank, at every point where they might attack. The Soviet-Chinese bloc does not lack manpower and spends it as something that is cheap. If an aggressor knew he could always prescribe the battle conditions that suited him and engage us in struggles mainly involving manpower, aggression might be encouraged. He would be tempted to attack in places and by means where his manpower superiority was decisive and where at little cost he could impose upon us great burdens. If the free world adopted that strategy, it could bankrupt itself and not achieve security over a sustained period.

The free world must devise a better strategy for its defense, based on its own special assets. Its assets include, especially, air and naval power and atomic weapons which are now available in a wide range, suitable not only for strategic bombing but also for extensive tactical use. The free world must make imaginative use of the deterrent capabilities of these new weapons and mobilities and exploit the full potential of collective security. Properly used, they can produce defensive power able to retaliate at once and effectively against any aggression.

To deter aggression, it is important to have the flexibility and the facilities which make various responses available. In many cases, any open assault by Communist forces could only result in starting a general war. But the free world must have the means for responding effectively on a selective basis when it chooses. It must not put itself in the position where the only response open to it is general war. The essential thing is that a potential aggressor should know in advance that he can and will be made to suffer for his aggression more than he can possibly gain by it. This calls for a system in which local defensive strength is reinforced by more mobile deterrent power. The method of doing so will vary according to the character of the various areas.

Some areas are so vital that a special guard should and can be put around them. Western Europe is such an area. Its industrial plant represents so nearly the balance of industrial power in the world that an aggressor might feel that it was a good gamble to seize it, even at the risk of considerable hurt to himself. In this respect, Western Europe is exceptional. Fortunately, the West European countries have both a military tradition and a large military potential, so that through a European Defense Community, and with support by the United States and Britain, they can create an adequate defense of the Continent.

Most areas within the reach of an aggressor offer less value to him than the loss he would suffer from well-conceived retaliatory measures. Even in such areas, however, local defense will always be important. In every endangered area there should be a sufficient military establishment to maintain order against subversion and to resist other forms of indirect aggression and minor satellite aggressions. This serves the indispensable need to demonstrate a purpose to resist, and to compel any aggressor to expose his real intent by such serious fighting as will brand him before all the world and promptly bring collective measures into operation. Potential aggressors have little respect for peoples who have no will to fight for their own protection or to make the sacrifices needed to make that fighting significant. Also, they know that such peoples do not attract allies to fight for their cause. For all of these reasons, local defense is important. But in such areas the main reliance must be on the power of the free community to retaliate with great force by mobile means at places of its own choice.

A would-be aggressor will hesitate to commit aggression if he knows in advance that he thereby not only exposes those particular forces which he chooses to use for his aggression, but also deprives his other assets of "sanctuary" status. That does not mean turning every local war into a world war. It does not mean that, if there is a Communist attack somewhere in Asia, atom or hydrogen bombs will necessarily be dropped on the great industrial centers of China or Russia. It does mean that the free world must maintain the collective means and be willing to use them in the way which most effectively makes aggression too risky and expensive to be tempting.

It is sometimes said that this system is inadequate because it assures an invaded country only that it will eventually be liberated and the invader punished. That observation misses the point. The point is that a prospective attacker is not likely to invade if he believes the probable hurt will outbalance the probable gain. A system which compels potential aggressors to face up to that fact indispensably supplements a local defensive system.

7. Nuclear Weapons and Foreign Policy

Henry A. Kissinger

Whatever the calculation, . . . whether it be based on the feasibility of a surprise attack with present weapons and delivery systems or on the impact of imminent technological trends, it is difficult to see how either side can count on achieving its objectives through all-out war. It is possible to calculate relative advantages in base structure or delivery capabilities, but they do not add up to a margin which would leave either side sufficient resources to impose its will. The essence of the nuclear stalemate is that it keeps the two superpowers from launching an all-out war because each can force the other to pay an exorbitant price for victory.

The speed and power of modern weapons has thus brought about a paradoxical consequence: henceforth the only outcome of an all-out war will be that *both* contenders must lose. Under almost any foreseeable circumstances, an upper limit of destruction will be reached before attrition of industrial potential can make itself felt and, long before that point is reached, the forces-in-being on both sides will have inflicted losses completely disproportionate to any objective which is likely to be the original purpose of the war. Nuclear stalemate should, therefore, not be confused with nuclear parity. It comes about because, after a certain point, superiority in destructive power no longer pays strategic returns.

In such a situation deterrence becomes a complex problem. From now on the decision between peace and war, never an easy one, will be complicated by the consciousness that all-out war entails the risk of national catastrophe. Obviously no power will start a war it thinks it is going to lose. But it will also be reluctant to start a war if the price may be its national substance. All-out war is therefore likely to turn into a last resort: an act of desperation to be invoked only if national survival is unambiguously threatened. And what constitutes an unambiguous threat will be interpreted with increasing rigidity as the risks of all-out war become better understood.

The capability for waging all-out war thus operates as a protection against a sudden onslaught on the territorial United States. It also poses risks which may make the decision to initiate war for any lesser objective increasingly difficult. The nuclear stalemate may prevent all-out war. It will not deter other forms of conflict; in fact, it may even encourage them. The side which can present its challenges in less than an all-out form thereby gains a psychological advantage. It can shift to its opponent the agonizing choice of whether a challenge which explicitly stops short of all-out war should be dealt with by total retaliation.

If the decision to engage in all-out war is going to be difficult for the United States, it will be next to impossible for most of our allies. We have, for a little while at least, the protection of distance. While this cannot avert a heavy attack, it can at least provide a measure of warning and it permits some degree of defense. Most of Europe, however, is only some forty minutes' flying time from advanced Soviet bases, and the greatest Soviet air strength is in light and medium jet bombers. Moreover, while the size of the United States makes it at least conceivable that our society could withstand a thermonuclear onslaught, both history and geography combine to cast doubt on Europe's ability to do so. Thermonuclear devastation, coming after the ravages of two world wars, might lead to the collapse of the European social structure. And the density of Europe's population would involve fearful casualties. A minister who knows that a small number of megaton weapons properly distributed could kill or injure over half the population of his country will not resort to all-out war except as a very last resort—and perhaps not even then.

Finally, the nuclear stalemate affects not only relations among the nuclear powers but their relations to powers which do not have a nuclear establishment. At a time when the relations between the two great powers are increasingly shaped by the awareness that their thermonuclear capability enables them to avoid defeat but not to achieve a meaningful victory, the role of the smaller nations is being strangely enhanced by the increased strength of the superpowers. . . .

Although the margin of superiority of the industrialized over the under-developed nations has never been greater, it has also never been less effective. In every crisis from Korea to Suez, the nonnuclear powers have behaved as if nuclear technology did not exist. They could do so because they knew that two considerations would inhibit the major powers: the consciousness that the employment of nuclear weapons would bring about a feeling of moral revulsion in the rest of the world, and the fear that the employment of nuclear weapons anywhere might set off a chain of events ending in an all-out war. Thus, the minor powers are in a sense insulated from the nuclear age by the incommensurability between the power of nuclear weapons and the objectives for which they might be employed, as well as by the inhibitions which are generated by a major reliance on an all-out strategy.

All-out war has therefore ceased to be a meaningful instrument of policy. It cannot be used against the minor powers for fear of the reaction of world opinion and also because its intricate strategy is not appropriate to wars of limited objectives. And it cannot be used against a major power for anything except negative ends: to prevent the opponent's victory. Thus an all-out war which starts as an all-out war is the least likely contingency, although it is the only one for which we have an adequate doctrine. . . . The fact remains that under present or foreseeable conditions it is difficult to think of a national purpose that could be advanced by all-out war.

This is not to say that we can afford to be without a capability for fighting an all-out war, or that it will be easy to maintain the conditions which will make such a war seem unattractive to an opponent. Obviously, if we do not retain a well-protected capability for massive retaliation, the calculus of risks described in this chapter with respect to all-out war would shift. An opponent might then consider it worth the gamble to launch a surprise attack against us. It does mean, however, that, if we behave effectively, we can always make the risks of all-out war seem prohibitive to an adversary. The control over the conditions which will determine whether there is going to be an all-out war will depend to a large extent on us. Whether we keep up in the technological race, whether our retaliatory force is well dispersed, whether our air defense exacts the maximum attrition and our civil defense is capable of preventing panics—all these decisions are within our exclusive control. But although these conditions are within our control they will not be easy to achieve. At the current rate of technological change the side which has conceded the first blow to its enemy will always live on the verge of catastrophe, for an adverse technological breakthrough is always possible. Thus the stalemate for all-out war is inherently precarious. It will impel a continuous race between offense and defense, and it will require a tremendous effort on our part simply to stay even.

Because all-out war is so intricate and so sophisticated, and because its risks are so fearful, the necessary concern with it leads to a psychological distortion: it tends to transform the modes of a war which can only be a last resort into a doctrine for the only feasible strategy; it leads our military and political leaders to identify each technological advance with strategically significant progress. In every crisis, short of direct Soviet attack, it furnishes arguments for a policy of minimum risk, not only because of the dangers of all-out war but also to reserve our all-out capability either for a principal enemy or to meet a less ambiguous challenge. . . .

We thus return to the dilemma which has plagued all our postwar military thinking. Does the nuclear age permit the establishment of a relationship between force and diplomacy? Is it possible to imagine applications of power less catastrophic than all-out thermonuclear war?

. . . The key problem of present-day strategy is to devise a spectrum of capabilities with which to resist Soviet challenges. These capabilities should enable us to confront the opponent with contingencies from which he can extricate himself only by all-out war, while deterring him from this step by a superior retaliatory capacity. Since the most difficult decision for a statesman is whether to risk the national substance by unleashing an all-out war, the psychological advantage will always be on the side of the power which can shift to its opponent the decision to initiate all-out war. All Soviet moves in the postwar period have had this character. They have faced us with problems which by themselves did not seem worth an all-out war, but with which we could not deal by an alternative capability. We refused to defeat the Chinese in Korea because we were unwilling to risk an all-out conflict. We saw no military solution to the Indochinese crisis without accepting risks which we were reluctant to confront. We recoiled before the suggestion of intervening in Hungary lest it unleash a thermonuclear holocaust. A strategy of limited war might reverse or at least arrest this trend. Limited war is thus not an alternative to massive retaliation, but its complement. It is the capability for massive retaliation which provides the sanction against expanding the war.

The conduct of limited war has two prerequisites: a doctrine and a capability. So long as we consider limited war as an aberration from the "pure" case of all-out war we will not be ready to grasp its opportunities, and we will conduct the wars we do fight hesitantly and ambiguously, oscillating between the twin temptations to expand them (that is, to bring them closer to our notion of what war should be like), or to end them at the first enemy overture.

A doctrine for limited war will have to discard any illusions about what can be achieved by means of it. Limited war is not a cheaper substitute for massive retaliation. On the contrary, it must be based on the awareness that with the end of our atomic monopoly it is no longer possible to impose unconditional surrender at an acceptable cost.

The purpose of limited war is to inflict losses or to pose risks for the enemy out of proportion to the objectives under dispute. The more moderate the objective, the less violent the war is likely to be. This does not mean that military operations cannot go beyond the territory or the objective in dispute; indeed, one way of increasing the enemy's willingness to settle is to deprive him of something he can regain only by making peace. But the result of a limited war cannot depend on military considerations alone; it reflects an ability to harmonize political and military objectives. An attempt to reduce the enemy to impotence would remove the psychological balance which makes it profitable for both sides to keep the war limited. Faced with the ultimate threat of complete defeat, the losing side may seek to deprive its opponent of the margin to impose his will by unleashing a thermonuclear holocaust.

Nevertheless, a strategic doctrine which renounces the imposition of unconditional surrender should not be confused with the acceptance of a stalemate. The notion that there is no middle ground between unconditional surrender and the *status quo ante* is much too mechanical. To be sure, a restoration of the *status quo ante* is often the simplest solution, but it is not the only possible one.

. . . It simply does not follow that because one side stands to lose from a limited war, it could gain from an all-out war. On the contrary, both sides face the same dilemma: that the power of modern weapons has made all-out war useless as an instrument of policy, except for acts of desperation.

There exist three reasons, then, for developing a strategy of limited war. First, limited war represents the only means for preventing the Soviet bloc, at an acceptable cost, from overrunning the peripheral areas of Eurasia. Second, a wide range of military capabilities may spell the difference between defeat and victory even in an all-out war. Finally, intermediate applications of our power offer the best chance to bring about strategic changes favorable to our side. For while a balance can be maintained along existing lines on the Eurasian continent, it will always be tenuous. So long as Soviet armies are poised on the Elbe, Western Europe will be insecure. So long as Chinese might presses upon free Asia, the uncommitted powers will seek safety in neutralism. To the outside world the Soviet bloc presents a picture of ruthless strength allied with artful cunning, of constant readiness to utilize force coupled with the diplomatic skill to secure the fruits of its use. The United States, therefore, faces the task not only of stemming the Soviet pressures, but also of reducing the Soviet sphere and demonstrating the limitations of Soviet power and skills. The last is almost as important as the reduction of the Soviet sphere. For the resolution of the free world, now assailed by a sense of its impotence, will improve to the extent that it realizes that the Soviet bloc, behind its façade of monolithic power, also shrinks from certain consequences.

A strategy of limited war is more likely to achieve this objective than the threat of a total nuclear war. . . .

Is limited nuclear war an advantageous strategy for us? It is important to define what is meant by an advantageous strategy in the nuclear age. It emphatically does not mean that limited nuclear war should be our *only* strategy. We must maintain at all times an adequate retaliatory force and not shrink from using it if our survival is threatened. Even against less than all-out challenges, limited nuclear war may not always be the wisest course. In a police action against a nonnuclear minor power, in a civil war in which the population must be won over, the use of nuclear weapons may be unnecessary or unwise for either political or psychological reasons. As a general rule, in a limited war the smallest amount of force consistent with achieving the objective should be used. The problem of limited nuclear war arises primarily in actions against nuclear powers or against powers with vast resources of manpower which are difficult to overcome with conventional technology.

The decision of whether it is to our strategic advantage to fight nuclear or nonnuclear wars must be made well in advance of hostilities. We cannot utilize the threat of nuclear war for purposes of deterrence and at the same time keep open the option of waging a conventional war should deterrence fail; at least we cannot do so against a nuclear power. For an aggressor may not guess our intentions correctly. In the absence of an unmistakable indication on our part, he may assume the worst contingency and conduct nuclear operations.

In a war against a nuclear power, the decision to rely on conventional war will be ineffective by itself unless it is understood in advance by the opponent. . . .

We thus return to the basic problem of limited war in the nuclear age: where to strike the balance between the desire of posing the maximum threat and the need for a strategy which does not paralyze the will. The quest for the greatest physical threat may create a psychological vulnerability, for the risks may always seem greater than the goals to be achieved. An overemphasis on a strategy which maximizes the readiness to act may reduce the physical sanctions of deterrence below the level of safety. In this task of posing the maximum *credible* threat, limited nuclear war seems a more suitable deterrent than conventional war. From the point of view of deterrence, the availability of a wide spectrum of nuclear weapons increases the aggressor's risks. It puts him on notice that any additional increment of power which he commits to the war can be matched by a similar increase in the power of the other side. And, limited nuclear war greatly complicates the problem of controlling territory, which is one of the purposes of aggression.

As compared to conventional war, does limited nuclear war reduce the credibility of the threat? Will it lower our willingness to resist? To be sure, limited nuclear war poses greater risks for both sides than conventional war. But it is a mistake to assume that the risks of nuclear war can be avoided by a decision to resist aggression with conventional weapons. Against a nuclear power, conventional war carries with it almost the same risks as nuclear war, for the side which engages in a conventional war against a nuclear power, without being willing to accept the risks of nuclear war, is at a hopeless disadvantage. Such an effort to hedge the risks would enable the opponent to gain his ends either by the threat or by the reality of nuclear war. Against a nuclear power, the decision to fight a conventional war can be justified only on the grounds that it represents an advantageous strategy; the over-all risks are not substantially less.

Does a strategy of limited nuclear war increase or decrease the risks of all-out war? The previous discussion has shown that there is no inevitable progression from limited nuclear war to all-out thermonuclear conflict. It remains to demonstrate that under most circumstances limited nuclear war may actually be less likely than conventional war to produce an all-out showdown.

Whether a limited war, nuclear or otherwise, may remain limited will depend on the working out of a subtle equation between the willingness of the contenders to assume risks and their ability to increase their commitments. By definition, a limited war between major powers involves the *technical* possibility that either side will be able to raise its commitment. If both sides are willing to do so in preference to accepting a limited defeat, a limited victory or a stalemate, the result will be all-out war. If one side is willing to run greater risks, or, what amounts to the same thing, is less reluctant to engage in an all-out war, it will have a decisive advantage. If both sides are willing to run the same risks and are able to make the same commitments short of all-out war, the result will be a stalemate or a victory for the side which develops the superior strategy.

Assuming that our determination is equal to that of our opponent— and no strategy can be productive without this—the crucial question is whether, from the point of view of the desirability of avoiding all-out war, a conventional war will be "safer" than a nuclear war. Obviously, a nuclear war involves a larger initial commitment than a conventional war. Is it safer then to begin a limited war with an initial commitment so large that any addition involves the danger of merging into all-out war or is it wiser to begin it with a commitment which it is possible to raise at smaller risk?

Paradoxically, in a war which begins with a smaller investment it may prove much more difficult to establish an equilibrium. The consciousness that the opponent is able at any moment to increase his commitment will insert an element of instability into the psychological equation of limited war. The temptation to anticipate the other side may lead to an increasingly explosive situation and to a cycle of gradually expanding commitments. Moreover, if reliance is placed on conventional war, it follows almost inevitably that the nature of limited nuclear conflict will not be fully explored either in staff planning or in diplomacy. Because the two sides will be less clear about each other's contentions, the detonation of any nuclear device could then set off an all-out holocaust. The fact that there exists a clear cut-off point between conventional and nuclear war may turn into a double-edged sword: the existence of two families of weapons may serve to limit the war as long as the limitation holds. Once breached, however, it may set off a vicious spiral, difficult to control. At best it will force the contenders to confront the problem of limited nuclear war under the most difficult circumstances; at worst it might unleash a thermonuclear holocaust.

A war which began as a limited nuclear war would have the advantage that its limitations could have been established—and, what is more important, understood—well in advance of hostilities. In such a conflict, moreover, the options of the aggressor are reduced in range. Whereas in a conventional war the choice is between continuing the war with its existing restrictions or risking an expanded *limited* war, in a nuclear war the choice is the much more difficult one between the existing war and all-out conflict. To be sure, even in a nuclear war it is possible to step up the commitment by resorting to higher yield weapons. But, given the proper tactics, such a course may not drastically alter the outcome and, if carried beyond a certain point, it will unleash all-out war. As long as both sides are eager to avoid a final showdown, a nuclear war which breaks out after diplomacy has established a degree of understanding of the possibilities of the new technology would probably stand a better chance of remaining limited than would a conflict that began as a limited conventional war in an international environment which was unsure about the significance of nuclear weapons and which has come to identify any explosion of a nuclear device with total war.

The choice between conventional and nuclear war then becomes an essentially practical one: which side is likely to gain from adopting limited nuclear war? Here our superior industrial potential, the broader range of our technology and the adaptability of our social institutions should give us the advantage. When the destructiveness of individual weapons is too small, manpower can substitute for technology, as was the case with Communist China in Korea. If weapons are too destructive, the importance of industrial potential is reduced because a very few weapons suffice to establish an equilibrium. For a nation with a superior industrial potential and a broader base of technology, the strategically most productive form of war is to utilize weapons of an intermediary range of destructiveness, sufficiently complex to require a substantial productive effort, sufficiently destructive so that manpower cannot be substituted for technology, yet discriminating enough to permit the establishment of a significant margin of superiority. . . .

When manpower can no longer be substituted for materiel, the strategic significance of Communist China may be much reduced, and in certain circumstances it may even constitute a drain on the resources of the U.S.S.R.

Even should the Soviet Union overcome its difficulties in producing the required spectrum of weapons—and over a period of time it undoubtedly can do so—it will still be handicapped by the nature of its institutions and by its historical experience. For just as the growth of the Soviet Long-Range Air Force and nuclear stockpile should force us to reassess our traditional reliance on all-out war, the introduction of nuclear weapons on the battlefield will shake the very basis of Soviet tactical doctrine. No longer will the Soviet bloc be able to rely on massed manpower as in World War II and in Korea. In a limited nuclear war dispersal is the key to survival and mobility the prerequisite to success. Everything depends on leadership of a high order, personal initiative and mechanical aptitude, qualities more prevalent in our society than in the regimented system of the U.S.S.R. To be sure, the Soviet forces can train and equip units for nuclear war. But self-reliance, spontaneity and initiative cannot be acquired by training; they grow naturally out of social institutions or they do not come into being. And a society like that of the Soviet Union, in which everything is done according to plan and by government direction, will have extraordinary difficulty inculcating these qualities.

While it may be true, as many advocates of the conventional war thesis maintain, that nuclear weapons do not permit economies of manpower, the significant point is how that manpower is used and on what qualities it places a premium. In conventional war, manpower is required to establish an equilibrium on the battlefield; its training is in the handling of a few relatively simple weapons; it can substitute discipline for conception. To conduct a nuclear war, manpower must be trained in a wide spectrum of abilities; here rewards go to initiative and technical competence at all levels. Such a utilization of manpower would seem to take advantage of the special qualities of our society. It is not for nothing that Soviet propaganda has been insistent on two themes: there is "no such thing" as limited nuclear war, and "ban the bomb." Both themes, if accepted, deprive us of flexibility and undermine the basis of the most effective United States strategy.

It may be objected that if a strategy of limited nuclear war is to our advantage it must be to the Soviet disadvantage, and the Kremlin will therefore seek to escape it by resorting to all-out war. But the fact that the Soviet leadership may stand to lose from a limited nuclear war does not mean that it could profit from all-out war. On the contrary, if our retaliatory force is kept at a proper level and our diplomacy shows ways out of a military impasse short of unconditional surrender, we should always be able to make all-out war seem an unattractive course. . . .

The discussion in this chapter has led to these conclusions: War between nuclear powers has to be planned on the assumption that it is likely to be a nuclear war. Nuclear war should be fought as something less than an all-out war. Limited nuclear war represents our most effective strategy against nuclear powers or against a major power which is capable of substituting manpower for technology.

Such a strategy is not simple or easy to contemplate. It requires an ability to harmonize political, psychological and military factors and to do so rapidly enough so that the speed of war waged with modern weapons does not outstrip the ability of our diplomacy to integrate them into a framework of limited objectives. It presupposes a careful consideration of the objectives appropriate for a limited war and of the weapons systems which have a sufficient degree of discrimination so that limited war does not merge insensibly into an all-out holocaust. More than ever the test of strategy will be its ability to relate military capability to psychological readiness. It cannot strive to combine the advantages of every course of action, of deterrence based on a maximum threat and of a strategy of minimum risk. It must decide on the price to be paid for deterrence, but then make certain that this price will be paid if deterrence should fail. It is a paradoxical consequence of a period when technology has never been more complicated that its effectiveness depends to such an extent on intangibles: on the subtlety of the leadership and on its conception of alternatives.

Nor should a policy of limited nuclear war be conceived as a means to enable us to reduce our readiness for all-out war. None of the measures described in this chapter is possible without a substantial retaliatory force; it is the fear of thermonuclear devastation which sets the bounds of limited war. . . .

The American strategic problem can, therefore, be summed up in these propositions:

1. Thermonuclear war must be avoided, except as a last resort.

2. A power possessing thermonuclear weapons is not likely to accept unconditional surrender without employing them, and no nation is likely to risk thermonuclear destruction except to the extent that it believes its survival to be directly threatened.

3. It is the task of our diplomacy to make clear that we do not aim for unconditional surrender, to create a framework within which the question of national survival is not involved in every issue. But equally, we must leave no doubt about our determination to achieve intermediary objectives and to resist by force any Soviet military move.

4. Since diplomacy which is not related to a plausible employment of force is sterile, it must be the task of our military policy to develop a doctrine and a capability for the graduated employment of force.

5. Since a policy of limited war cannot be implemented except behind the shield of a capability for all-out war, we must retain a retaliatory force

sufficiently powerful and well protected so that by no calculation can an aggressor discern any benefit in resorting to all-out war. . . .

8. Recommendations Relating to Our Defense Budget

John F. Kennedy

. . . It would not be appropriate at this time or in this message to either boast of our strength or dwell upon our needs and dangers. It is sufficient to say that the budgetary recommendations which follow, together with other policy, organizational and related changes and studies now under way administratively, are designed to provide for an increased strength, flexibility, and control in our Defense Establishment in accordance with the above policies.

II. STRENGTHENING AND PROTECTING OUR STRATEGIC DETERRENT AND DEFENSES

A. Improving our missile deterrent: As a power which will never strike first, our hopes for anything close to an absolute deterrent must rest on weapons which come from hidden, moving, or invulnerable bases which will not be wiped out by a surprise attack. A retaliatory capacity based on adequate numbers of these weapons would deter any aggressor from launching or even threatening an attack—an attack he knew could not find or destroy enough of our force to prevent his own destruction.

1. Polaris: The ability of the nuclear-powered Polaris submarine to operate deep below the surface of the seas for long periods and to launch its ballistic, solid-fuel nuclear-armed missiles while submerged gives this weapons system a very high degree of mobility and concealment, making it virtually immune to ballistic missile attack.

In the light of the high degree of success attained to date in its development, production, and operation, I strongly recommend that the Polaris program be greatly expanded and accelerated. I have earlier directed the Department of Defense, as stated in my State of the Union message, to increase the fiscal year 1961 program from 5 submarine starts to 10, and to accelerate the delivery of these and other Polaris submarines still under construction. This action will provide five more operational submarines about 9 months earlier than previously planned.

For fiscal year 1962, I recommend the construction of 10 more Polaris submarines, making a total of 29, plus 1 additional tender. These 10 submarines, together with the 10 programed for fiscal year 1961, are scheduled to be delivered at the rate of 1 a month or 12 a year, beginning in June

1963, compared with the previous rate of 5 a year. Under this schedule, a force of 29 Polaris submarines can be completed and at sea 2 months before the present program called for 19 boats, and 2 years earlier than would be possible under the old 5-a-year rate. These 29 submarines, each with a full complement of missiles, will be a formidable deterrent force. The sooner they are on station, the safer we will be. And our emphasis upon a weapon distinguished primarily for its invulnerability is another demonstration of the fact that our posture as a nation is defensive and not aggressive.

I also recommend that the development of the long-range Polaris A-3 be accelerated in order to become available a year earlier, at an eventual savings in the procurement of the A-2 system.

This longer range missile with improved penetration capability will greatly enhance the operational flexibility of the Polaris force and reduce its exposure to shore-based antisubmarine warfare measures. Finally, we must increase the allowance of Polaris missiles for practice firing to provide systematic "proving ground" data for determining and improving operational reliability.

The increases in this program, including $15 million in new obligational authority for additional crews, constitute the bulk of the budget increases—$1.34 billion in new obligational authority on a full funded basis, over a 4-year period though only $270 million in expenditures in fiscal 1962. I consider this a wise investment in our future.

2. Minuteman: Another strategic missile system which will play a major role in our deterrent force, with a high degree of survivability under ballistic missile attack, is the solid-fuel Minuteman. This system is planned to be deployed in well-dispersed hardened sites and, eventually, in a mobile mode on railroad cars. On the basis of the success of tests conducted to date and the importance of this system to our overall strategy, I recommend the following steps:

(1) Certain design changes to improve the reliability, guidance accuracy, range, and reentry of this missile should be incorporated earlier than previously planned, by additional funding for research and development.

(2) A more generous allotment of missiles for practice firing should, as in the case of the Polaris, be provided to furnish more operational data sooner.

(3) The three mobile Minuteman squadrons funded in the January budget should be deferred for the time being and replaced by three more fixed-base squadrons (thus increasing the total number of missiles added by some two-thirds). Development work on the mobile version will continue.

(4) Minuteman capacity production should be doubled to enable us to move to still higher levels of strength more swiftly should future conditions warrant doubling our production. There are great uncertainties as to the future capabilities of others; as to the ultimate outcome of struggles now going in many of the world's trouble spots; and as to future technological breakthroughs either by us or any other nation. In view of these major uncertainties, it is essential that, here again, we adopt an insurance philosophy and hedge our risks by buying options on alternative courses of action.

We can reduce leadtime by providing, now, additional standby production capacity that may never need to be used, or used only in part, and by constructing additional bases which events may prove could safely have been postponed to the next fiscal year. But that option is well worth the added cost.

Together, these recommendations for Minuteman will require the addition of $96 million in new obligational authority to the January budget estimate.

3. Skybolt: Another type of missile less likely to be completely eliminated by enemy attack is the air-to-ground missile carried by a plane that can be off the ground before an attack commences. Skybolt is a long-range (1,000 miles), air-launched, solid-fuel, nuclear-warhead ballistic missile designed to be carried by the B-52 and the British V bombers. Its successful development and production may extend the useful life of our bombers into the missile age—and its range is far superior to the present Hound Dog missiles.

I recommend that an additional $50 million in new obligational authority be added to the 1962 budget to enable this program to go forward at an orderly rate.

B. Protecting our bomber deterrent: The considerably more rapid growth projected for our ballistic missile force does not eliminate the need for manned bombers—although no funds were included in the January budget for the further procurement of B-52 heavy bombers and B-58 medium bombers, and I do not propose any. Our existing bomber forces constitute our chief hope for deterring attack during this period prior to the completion of our missile expansion. However, only those planes that would not be destroyed on the ground in the event of a surprise attack striking their base can be considered sufficiently invulnerable to deter an aggressor.

I therefore recommend the following steps to protect our bomber deterrent:

1. Airborne alert capacity: That portion of our force which is constantly in the air is clearly the least vulnerable portion. I am asking for the funds to continue the present level of indoctrination training flights, and to complete the standby capacity and materials needed to place one-eighth of our entire heavy bomber force on airborne alert at any time. I also strongly urge the reenactment of section 512(b) of the Department of Defense Appropriation Act for 1961, which authorizes the Secretary of Defense, if the President determines it is necessary, to provide for the cost of a full airborne alert as a deficiency expense approved by the Congress.

2. Increased ground alert force and bomb alarms: Strategic bombers standing by on a ground alert of 15 minutes can also have a high degree of survivability provided adequate and timely warning is available. I therefore recommend that the proportion of our B-52 and B-47 forces on ground alert should be increased until about half of our total force is on alert. In addition, bomb alarm detectors and bomb alarm signals should be installed at key warning and communication points and all SAC bases, to make certain that a dependable notification of any surprise attack cannot be eliminated. Forty-five million dollars in new obligational authority will pay for all of these measures.

C. Improving our continental defense and warning systems: Because of the speed and destructiveness of the intercontinental ballistic missile and the secrecy with which it can be launched, timely warning of any potential attack is of crucial importance not only for preserving our population but also for preserving a sufficient portion of our military forces—thus deterring such an attack before it is launched. For any attacker knows that every additional minute gained means that a larger part of our retaliatory force can be launched before it can be destroyed on the ground. We must assure ourselves, therefore, that every feasible action is being taken to provide such warning. . . .

III. STRENGTHENING OUR ABILITY TO DETER OR CONFINE LIMITED WARS

The free world's security can be endangered not only by a nuclear attack, but also by being slowly nibbled away at the periphery, regardless of our strategic power, by forces of subversion, infiltration, intimidation, indirect or nonovert aggression, internal revolution, diplomatic blackmail, guerrilla warfare, or a series of limited wars.

In this area of local wars, we must inevitably count on the cooperative efforts of other peoples and nations who share our concern. Indeed, their interests are more often directly engaged in such conflicts. The self-reliant are also those whom it is easiest to help—and for these reasons we must continue and reshape the military assistance program which I have discussed earlier in my special message on foreign aid.

But to meet our own extensive commitments and needed improvements in conventional forces, I recommend the following:

A. Strengthened capacity to meet limited and guerrilla warfare—limited military adventures and threats to the security of the free world that are not large enough to justify the label of "limited war." We need a greater ability to deal with guerrilla forces, insurrections, and subversion. Much of our effort to create guerrilla and antiguerrilla capabilities has in the past been aimed at general war. We must be ready now to deal with any size of force, including small, externally supported bands of men; and we must help train local forces to be equally effective.

B. Expanded research on nonnuclear weapons: A few selected high priority areas—strategic systems, air defense and space—have received the overwhelming proportion of our defense research effort. Yet, technology promises great improvements in nonnuclear armaments as well; and it is important that we be in the forefront of these developments. What is needed are entirely new types of nonnuclear weapons and equipment— with increased firepower, mobility, and communications, and more suited to the kind of tasks our limited war forces will most likely be required to perform. I include here antisubmarine warfare as well as land and air operations. I recommend, therefore, an additional $122 million in new obligational authority to speed up current limited warfare research and development programs and to provide for the initiation of entirely new programs.

C. Increased flexibility of conventional forces: Our capacity to move forces in sizable numbers on short notice and to be able to support them in one or more crisis areas could avoid the need for a much larger commitment later. Following my earlier direction, the Secretary of Defense has taken steps both to accelerate and increase the production of airlift aircraft. A total of 129 new, longer range, modern airlift aircraft will be procured through fiscal year 1962, compared with the 50 previously programed. An additional $172 million in new obligational authority will be required in the 1962 budget to finance this expanded program.

These additional aircraft will help to meet our airlift requirements until the new specially designed, long-range, jet-powered C-141 transport becomes available. A contractor for this program has been selected and active development work will soon be started. Adequate funds are already included in the January budget to finance this program through the coming fiscal year.

I am also recommending in this message $40 million in new obligational authority for the construction of an additional amphibious transport of a new type, increasing both the speed and the capability of Marine Corps sealift capacity; and $84 million in new obligational authority for an increase in the Navy's ship rehabilitation and modernization program, making possible an increase in the number of ship overhauls (as well as a higher level of naval aircraft maintenance).

But additional transport is not enough for quick flexibility. I am recommending $230 million in new obligational authority for increased procurement of such items as helicopters, rifles, modern nonnuclear weapons, electronics and communications equipment, improved ammunition for artillery and infantry weapons, and torpedoes. Some important new advances in ammunition and bombs can make a sizable qualitative jump in our limited war capabilities.

D. Increased nonnuclear capacities of fighter aircraft: Manned aircraft will be needed even during the 1965–75 missile era for various limited war missions. Target recognition, destruction of all types of targets when extreme accuracy is required, and the control of airspace over enemy territory will all continue to be tasks best performed by manned aircraft. . . .

9. Defense Arrangements of the North Atlantic Community

Robert S. McNamara

NUCLEAR STRATEGY

A central military issue facing NATO today is the role of nuclear strategy. Four facts seem to us to dominate consideration of that role. All of them point in the direction of increased integration to achieve our common defense. First, the alliance has overall nuclear strength adequate to any challenge confronting it. Second, this strength not only minimizes the likelihood of major nuclear war but makes possible a strategy designed to preserve the fabric of our societies if war should occur. Third, damage to the civil societies of the alliance resulting from nuclear warfare could be very grave. Fourth, improved nonnuclear forces, well within alliance resources, could enhance deterrence of any aggressive moves short of direct, all-out attack on Western Europe.

Let us look at the situation today. First, given the current balance of nuclear power, which we confidently expect to maintain in the years ahead, a surprise nuclear attack is simply not a rational act for any enemy. Nor would it be rational for an enemy to take the initiative in the use of nuclear weapons as an outgrowth of a limited engagement in Europe or elsewhere. I think we are entitled to conclude that either of these actions has been made highly unlikely.

Second, and equally important, the mere fact that no nation could rationally take steps leading to a nuclear war does not guarantee that a nuclear war cannot take place. Not only do nations sometimes act in ways that are hard to explain on a rational basis, but even when acting in a "rational" way they sometimes, indeed disturbingly often, act on the basis of misunderstandings of the true facts of a situation. They misjudge the way others will react and the way others will interpret what they are doing.

We must hope—indeed I think we have good reason to hope—that all sides will understand this danger and will refrain from steps that even raise the possibility of such a mutually disastrous misunderstanding. We have taken unilateral steps to reduce the likelihood of such an occurrence. We look forward to the prospect that through arms control the actual use of these terrible weapons may be completely avoided. It is a problem not just for us in the West but for all nations that are involved in this struggle we call the cold war.

For our part we feel we and our NATO allies must frame our strategy with this terrible contingency, however remote, in mind. Simply ignoring the problem is not going to make it go away.

The United States has come to the conclusion that, to the extent feasible, basic military strategy in a possible general nuclear war should be approached in much the same way that more conventional military operations have been regarded in the past. That is to say, principal military objectives, in the event of a nuclear war stemming from a major attack on the alliance, should be the destruction of the enemy's military forces, not of his civilian population.

The very strength and nature of the alliance forces make it possible for us to retain, even in the face of a massive surprise attack, sufficient reserve striking power to destroy an enemy society if driven to it. In other words, we are giving a possible opponent the strongest imaginable incentive to refrain from striking our own cities.

The strength that makes these contributions to deterrence and to the hope of deterring attack upon civil societies even in wartime does not come cheap. We are confident that our current nuclear programs are adequate and will continue to be adequate for as far into the future as we can reasonably foresee. During the coming fiscal year the United States plans to spend close to $15 billion on its nuclear weapons to assure their adequacy. For what this money buys, there is no substitute.

In particular, relatively weak national nuclear forces with enemy cities as their targets are not likely to be sufficient to perform even the function of deterrence. If they are small, and perhaps vulnerable on the ground or in the air, or inaccurate, a major antagonist can take a variety of measures to counter them. Indeed, if a major antagonist came to believe there was a substantial likelihood of its being used independently, this force would be inviting a preemptive first strike against it. In the event of war, the use of such a force against the cities of a major nuclear power would be tantamount to suicide, whereas its employment against significant military targets would have a negligible effect on the outcome of the conflict. Meanwhile the creation of a single additional national nuclear force encourages the proliferation of nuclear power with all of its attendant dangers.

In short, then, limited nuclear capabilities, operating independently, are dangerous, expensive, prone to obsolescence, and lacking in credibility as a deterrent. Clearly, the United States nuclear contribution to the alliance is neither obsolete nor dispensable. . . .

10. The Dynamics of Nuclear Strategy

Robert S. McNamara

. . . No sane citizen, no sane political leader, no sane nation, wants thermonuclear war. But merely not wanting it is not enough. We must understand the difference between actions which increase its risk, those which reduce it, and those which, while costly, have little influence one way or another.

Now this whole subject matter tends to be psychologically unpleasant. But there is an even greater difficulty standing in the way of constructive and profitable debate over the issues. And that is that nuclear strategy is exceptionally complex in its technical aspects. Unless these complexities are well understood, rational discussion and decisionmaking are simply not possible. . . .

"ASSURED DESTRUCTION CAPABILITY"

One must begin with precise definitions. The cornerstone of our strategic policy continues to be to deter deliberate nuclear attack upon the United States, or its allies, by maintaining a highly reliable ability to inflict an unacceptable degree of damage upon any single aggressor, or combination of aggressors, at any time during the course of a strategic nuclear exchange—even after our absorbing a surprise first strike. This can be defined as our "assured destruction capability."

Now, it is imperative to understand that assured destruction is the very essence of the whole deterrence concept.

We must possess an actual assured destruction capability. And that actual assured destruction capability must also be credible. Conceivably, our assured destruction capability could be actual without being credible—in which case it might fail to deter an aggressor. The point is that a potential aggressor must himself believe that our assured destruction capability is in fact actual and that our will to use it in retaliation to an attack is in fact unwavering.

The conclusion, then, is clear: If the United States is to deter a nuclear attack on itself or on its allies, it must possess an actual and a credible assured destruction capability.

When calculating the force we require, we must be "conservative" in all our estimates of both a potential aggressor's capabilities and his intentions. Security depends upon taking a "worst plausible case"—and having the ability to cope with that eventuality.

In that eventuality we must be able to absorb the total weight of nuclear attack on our country—on our strike-back forces; on our command and control apparatus; on our industrial capacity; on our cities; and on our population—and still be fully capable of destroying the aggressor to the point that his society is simply no longer viable in any meaningful 20th-century sense.

That is what deterrence to nuclear aggression means. It means the certainty of suicide to the aggressor—not merely to his military forces but to his society as a whole.

"FIRST-STRIKE CAPABILITY"

Now let us consider another term: "first-strike capability." This, in itself, is an ambiguous term, since it could mean simply the ability of one nation to attack another nation with nuclear forces first. But as it is normally used, it connotes much more: the substantial elimination of the attacked nation's retaliatory second-strike forces. This is the sense in which "first-strike capability" should be understood.

Now, clearly, such a first-strike capability is an important strategic concept. The United States cannot—and will not—ever permit itself to get into the position in which another nation or combination of nations would possess such a first-strike capability, which could be effectively used against it.

To get into such a position vis-à-vis any other nation or nations would not only constitute an intolerable threat to our security, but it would obviously remove our ability to deter nuclear aggression—both against ourselves and against our allies.

Now, we are not in that position today—and there is no foreseeable danger of our ever getting into that position.

Our strategic offensive forces are immense: 1,000 Minuteman missile launchers, carefully protected below ground; 41 Polaris submarines, carrying 656 missile launchers—with the majority of these hidden beneath the seas at all times; and about 600 long-range bombers, approximately 40 percent of which are kept always in a high state of alert.

Our alert forces alone carry more than 2,200 weapons, averaging more than 1 megaton each. A mere 400 1-megaton weapons, if delivered on the Soviet Union, would be sufficient to destroy over one-third of her population and one-half of her industry.

And all of these flexible and highly reliable forces are equipped with devices that insure their penetration of Soviet defenses.

Now, what about the Soviet Union? Does it today possess a powerful nuclear arsenal?

The answer is that it does.

Does it possess a first-strike capability against the United States?

The answer is that it does not.

Can the Soviet Union, in the foreseeable future, acquire such a first-strike capability against the United States?

The answer is that it cannot. It cannot because we are determined to remain fully alert and we will never permit our own assured destruction capability to be at a point where a Soviet first-strike capability is even remotely feasible.

Is the Soviet Union seriously attempting to acquire a first-strike capability against the United States?

Although this is a question we cannot answer with absolute certainty, we believe the answer is "No." In any event, the question itself is, in a sense, irrelevant. It is irrelevant since the United States will so continue to maintain—and where necessary strengthen—our retaliatory forces that, whatever the Soviet Union's intentions or actions, we will continue to have an assured destruction capability vis-à-vis their society in which we are completely confident.

But there is another question that is most relevant. And that is: Do we—the United States—possess a first-strike capability against the Soviet Union?

The answer is that we do not.

And we do not, not because we have neglected our nuclear strength. On the contrary, we have increased it to the point that we possess a clear superiority over the Soviet Union. . . .

U.S. NUCLEAR SUPERIORITY

The more frequent question that arises in this connection is whether or not the United States possesses nuclear superiority over the Soviet Union.

The answer is that we do.

But the answer is—like everything else in this matter—technically complex. The complexity arises in part out of what measurement of superiority is most meaningful and realistic.

Many commentators on the matter tend to define nuclear superiority in terms of gross megatonnage or in terms of the number of missile launchers available.

Now, by both these two standards of measurement, the United States does have a substantial superiority over the Soviet Union in the weapons targeted against each other.

But it is precisely these two standards of measurement that are themselves misleading. For the most meaningful and realistic measurement of nuclear capability is neither gross megatonnage nor the number of available missile launchers, but rather the number of separate warheads that are capable of being *delivered* with accuracy on individual high-priority targets with sufficient power to destroy them.

Gross megatonnage in itself is an inadequate indicator of assured destruction capability, since it is unrelated to survivability, accuracy, or penetrability and poorly related to effective elimination of multiple high-priority targets. There is manifestly no advantage in overdestroying one target at the expense of leaving undamaged other targets of equal importance.

Further, the number of missile launchers available is also an inadequate indicator of assured destruction capability, since the fact is that many of our launchers will carry multiple warheads.

But by using the realistic measurement of the number of warheads available, capable of being reliably delivered with accuracy and effectiveness on the appropriate targets in the United States or Soviet Union, I can tell you that the United States currently possesses a superiority over the Soviet Union of at least three or four to one.

Furthermore, we will maintain a superiority—by these same realistic criteria—over the Soviet Union for as far ahead in the future as we can realistically plan.

I want, however, to make one point patently clear: Our current numerical superiority over the Soviet Union in reliable, accurate, and effective warheads is both greater than we had originally planned and in fact more than we require.

Moreover, in the larger equation of security, our "superiority" is of limited significance; since even with our current superiority, or indeed with any numerical superiority realistically attainable, the blunt, inescapable fact remains that the Soviet Union could still—with its present forces—effectively destroy the United States, even after absorbing the full weight of an American first strike. . . .

I have noted that our present superiority is greater than we had planned. Let me explain to you how this came about; for I think it is a significant illustration of the intrinsic dynamics of the nuclear arms race.

In 1961, when I became Secretary of Defense, the Soviet Union possessed a very small operational arsenal of intercontinental missiles. However, they did possess the technological and industrial capacity to enlarge that arsenal very substantially over the succeeding several years.

Now, we had no evidence that the Soviets did in fact plan to fully use that capability. But as I have pointed out, a strategic planner must be "conservative" in his calculations; that is, he must prepare for the worst plausible case and not be content to hope and prepare merely for the most probable.

Since we could not be certain of Soviet intentions—since we could not be sure that they would not undertake a massive buildup—we had to insure against such an eventuality by undertaking ourselves a major buildup of the Minuteman and Polaris forces.

Thus, in the course of hedging against what was then only a theoretically possible Soviet buildup, we took decisions which have resulted in our current superiority in numbers of warheads and deliverable megatons.

But the blunt fact remains that if we had had more accurate information about planned Soviet strategic forces we simply would not have needed to build as large a nuclear arsenal as we have today.

Now let me be absolutely clear. I am not saying that our decision in 1961 was unjustified. I am simply saying that it was necessitated by a lack of accurate information. Furthermore, that decision itself—as justified as it was—in the end could not possibly have left unaffected the Soviet Union's future nuclear plans.

What is essential to understand here is that the Soviet Union and the United States mutually influence one another's strategic plans. Whatever be their intentions, whatever be our intentions, actions—or even realistically potential actions—on either side relating to the buildup of nuclear forces, be they either offensive or defensive weapons, necessarily trigger reactions on the other side. It is precisely this action-reaction phenomenon that fuels an arms race. . . .

11. Hearings on Military Posture Before the U.S. Congress

Robert S. McNamara

. . . It seemed to us in 1961 that one of the first things we had to do was to separate the problem of strategic nuclear war from that of all other kinds of war. Although the matter had long been debated, the fact that strategic nuclear forces, no matter how versatile and powerful they may be, do not by themselves constitute a credible deterrent to all kinds of aggression had still to be squarely faced.

There was, of course, a deep and vivid awareness from the very beginning of the nuclear era that a war in which large numbers of atomic bombs were employed would be far different, not only in degree but in kind, from any ever fought before. In such a war the potential battlefield would be the entire homelands of the participants.

Furthermore, because of the enormous destructive power of nuclear weapons and the great speed and diverse ways in which they can be delivered, nothing short of a virtually perfect defensive system would provide anything approaching complete protection for populations and cities against a determined, all-out attack by a major nuclear power. This is not simply a matter of technology, it is inherent in the offensive-defensive problem. A nuclear-armed offensive weapon which has a 50/50 chance of destroying its target would be highly effective. But a defensive weapon with the same probability of destroying incoming nuclear warheads would be of little value.

This point was well understood by many who had closely studied the problem, even at the beginning of the nuclear era. In late 1945, for example, General Arnold noted that ". . . measures intended for protection against an atom bomb attack must be highly efficient from the very start of a war if they are to be any good at all. Our experience in this war has shown that it is most difficult to attain this goal." I might add, all of our experience since that time has conclusively demonstrated that a defense of such a high order of perfection is still technically unobtainable.

But the point to note here is that throughout the 1950s, and indeed since the end of World War II, it has always been our capacity to retaliate with massive nuclear power which was considered to be the deterrent against Soviet attack. It was this tendency to rely on nuclear weapons as the "universal deterrent" that helped contribute to the decline in our non-nuclear limited war forces, first during the late 1940s, and then during the second half of the 1950s. And yet by 1961, it was becoming clear that large scale use of nuclear weapons by the West as a response to Soviet aggression, other than an all-out attack, was not desirable. Therefore, other types of forces would have to be provided both to deter and, in the event deterrence failed, to cope with conflicts at the middle and lower end of the spectrum.

Thus, the time was ripe for a major reassessment of our military forces in relation to our national security policies and objectives.

With regard to our strategic nuclear war capabilities as such, our initial analysis impressed us with the need for prompt action in three related areas. First, while our strategic offensive forces were then fully adequate for their mission, it was apparent that our soft missiles and bombers would become exceedingly vulnerable to a nuclear surprise attack once our opponent had acquired a large number of operational ICBMs. Second, when that potential threat became a reality, reliable warning and timely response to warning of a missile attack would be of crucial importance to the survival of our bomber forces. Third, considerable improvements would have to be made in our command and communication systems if the strategic offensive forces were to be kept continuously under the control of the constituted authorities—before, during, and after a nuclear attack.

Essentially, there appeared to be two approaches available to us at the time: (1) we could provide offensive forces which could be launched within the expected period of tactical warning from the Ballistic Missile Early Warning System which was then still under construction, or (2) we could provide forces which would be able to survive a massive ICBM attack and then be launched in retaliation. As a long-term solution for the protection of our missiles, the first approach was rejected because of its great dependence on timely and unambiguous warning. While the timeliness of warning was reasonably assured, we could not be completely certain that the warning would be unambiguous. In the case of the manned bombers, this uncertainty presented serious, but not necessarily critical, problems. The bombers could be launched upon warning and ordered to proceed to their targets only after the evidence of an attack was unmistakable. But once launched, a ballistic missile could not be recalled. Yet, unless it is deployed in a mode which gives it a good chance of surviving an attack, it, too, would have to be launched before the enemy's missiles strike home, or risk destruction on the ground.

Obviously, it would be extremely dangerous for everyone involved if we were to rely on a deterrent missile force whose survival depended on a hair-trigger response to the first indications of an attack. Accordingly, we decided to accelerate the shift from the first generation ICBMs, the liquid fuel Atlas and Titan, to the second generation solid fuel missiles, Polaris and Minuteman, the former types being very costly and difficult to deploy in hardened underground sites and maintain on a suitable alert status. We knew that the Minuteman would not only be less expensive to produce and deploy in protected sites (and, thereby, provide more aim points per dollar expended), but would also be considerably easier and less costly to keep on alert. Because of its unique launching platform, the submarine-carried Polaris missile inherently promised a high likelihood of surviving a surprise attack, due to its mobility and concealment.

As these more survivable and effective Polaris and Minuteman missiles entered the operational forces in large numbers during FY 1964-65, the older Regulus, Atlas and Titan I types were phased out. And over the years as advancing technology produced new models of the Minuteman and Polaris ("models" which represented as great an advance over their predecessors as the B-52 over the B-47), these too have been promptly introduced. Concurrently with the deployment of the strategic missile force, we concluded an unprecedented testing program in order to assure ourselves that they could be relied upon to perform their mission. Finally, a very large missile penetration aids effort was undertaken to make certain that we could overcome any enemy defensive measures designed to stop our missiles. Yet, notwithstanding the retirement of all of the Atlas and Titan Is, the number of land-based ICBMs increased from 28 at end FY 1961 to 1,054 by end FY 1967. And, all of the planned 41 Polaris submarines have now become operational, most with advanced model Polaris missiles.

With regard to the manned bombers, it was clearly evident in 1961 that the number that could be maintained on alert status was far more important than the total in the inventory, which was then very sizable. Accordingly, until the Minuteman and Polaris forces could be deployed, we increased by 50 percent the proportion of the force being maintained on a 15-minute ground alert, the warning time we could expect from BMEWS.

The build-up of the strategic bomber force to 14 wings of B-52s and two wings of B-58s was completed in FY 1963. During this same period the medium bomber force of older B-47s was phased down, eventually being retired completely in 1966 on essentially the same schedule planned by the previous Administration. In addition, a large and very expensive B-52 modification program was placed under way in order to extend the useful life of the later models of these aircraft well into the 1970s and to enable them to employ low-altitude tactics in order to improve their penetration capabilities against enemy defenses.

As a result of these changes, and notwithstanding the retirement of the Atlas, Titan I and B-47s, the number of nuclear weapons in the alert force increased over threefold during the period. Now that the Minuteman and Polaris forces have been deployed, we can reduce somewhat the proportion of the bomber force on alert.

Not much could be done in 1961 to improve the continental air defense system which had been designed against bomber attack. However, recognizing the vulnerability of the SAGE ground control system sites to missile attack, we did start deployment of a backup system which has since been greatly expanded and made more effective. And because adequate warning of ballistic missile attack was so important to the survival and ultimate effectiveness of our strategic bomber force, we pressed forward the construction of BMEWS and somewhat later began the deployment of Over-the-Horizon radars. As the weight of the threat continued to shift from bombers to missiles, we began to modify the air defense system, phasing out those elements which became obsolete or excess to our needs.

We also considered in 1961 the advisability of deploying an active defense against ballistic missile attack. However, there were widespread doubts even then as to whether the Nike-Zeus system, which had been under development since 1956, should ever be deployed. Aside from outstanding questions as to its technical feasibility and our concern over operating problems which might be encountered, we were convinced that its effectiveness could be critically degraded by the use of more sophisticated warheads screened by multiple decoys or chaff. Weighing all the pros and cons, we concluded in 1962 that the best course was to shift the development of the system to a more advanced approach and to take no action to produce and deploy it at that time. We stepped up the pace and scope of our efforts to expand our knowledge of the entire problem of detecting, tracking, intercepting and destroying ballistic missiles. It was from these efforts that we have since drawn much of the technology incorporated in our present ballistic missile defense concepts.

Finally, we undertook an extensive program to improve and make more secure the command and control of our strategic offensive forces. Among the measures taken was the establishment of a number of alternate national command centers, including some which would be maintained continuously in the air so that the direction of all our forces would not have to depend upon the survival of a single center. Steps were also taken to enhance the survivability, reliability and effectiveness of the various command and communications systems, including, for example, provision for the airborne control of bomber, Minuteman and Polaris launchings. These were all forged into a new integrated National Military Command System. To guard against accidental or unauthorized firings, new procedures, equipment and command arrangements were introduced to ensure that all nuclear weapons could be released only on the positive command of the national authorities.

Many of the tasks we set out for ourselves seven years ago have been successfully accomplished. But, the situation which we foresaw then is now well upon us. The Soviets have, in fact, acquired a large force of ICBMs installed in hardened underground silos. To put it bluntly, neither the Soviet Union nor the United States can now attack the other, even by complete surprise, without suffering massive damage in retaliation. This is so because each side has achieved, and will most likely maintain over the foreseeable future, an actual and credible second strike capability against the other. It is precisely this mutual capability to destroy one another, and, conversely, our respective inability to prevent such destruction, that provides us both with the strongest possible motive to avoid a strategic nuclear war.

That we would eventually reach such a stage had been clearly foreseen for many years. Five years ago I pointed out to this Committee that: "We are approaching an era when it will become increasingly improbable that either side could destroy a sufficiently large portion of the other's strategic nuclear force, either by surprise or otherwise, to preclude a devastating retaliatory blow."

In January 1956, Secretary of Defense Wilson noted that, ". . . independent of what year it might happen, within a reasonable number of years we are almost bound to get into a condition sometimes described as 'atomic plenty' or a condition where the two parties could, as a practical matter, destroy each other." In the following month, Secretary of the Air Force Quarles was even more explicit. He said, "I believe it will mean that each side will possess an offensive capability that is so great and so devastating that neither side will have a knockout capability, and, therefore, a situation in which neither side could profitably initiate a war of this kind. . . . This has been frequently referred to as a position of mutual deterrence, and I believe we are moving into that kind of a situation."

Indeed, as far back as February 1955, a distinguished group of scientists and engineers, frequently referred to as the Killian Committee, had concluded on the basis of a comprehensive study of our continental air defense that within probably less than a decade a nuclear attack by either the United States or the Soviet Union would result in mutual destruction. "This is the period," the Committee's report stated, "when both the U.S. and Russia will be in a position from which neither country can derive a winning advantage, because each country will possess enough multimegaton weapons and adequate means of delivering them, either by conventional or more sophisticated methods, through the defenses then existing. The ability to achieve surprise will not affect the outcome because each country will have the residual offensive power to break through the defenses of the other country and destroy it regardless of whether the other country strikes first."

Clearly, nothing short of a massive pre-emptive first strike on the Soviet Union in the 1950s could have precluded the development of the situation in which we now find ourselves. This point, too, was noted by Secretary McElroy in 1958. Indeed, the hearings of the Congressional Committees concerned with national defense during that period are replete with references to this crucial issue.

Be that as it may, the problem now confronting the Nation is how best to ensure our safety and survival in the years ahead, in an era when both we and the Soviet Union will continue to have large and effective second strike strategic offensive forces and when the Red Chinese may also acquire a strategic nuclear capability.

I believe we can all agree that the cornerstone of our strategic policy must continue to be the deterrence of a deliberate nuclear attack against either the United States or its allies. But this immediately raises the question, what kind and level of forces do we need to ensure that we have such a deterrent, now and in the foreseeable future?

Having wrestled with this problem for the last seven years, I am convinced that our forces must be sufficiently large to possess an "Assured Destruction" capability. By this I mean an ability to inflict at all times and under all foreseeable conditions an unacceptable degree of damage upon any single aggressor or combination of aggressors—even after absorbing a surprise attack. One can add many refinements to this basic concept, but the fundamental principle involved is simply this: it is the clear and present ability to destroy the attacker as a viable 20th Century nation and an unwavering will to use these forces in retaliation to a nuclear attack upon ourselves or our allies that provides the deterrent, and not the ability partially to limit damage to ourselves.

This is not to say that defense measures designed to significantly limit damage to ourselves (which is the other major objective of our strategic forces) might not also contribute to the deterrent. Obviously, they might—if an increase in our "Damage Limiting" capability could actually undermine our opponent's confidence in his offensive capability. But for a "Damage Limiting" posture to contribute significantly to the deterrent in this way, it would have to be extremely effective, i.e., capable of reducing damage to truly nominal levels—and as I will explain later, we now have no way of accomplishing this.

As long as deterrence of a deliberate Soviet (or Red Chinese) nuclear attack upon the United States or its allies is the vital first objective of our strategic forces, the capability for "Assured Destruction" must receive the first call on all of our resources and must be provided regardless of the costs and the difficulties involved. That imperative, it seems to me, is well understood and accepted by all informed Americans. What is not so well understood, apparently, is the basis upon which our force requirements must logically be determined—in other words, how much "Assured Destruction" capability do we need and what is the proper way to measure that need?

The debate on how much is enough, I suspect, is as old as war itself, but it acquired a new and very special significance with the advent of the atomic bomb. As one observer, Bernard Brodie, noted in 1946, at the very beginning of the nuclear era:

> Superiority in numbers of bombs is not in itself a guarantee of strategic superiority in atomic bomb warfare . . . it appears that for any conflict a specific number of bombs will be useful to the side using it, and anything beyond that will be luxury. What that specific number would be for any given situation it is now wholly impossible to determine. But we can say that if 2,000 bombs in the hands of either party is enough to destroy entirely the economy of the other, the fact that one side has 6,000 and the other 2,000 will be of relatively small significance . . . the actual critical level could never be precisely determined in advance and all sorts of contingencies would have to be provided for. Moreover, nations will be eager to make whatever political capital (in the narrowest sense of the term) can be made out of superiority in numbers. But it nevertheless remains true that superiority in numbers of bombs does not endow its possessor with the kind of military security which formerly resulted from superiority in armies, navies, and air forces.

A decade later, in a speech appropriately entitled "How Much Is Enough," Secretary of the Air Force Quarles took up the same theme in a somewhat more elaborate and sophisticated manner. He presented the case as follows:

> The advent of atomic weapons in great numbers and variety, and now in megaton yields, has brought us to the point where the airpower we now hold poised is truly powerful beyond the imagination of man. But there comes a time in the course of increasing our airpower when we must make a determination of sufficiency. . . . Sufficiency of air power, to my mind, must be determined period by period on the basis of the force required to accomplish the mission assigned. Because technological changes are constantly occurring, which alter the power of any force to execute its mission . . . we must constantly review our mission requirements and tailor our concept of sufficiency to the current and foreseeable needs.
> . . . [T]he build-up of atomic power in the hands of the two opposed alliances of nations makes total war an unthinkable catastrophe for both sides. Neither side can hope by a mere margin of superiority in airplanes or other means of delivery of atomic weapons to escape the catastrophe of such a war. Beyond a certain point, this prospect is not the result of *relative* strength of the two opposed forces. It is the *absolute* power in the hands of each, and in the substantial invulnerability of this power to interdiction.
> Under such circumstances, each potential belligerent in total war could possess what might be called a "mission capability" relative to the other. So great is the destructive power of even a single weapon that these capabilities can exist even if there is a wide disparity between the offensive or defensive strengths of the opposing forces. . . . It is crucially important that we maintain the level of strength constituting a "mission capability." It is neither necessary nor desirable in my judgment to maintain strength above that level.

Although the technology of strategic nuclear war has undergone dramatic changes since 1956, the general principle laid down by Secretary Quarles is as valid today as it was then. The requirement for strategic forces must still be determined on the basis of the "mission capability" we are seeking to achieve. That, in turn, must be related to our overall policy objective, i.e., deterrence of a deliberate nuclear attack on ourselves or our allies. Thus, the first quantitative question which presents itself is: What kind and amount of destruction must we be able to inflict upon the attacker in retaliation to ensure that he would, indeed, be deterred from initiating such an attack?

As I have explained to the Committee in previous years, this question cannot be answered precisely. Some people have argued that the Soviet or Red Chinese tolerance of damage would be much higher than our own. Even if this were true (which is debatable), it would simply mean that we must maintain a greater "Assured Destruction" capability. For example, if we believe that a 10 percent fatality level would not deter them, then we must maintain a capability to inflict 20 or 30 percent, or whatever level is deemed necessary. In the case of the Soviet Union, I would judge that a capability on our part to destroy, say, one-fifth to one-fourth of her population and one-half of her industrial capacity would serve as an effective deterrent. Such a level of destruction would certainly represent intolerable punishment to any 20th Century industrial nation.[1]

The next question which has to be answered is: What kind and how large a force do we need to ensure at all times and under all foreseeable conditions that we can inflict the desired level of damage on the attacker? Obviously, the number of strategic missiles and aircraft we need cannot be determined solely on the basis of some fixed ratio to the number our opponents might have, or for that matter, to the number of nuclear warheads or the gross megatonnage those weapons could carry. Certainly, these are very important factors, each in its own right, and they must be and are taken into account in our calculations. But these are not the only or even most important factors. The requirement for "Assured Destruction" forces can be determined logically only on the basis of the size and character of the target system they may be called upon to destroy, taking account of all the other relevant factors involved. Among these are: the number of our weapons which at any given time are ready to be launched toward their targets; the number of these which could be expected to survive a Soviet surprise first attack; and the number of the "ready," "surviving" weapons which can reasonably be expected to reach the objective area, survive the enemy defenses and detonate over or on their intended targets.

Thus, a logical determination of strategic force requirements involves a rather complex set of calculations. You may recall that when I appeared here six years ago in support of our first Five Year Defense Program, I described the steps of this process in some detail. In view of the misunderstandings which have arisen over the issue, I believe it might be useful to restate them here.

The first step is to determine the number, types, and locations of the aiming points in the target system.

The second step is to determine the numbers and explosive yields of weapons which must be delivered on the aiming points to ensure the destruction or substantial destruction of the target system.

The third step involves a determination of the size and character of the forces best suited to deliver these weapons, taking into account such factors as: size of warhead, system reliability, delivery accuracy, ability to penetrate enemy defenses, and cost.

Since we must be prepared for a first strike by the enemy, allowances must also be made in our calculations for the losses which our own forces would suffer from the initial enemy attack. This, in turn, introduces additional factors:

1. The size, weight, and effectiveness of a possible enemy attack.
2. The degree of vulnerability of our own strategic weapon systems to such an attack.

Clearly, each of these factors involves various degrees of uncertainty. But these uncertainties are not unmanageable. By postulating various sets of assumptions, ranging from optimistic to pessimistic, it is possible to introduce into our calculations reasonable allowances for these uncertainties. For example, we can use in our analysis both the higher and lower limits of the range of estimates of the number of enemy ICBMs and long-range bombers. We can assign to these forces a range of capabilities as to warhead yield, accuracy, reliability, etc.

With respect to our own forces, we can establish, within reasonable limits, the degree of reliability, accuracy and vulnerability of each type of offensive weapon system and its ability to penetrate the enemy defenses under various modes of operation. The last factor also involves an estimate of the size and character of the enemy's defenses.

Obviously, a change in any major element of the problem necessitates changes in many other elements. For example, the Soviets' deployment of a very extensive air defense system during the 1950s forced us to make some very important changes in our strategic bomber forces. The B-52s had to be provided with penetration aids—i.e., standoff missiles, decoys, electronic countermeasure equipment, etc. In addition, the B-52's airframe had to be substantially strengthened to permit sustained low-altitude operations.

Now, in the late 1960s, because the Soviet Union *might* deploy extensive ABM defenses, we are making some very important changes in our strategic missile forces. Instead of a single large warhead, our missiles are now being designed to carry several small warheads and penetration aids, because it is the number of warheads, or objects which appear to be warheads to the defender's radars, that will determine the outcome in a contest with an ABM defense.

Gross megatonnage is not a reliable indicator of the destructive power of an offensive force. For example, one missile carrying ten 50-kiloton warheads (a total yield of ½-megaton) would be just as effective against a large city (2,000,000 people) as a single 10-megaton warhead with 20 times the total yield. Against smaller cities (100,000 people) ten 50-kiloton warheads would be 3½ times as effective as the single 10-megaton warhead, and against airfields 10 times as effective. Even against hard ICBM sites, the ten 50-kiloton warheads would (given the accuracy we anticipate) be more effective than a single 10-megaton warhead. And, of course, it would take 10 times as many ABM interceptors to defend a city against ten 50-kiloton warheads as it would against a single 10-megaton warhead.

It is clear, therefore, that gross megatonnage is an erroneous basis on which to compare the destruction capability of two forces. And as I pointed out to the Committee last year, the number of missiles on launchers alone is not a much better measure. Far more important is the surviving number of separately targetable, serviceable, accurate, reliable warheads. But the only true measure of relative effectiveness of two "Assured Destruction" forces is their ability to survive and to destroy the target systems they are designed to take under attack.

In terms of numbers of separately targetable, survivable, accurate, reliable warheads, our strategic forces are superior to those of the Soviet Union. But I must caution that in terms of national security, such "superiority" is of little significance. For even with that "superiority," or indeed with any "superiority" realistically attainable, the blunt, inescapable fact remains that the Soviet Union could still effectively destroy the United States, even after absorbing the full weight of an American first strike.

We should be under no illusion that "Damage Limiting" measures, regardless of how extensive they might be, could, by themselves, change that situation. This is so far the same reason that the deployment by the Soviets of a ballistic missile defense of their cities will not improve their situation. We have already taken the necessary steps to guarantee that our strategic offensive forces will be able to overcome such a defense. Should the Soviets persist in expanding what now appears to be a light and modest ABM deployment into a massive one, we will be forced to take additional steps. We have available the lead time and the technology to so increase both the quality and the quantity of our strategic offensive forces—with particular attention to more sophisticated penetration aids—so that this expensive "Damage Limiting" effort would give them no edge in the nuclear balance whatsoever. By the same token, however, we must realistically assume that the Soviet Union would take similar steps to offset any threat to their deterrent that might result from our deploying an ABM defense of our own cities.

Under these circumstances, surely it makes sense for us both to try to halt the momentum of the arms race which is causing vast expenditures on both sides and promises no increase in security. The logic of discussions to limit offensive and defensive strategic weapons is even more compelling than it was a year ago when the President proposed such discussions to the Soviet Union. We are continuing our attempt to persuade the Soviets to agree to our proposal for discussions.

It is important to distinguish between an ABM system designed to protect against a Soviet attack on our cities and an ABM system designed for other purposes. One such purpose would be to provide greater protection for our strategic offensive forces; another would be to protect our cities against an attack by Red China. The first is not a "Damage Limiting" measure, but rather an action designed to strengthen our "Assured Destruction" capability by ensuring the survival of a larger proportion of our retaliatory forces. The second is a "Damage Limiting" measure, but one against a small force—because of the size and character of the attacks involved, a good defense becomes feasible.

As I noted last year, Red China may achieve an initial ICBM operational capability in the early 1970s and a modest force in the mid-1970s. Depending upon the rate of growth thereafter, a thin ABM deployment, with some additions and improvements, could be highly effective through the mid-1980s. The ability of the thin ABM to limit damage to our Nation in the event our offensive force failed to deter an "irrational" aggressor was the basis for our decision to deploy such a force. . . .

NOTES

1. Red China represents a somewhat different problem. Today Red China is still far from being an industrial nation. What industry it has is heavily concentrated in a comparatively few cities. We estimate, for example, that a relatively small number of warheads detonated over 50 Chinese cities would destroy half of the urban population (more than 50 million people) and more than one-half of the industrial capacity. And, as I noted last year, such an attack would also destroy most of the key governmental, technical, and managerial personnel, as well as a large proportion of the skilled workers. Since Red China's capacity to attack the U.S. with nuclear weapons will be very limited at least through the 1970s, the ability of even so small a portion of our strategic forces to inflict such heavy damage upon them should serve as a major deterrent to a deliberate attack on us by that country.

CHAPTER 3

The Effects
of Vulnerability
and Parity

Contrary to the hopes of the Kennedy and Johnson administrations, the Soviets declined to follow the U.S. lead and to stabilize their missile forces at relatively low levels, which would have allowed for early and deep reductions and the mutual adoption of a nuclear strategy based on Assured Destruction. Starting several years later than the United States, the Soviet Union launched a massive buildup of its own. From a modest missile force of 50 intercontinental ballistic missiles (ICBM) and a few submarine-launched ballistic missiles (SLBM) in 1961, the Soviet Union deployed 320 ICBMs and SLBMs in 1964 and 930 in 1968. Soviet ICBMs and SLBMs reached a peak of 2,428 in 1978 before stabilizing at 2,426 in 1979 (see Table 1 in "The Fundamentals of Nuclear Strategy and Arms Control").

The pace and magnitude of the Soviet buildup troubled U.S. leaders for strategic and political reasons. Strategically, the problem was that the growing Soviet strength in missile launchers would sooner or later be teamed with an increasingly sophisticated technology to produce thousands of multiple independently targetable reentry vehicles (MIRV) with sufficient accuracy to threaten the survivability and therefore credibility of the U.S. land-based deterrent. Because of a nationwide antidefense psychology engendered by the Vietnam War, the United States cut arms spending for a decade. In order to avoid a first-strike posture, the United States had not developed a land-based force with the capabilities the Soviets had achieved. For these reasons, as the Soviets persisted in their deployments and modernization, it became clear that a strategic response would be required of the United States. Politically, the uncertainty and fear spread by the Soviet buildup and by bold Soviet initiatives in Africa, Vietnam, and Afghanistan undercut the efforts of the Nixon and Carter administrations to stabilize the arms competition and to sustain a movement toward lower levels.

Together, the adverse political and strategic developments made the arguments in favor of a major U.S. buildup more and more convincing in the bump and run of presidential and congressional elections in the United States. Ultimately, the doubters won, and despite initial steps by the Carter administration, the Reagan administration entered office pledged to dramatic increases in U.S. strategic nuclear forces.

The initial strategic response came as a change in targeting and control of nuclear weapons in the Nixon and Carter administrations. Beginning in 1960 the United States coordinated the strategic nuclear targeting plans of its military services in a Single Integrated Operational Plan (SIOP). Made public in speeches by two secretaries of defense—James Schlesinger in 1974 and Harold Brown in 1980—the new targeting policy called for planning to use nuclear weapons much more selectively than in the past. Schlesinger made clear that the strikes envisioned by existing war plans, although they were not aimed at civilian populations as such, would cause such immense collateral damage that the Soviet leaders might miss the distinction, if one were intended, between countercity and counterforce attack. Secretary Brown held that because the Soviets had developed a capability to destroy large numbers of U.S. land-based missiles in a first counterforce strike, the United States must acquire a comparable capability in order to maintain the credibility of its deterrent. Because both sides were acquiring large numbers of extremely accurate warheads, thereby limiting collateral damage to population, it followed that nuclear exchanges might occur over a relatively long period. It would therefore be necessary, if the conflict were to remain limited, to make sure that the United States could know what damage its forces had suffered, could launch some missiles and withhold others, and could assess damage to the Soviet Union. The military jargon for this is command, control, communication, and intelligence—C^3I. Among strategists these changes produced two reactions. The first was to seriously challenge the adequacy of Assured Destruction and its targeting of civil populations as a last resort. The second was to assail the notion that a limited nuclear war could be kept limited.

Those who challenged Assured Destruction began by talking to one another, but as the political opposition to nuclear orthodoxy increased in the 1970s, they soon came to speak for many Americans and, ultimately, for many in the Reagan administration. Their argument was this: Now that the Soviets had the capability to devastate the United States and had formulated a strategy to attack U.S. missiles while keeping collateral damage relatively low, the threat to retaliate against Soviet civilians was no longer adequate either to deter Soviet aggression or to defend the United States in case of war. Colin Gray became a prominent spokesman for this school of thought, as did Senator Henry Jackson and his assistant, Richard Perle, who was to serve as an assistant secretary of defense in the Reagan administration. Gray's "Nuclear Strategy: The Case for a Theory of Victory" builds on the reservations about Assured Destruction to suggest the need to formulate a war-fighting strategy that threatens the survival of the Soviet regime, something one could be certain that the Soviets valued. The pull toward symmetry with the Soviet Union could be found throughout the works of Gray and other critics of Assured Destruction.

In 1981 Desmond Ball captured the doubts about controlling nuclear war that plagued even the advocates of flexible nuclear targeting. Ball's provocative question "Can Nuclear War Be Controlled?" touched the heart of the issue. If the pressures for escalation were irresistible and the environment of nuclear war incompatible with the finely tuned responses apparently envisioned by Schlesinger and Brown, there were no grounds for departing from Assured Destruction or for adopting a new targeting scheme. One ought, instead, to get on with arms reductions.

The changes intended to add flexibility to U.S. nuclear targeting and to prepare for an extended period of nuclear war were endorsed by the Reagan administration as National Security Decision Directive 13 (NSDD-13). If the dissent by Ball and others failed to halt the move toward strategic symmetry, this was in good part because of a growing public awareness and expert concern with the nature of Soviet nuclear strategy and capabilities.

12. Flexible Strategic Options and Deterrence

James R. Schlesinger

. . . There is in prospect—or there has taken place, to be more precise— a change in the strategies of the United States with regard to the hypothetical employment of central strategic forces. A change in targeting strategy as it were.

This intersects with our position at SALT, but it is quite distinct from that position. In order to bring about alterations in our targeting strategies, we do not require to increase the numbers or the throw-weight of what we have in our strategic arsenal. The sizing of the strategic arsenal will depend upon SALT and will depend upon the position taken by the Soviets at SALT.

As I've indicated, we must be in a position [in] which the Soviets fully understand that if they are prepared—if they insist on racing—that we are prepared to match them, but that it is better by far, for both sides, to agree on essential equivalence. Consequently, the sizing of our strategic forces depends on SALT. The change in targeting doctrine is separable from that and does not impact necessarily the sizing of our strategic forces.

Q. Could you amplify on the changing in targeting strategy?

A. I think that this has been discussed over the years, that to a large extent the American doctrinal position has been wrapped around something called "assured destruction," which implies a tendency to target Soviet cities initially and massively and that this is the only option that the President of the United States or the national command authorities would have in the event of a possible recourse to strategic weapons. It is our intention that this not be the only option and possibly not the principal option open to the national command authorities.

Q. Can you just go a little further and tell me what you're talking about? What are you trying to do that is different? What is the change? What is the other option?

A. I'm not going to spell that out to you right now.

Q. Well could you put it in English then, so that a layman can understand what you're driving at?

A. The main point that should be understood is that both sides now have, and will continue to have, invulnerable second-strike forces, and that with those invulnerable second-strike forces it is inevitable, or virtually inevitable, that the employment by one side of its forces against the cities of the other side in an all-out strike will immediately bring a counter-strike against its own cities. Consequently, the range of circumstances in which an all-out strike against an opponent's cities can be contemplated has narrowed considerably and one wishes to have alternatives for employment of strategic forces other than what would be, for the party initiating, a suicidal strike against the cities of the other side.

Q. Mr. Secretary, a related question. Are you not saying, sir, that we would be shifting our targets to Soviet weapons rather than Soviet cities, and we will be perhaps improving the accuracy of our missiles so that we can knock out the Soviet missile sites if we have to, rather than going for the Soviet cities? Is that what you're saying?

A. No. I think parts of that are quite appropriate, but not entirely. As I indicated a moment ago, the forces on both sides are now so large and so secure that both sides have invulnerable second-strike capabilities, and therefore it is not possible to achieve a disarming first-strike which I think you were referring to.

Now, what I was referring to is a set of selective options against different sets of targets. We would not necessarily specify any particular set of targets, military targets, whether silos or other. Military targets are, of course, one of the possible target sets, but it is necessary to maintain a set of options which goes beyond the inherent attack—all-out attack—against enemy cities in the event of nuclear exchanges.

Q. Are you going to try to permit the improvement of nuclear accuracy?

A. Yes.

Q. When was that decision made? . . .

A. There is no single point in time at which such a decision can be described as having taken place. I have discussed these matters with the Joint Chiefs over a period of months and, consequently, at any point in time they were fully cognizant of my position. Now, if you are referring to when a piece of paper went forward, I believe that the pieces of paper went forward during the last summer. . . .

Q. How can we be assured—or how can the Russians be assured—that we are not seeking a first-strike capability when you enunciate this new targeting doctrine? How can we differentiate? Doesn't this have a momentum of its own?

A. The answer to that I think is quite clear. In the first place . . . the Soviets are allowed under the Interim Agreement 62 submarines and 950 missiles. They are submarine-launched ballistic missiles. In addition they have other forces. Any calculation would demonstrate, I believe, it is not possible for us even to begin to eliminate the city destruction capabilities embodied in their ICBM forces alone, so that . . . the attaining of a full disarming capability is not an option that is open either to the United States or to the Soviet Union.

Now, with the regard to the development of these kinds of capabilities, what we are saying at the present time is that the overall power represented by our arsenal must be a match for the overall power represented by the Soviet strategic arsenal.

The Soviets have had in recent years a highly vigorous research and development effort. They have four new ICBM in development, as you are aware. Three of those ICBM now have been tested with MIRV, and it is quite plausible to believe that the fourth will also be tested with MIRV. They have three submarine programs in operation—three ballistic submarine programs. Therefore, their program shows a great deal of vigor in the R & D stage. They have an immense amount of throw-weight presently coupled with technology inferior to that possessed by the United States.

Our concern is that, if they marry the technologies that are now emerging in their R & D program to the throw-weight and numbers that they have been allowed under the Interim Agreement, they would develop a capability that was preponderant relative to that of the United States, and this is impermissible from the standpoint of the American Government so long as we bear the obligation to carry the nuclear shield not only for ourselves but our allies.

If the Soviets were able to develop these improved technologies presently available to the United States in the form of guidance, MIRV, warhead technology, at some point around 1980 or beyond, they would be in a position in which they had a major counterforce option against the United States and we would lack a similar option.

Consequently, in the pursuit of symmetry, meaningful symmetry, for the two forces—and by meaningful symmetry, I do not mean symmetry in every respect—in the pursuit of meaningful symmetry we cannot allow the Soviets unilaterally to obtain a counterforce option which we ourselves lack. We must have a symmetrical balancing of the strategic forces on both sides.

I think that you can say that the kinds of forces that we might put in the R & D stage are anticipating the possibility of Soviet deployments of the type that I have described, and that if SALT eventuates in an agreement which at a lower level does not require the introduction of these improved categories of equipment that we would not feel obliged to introduce them. But we cannot be in a position—given the world-wide responsibilities of the United States and the need to maintain a world-wide balance—we cannot be in a position in which a major option is open to the Soviet Union which we through a self-denying ordinance have precluded for the

United States. . . . I think that one must also recognize, with regard to the strategic forces, that if both sides operate intelligently and perceptively the likelihood of the employment of strategic forces is very low—approaching zero—and that probability will remain very low provided that we have the symmetry that I mentioned on both sides. . . .

If one is thinking only in terms of the ability to bash the cities on the other side, we have more than enough forces. There are, however, target sets far more numerous than that, and far more difficult to destroy selectively, and what I have suggested is that if there are two classes of targets . . . to put ourselves in a position, as upholders of the nuclear umbrella for most of the free world, in which one's hypothetical opponent is in a position to attack all target classes and we are restricted to attacking only one limited destructive and suicidal target class, is not maintenance of essential equivalence. . . . The point that I'm making is that one does not necessarily have to go after a large set of targets; one must be in a position that one can respond in the event—in the hypothetical event of hypothetical aggression—with the strategic forces preferably in a way that limits damage to both sides—to all sides—rather than in a way that hopefully reduces the possibility of that outcome by flamboyant advertising of the destructiveness of such a war.

Q. Are you not talking about the possibility of tactical nuclear warfare with tactical nuclear weapons again?

A. The answer to that is the observations that I've made about the need for selectivity and constraint, designed to . . . minimize collateral damage and to avoid to as great an extent as possible bystander fatalities, applies as well to tactical nuclear warfare as to strategic warfare. It applies fully as well.

I was not, however, confining my observations to the tactical area. Yes, indeed, the implicit answer to your question was embodied in what I stated before, to wit, that you can have tactical applications of nuclear weapons which I believe will both enhance deterrence and substantially reduce the consequences for the nations involved and for mankind in the highly improbable event that strategic warfare would actually break out.

Q. Mr. Secretary, earlier you said that we have more than enough forces to bash the Soviet cities. . . . Does it not follow from that, therefore, that we already have a counterforce capability, and if so what is [it] that you are seeking that we don't already have?

A. No, that does not follow that we have a counterforce capability if one is able to destroy countervalue targets in that way. It depends upon the kinds of military targets that you may be referring to. It is evident, for example, that this large number of weapons in our stockpile provides us with the discriminating ability to go after certain classes of military targets—airfields and the like. We are also perfectly content to put a lid on the total quantity of nuclear forces at our present level, or below, should we be able to persuade the Soviet Union to agree.

Now what is it that we want? We have a sufficient amount of force now. We were not in difficulty in 1973 and, if there was a willingness on the other side . . . to put a lid on this level, that would be quite satisfactory to the United States—or to put a lid on at a lower level, that would be satisfactory to the United States. . . . What we must preserve is what I referred to before as "essential equivalence." We cannot be in a position of inferiority with regard to the strategic options that might be employed against the United States or its allies, and consequently we must be prepared for the possibility of an improvement and expansion of Soviet forces, for which there has been some recent intelligence evidence, and we must be prepared to match them, so that there is no asymmetrical advantage with regard to those options that the Soviets have. . . .

As I mentioned before, if the Soviets employed their present amount of throw-weight with the improved technologies, they could have on the order of 7,000 one-megaton weapons in their ICBM force as opposed to a much smaller number of much, much lower yield weapons in our forces. In the event of any conflict of wills in which the forces available to both sides must be assessed, the hypothetical leaders of the West under those circumstances would have to have much tougher wills than would their opponents given the asymmetry of forces.

Now, I think that is quite consistent with the objective of the SALT I agreements in Moscow in 1972, in which both sides fully endorsed the proposition that neither side should seek strategic advantage.

So what we are seeking, if I may put it this way, is to forestall the development of an asymmetrical situation that would be beneficial to the Soviet Union. If we can forestall it through SALT negotiations and avoid the major expenditures, that is desirable from the standpoint of the United States.

Q. You've introduced one new element here, as I understand it—that is the targeting. Is that to be taken, whatever it exactly means, as an interim device to prevent the point of asymmetry that you're talking about?

A. No. Earlier, I attempted to distinguish sharply between the *sizing of our forces*, which is dependent upon the progress of the SALT negotiations and [the fact] that we are satisfied with the present sizing and character of our forces, by and large, should the Soviets be willing to level off at the present balance, and the *change in targeting doctrine*, which is separable from that sizing problem or the problem of asymmetry that you refer to in your question.

The point that must be emphasized is that a strategy which . . . appears to involve inevitably a massive strike against the cities of one's opponents—where that opponent has a secure, impressive second-strike capability—involves inevitably the destruction of one's own cities. Therefore a threat is not likely to be employed.

We must be in a position in which the President of the United States, if he's called on to use strategic forces, has an option other than the option that I have referred to, which concentrates on cities and which therefore carries in its wake the notion of inevitable destruction of American cities.

That is not an option lightly implemented. In fact one can say that these kinds of strategies are strategies historically that have [been] embraced in two sets of circumstances. One set of circumstances is when a nation, such as the United States in the 1950s, or indeed during the early 1960s, was in a position of such preponderance in terms of its forces . . . that it could well threaten a potential opponent with that kind of all-out strike, and the relative damage in the event of such an exchange would be such that no potential foe would care to test those circumstances.

The other set of circumstances in which the threat of all-out retaliation against cities seems to be appealing—doctrinally appealing—is when a nation possesses a relatively small and weak nuclear capability, in which its deterrence depends upon the convincing of its potential foes that it has the will and the determination under any circumstances, no matter [if] involving its own destruction to wreak devastation on a potential opponent.

The United States is no longer in the first position in which it has preponderance of nuclear forces that prevailed into the later 1960s. It is certainly not in the position of the second category of power with a relatively weak nuclear force that [it] must threaten to employ immediately and entirely. It has a large and sophisticated establishment that permits the application of pressure while maintaining intra-war deterrence to protect its own cities and consequently, is in a position to maintain a credible threat of the application of force for the foreseeable future, because of the relative sophistication of its force in a period in which there is another power that has roughly equivalent forces.

Q. The change in our targeting doctrine is necessary, essentially, because the Russians have a large second-strike capability?

A. Yes, I think it is necessary now. I think that it might well have been desirable before. You will all recall that in the 1960s, in 1962 in fact, Secretary McNamara discussed this type of issue at Athens and later in his Ann Arbor address in which the emphasis was put on something called city avoidance.

Now this is quite parallel. That was in a period of time in which the United States did retain close to a disarming capability against the Soviet Union and recognized it. Therefore, I think, the subject lapsed. But it is now, I think, incumbent upon us to emphasize a variety of options in this area available to the United States. As you know, this subject has been discussed in the President's foreign policy reports since 1970. . . .

Q. Mr. Secretary, you talked about a major change that would appear in our nuclear strategy. Can you describe that in concise terms for us?

A. The change is one that has been discussed in various foreign policy reports of the President. What the intent is, is more effective deterrence of the possibility of strategic warfare, because it is known by all parties that the President of the United States has options other than the devastating option of going against the cities on the other side.

Q. Mr. Secretary, are you saying that the United States has actually retargeted some of its missiles, missiles which were once targeted for cities and are now targeted for other areas?

A. I will not go into that. What I will say is that we have targeting options which are more selective and which do not necessarily involve major mass destruction on the other side, and that the purpose of this, of course, is to maintain the capability to deter any desire on the part of an opponent to inflict major damage on the United States or its allies.

Q. You seem to be saying on the SALT talks that, if the Russians are willing to sign an agreement as of the 1973 situation, we'll go along with that, but if they want to get back into a missile race, we'll go back into a missile race.

A. That's right. We must maintain essential equivalence between the forces available to the Soviet Union and the forces available to the United States. There should be no question in the minds of the Soviets as we negotiate with them of our willingness to achieve that essential equivalence. We are prepared to maintain a rough balance that we have today, but the present balance is dependent upon the technological advantages that the United States asymmetrically possesses at the present time, as opposed to the quantitative advantages that the Soviet Union asymmetrically possesses.

Consequently, if the Soviets insist on introducing major technological improvements in their force, this must be compensated for, in our view, by a reduction in their present quantitative advantages.

Q. Why, after all these years, do we now have to develop the capability to fight limited wars? We have never found that necessary in the past. Why is that necessary now?

A. I'm not sure what you're referring to.

Q. Isn't that what you're really saying. That now we have to have the ability to fight a limited war, we have to have limited options rather than just massive retaliation or assured destruction or whatever you want to call it?

A. I think you're referring to selective strikes as limited wars or of wider variety of options. I think the reason for that is that circumstances have now changed considerably since the middle 1960s and that the acquisition by the Soviet Union of a secure second-strike capability means that the circumstances under which we can contemplate—or the Soviets can believe that we can contemplate—the use of our nuclear forces are narrower than they were in the past, and therefore we must seek additional ways of maintaining adequate deterrence, so that we avoid the possibility of nuclear war.

13. Nuclear War Strategy, PD-59

Harold Brown

ADMINISTRATION RESPONSES TO CONGRESSIONAL
QUESTIONS SUBMITTED BEFORE A HEARING

Question 1. What are the basic strategic targeting priorities in PD [Presidential Directive]-59? How do these differ from previous targeting guidance, particularly that contained in NSDM [National Security Decision Memorandum] 242?

Answer. PD-59 specifies the development of plans to attack a comprehensive Soviet/Warsaw Pact target system, with the flexibility to employ these plans, should deterrence fail, in a deliberate manner consistent with the needs of the situation and in a way which will deny an aggressor any gain, or would impose costs which clearly exceed his expected gains. This could entail initial retaliation on military and control targets while retaining the capability either to withhold for a relatively prolonged period, or to execute, broad retaliatory attacks on the political control system and on general industrial capacity. These individual target systems, which we feel the Soviet leaders value most, include leadership and control, military forces both nuclear and conventional and the industrial/economic base. Highlights of targeting aspects include an increased number of situation-oriented options, and more flexibility for selectively attacking all categories of targets.

PD-59 requires the option to attack a full range of industrial/economic targets be retained. PD-59 also places more emphasis on how to improve the effectiveness of targeting retaliation against Warsaw Pact leadership and control, nuclear forces, and conventional forces in a wartime situation. In contrast to some pronouncements by the press, the United States has never had a doctrine based simply and solely on reflexive, massive attacks on Soviet cities. Instead, we have always planned both more selectively (options limiting industrial/economic damage) and more comprehensively (a range of military targets in addition to the industrial/economic base). Previous Administrations, going back well into the 1960s, recognized the inadequacy of a strategic doctrine that would give us too narrow a range of options. The fundamental premises of our countervailing strategy are a natural evolution of the conceptual foundations built over the course of a generation. PD-59 is not a new strategic doctrine; it is not a radical departure from past U.S. strategic policy. Our countervailing strategy, as formally stated in PD-59, is in fact, a refinement, a codification of previous statements of our strategic policy. PD-59 takes the same essential strategic doctrine, and restates it more clearly, more cogently, in the light of current conditions and current capabilities.

Question 2. What are the fundamental political and military objectives for strategic targeting in PD-59? Is it envisaged that the United States could, under certain circumstances, conduct limited nuclear war for foreign policy, political or military objectives? Does the PD-59 envision the possibility of U.S. nuclear retaliation for any provocation short of a nuclear attack on the United States or its allies?

Answer. Deterrence remains, as it has been historically, our fundamental strategic objective. The overriding objective of our strategic forces is to deter nuclear war. But deterrence must restrain an adversary from carrying out any of a far wider range of threats than just that of massive attacks of U.S. cities. We seek to deter any adversary from any course of action that could lead to general nuclear war. Our strategic forces also must deter nuclear attacks on smaller sets of targets in the United States or on U.S. military forces overseas, and deter the nuclear coercion of, or attack on, our friends and allies. Our strategic forces, in conjunction with theater conventional and nuclear forces, must also contribute to deterrence of conventional aggression as well. I say "contribute" because we recognize that neither nuclear forces nor the cleverest theory for their employment can eliminate the need for us—and our allies—to provide a capable conventional deterrent.

In our analysis and planning, we are necessarily giving greater attention to how a nuclear war would actually be fought by both sides if deterrence fails. There is no contradiction between this focus on how a war would be fought and what its results would be, and our purpose of insuring continued peace through deterrence. Nor is there a contradiction between this focus and a judgment that escalation of a "limited" to an "all-out" nuclear war is likely. Indeed, this focus helps us achieve deterrence and peace, by insuring that our ability to retaliate is fully credible. We must have forces, contingency plans, and command and control capabilities that will convince the Soviet leadership that no war and no course of aggression by them that led to use of nuclear weapons—on any scale of attack and at any stage of conflict—could lead to victory, however they may define victory.

Operationally, our countervailing strategy requires that our plans and capabilities be structured to put more stress on being able to employ strategic nuclear forces selectively, as well as by all-out retaliation in response to massive attacks on the United States. It is our policy—and we have increasingly the means and the detailed plans to carry out this policy—to ensure that the Soviet leadership knows that if they chose some intermediate level of aggression, we could, by selective, large (but still less than maximum) nuclear attacks, exact an unacceptably high price in the things the Soviet leaders appear to value most—their military forces both nuclear and conventional, their political and military control apparatus, and the industrial capability to sustain a war. In our planning we have not ignored the problem of ending the war, nor would we ignore it in the event of a war. And, of course, we have, and we will keep, a survivable and enduring capability to attack the full range of targets, including the Soviet economic base, if that is the appropriate response to a Soviet strike.

The United States already retains the option of using weapons in a limited way in response to a conventional attack on us or our allies if necessary. However, PD-59 does *not* propose a first strike strategy. We are talking about what we could and (depending on the nature of a Soviet attack) would do in response to a Soviet attack. Nothing in the policy contemplates that nuclear war can be a deliberate instrument of achieving our national security goals because it cannot be. But we cannot afford the risk that the Soviet leadership might entertain the illusion that nuclear war could be an option—or its threat a means of coercion—for them.

Question 3. What alternative targeting strategies were examined in the studies which preceded PD-59? On what grounds were such alternatives rejected? Was the President presented with alternatives to the targeting policy set forth in PD-59?

Answer. Alternative targeting strategies were addressed. The alternative strategies examined were: (a) strengthen existing policy; (b) focus more heavily on denying Soviets a favorable war outcome; (c) add higher confidence capability against some target systems; and (d) rely more heavily on assured destruction.

Under alternative (a) the forces and related C³I to accomplish this strategy would be given added endurance.

Alternative (b) placed more emphasis on targeting of Soviet (and non-Soviet Warsaw Pact) nuclear and conventional forces to assure that they could not expect to achieve a favorable outcome or a victory, however victory might be defined, while retaining an assured destruction capability.

Alternative (c) would require greater capabilities against certain Soviet forces than in alternative (b).

The last alternative, (d), also would avoid the need to make any improvements to the flexibility and endurance of strategic forces and C³I.

Each of the alternatives was considered in light of: (a) what flexibility in our nuclear posture (i.e., how broad a range of options) is desired; (b) how much endurance do our forces and C³I require; (c) how much capability is considered necessary; (d) costs of achieving these capabilities.

These considerations were weighed against the ability of each of the alternatives to deter the Soviets, taking into account Soviet attitudes toward concepts of nuclear war and perceptions of our capabilities and will, as well as the perceptions of our friends and allies. In the final analysis, a policy was selected which was judged to be most realistic considering the current relationship between the U.S. and the U.S.S.R., and the world situation, and considering the continued aggressive pursuit by the Soviets of comprehensive improvement in all aspects of military force capabilities, both nuclear and conventional.

What we have done in the past three and a half years is to look more closely at our capabilities, our doctrine and our plans in the light of what we know about Soviet forces, doctrine, and plans. The Soviet leadership appears to contemplate at least the possibility of a relatively prolonged exchange if a war comes, and in some circles at least, they seem to take seriously the theoretical possibility of victory in such a war. We cannot afford to ignore these views—even if we think differently, as responsible U.S. officials do. We need to have, and we do have, a posture—both forces and doctrine—that makes it clear to the Soviets, and to the world, that any notion of victory in nuclear war is unrealistic.

Question 4. What is the Defense Department's assessment of Soviet strategic targeting doctrine and capabilities, as well as the willingness of Soviet political and military leadership to conduct limited nuclear war for political or military objectives? What is the basis for this assessment?

Answer. Our understanding of Soviet concepts of the role and possible results of nuclear war is uncertain. This is partly because our evidence is ambiguous and our analysis clouded by that ambiguity, and partly, no doubt, because different Soviet leaders address these inherently uncertain issues from different perspectives. We do study Soviet literature and information on force posture and force characteristics to develop our assessments.

Soviet leaders acknowledge that nuclear war would be destructive beyond even the Russian historical experience of the horrors of war. But at the same time some things the Soviet spokesmen say—as well as the things they do in their military preparations—suggest they take the prospect that a nuclear war might actually be fought more seriously than we do. In their discussion of that prospect, there are suggestions also that if nuclear war occurred, national survival and dominant military position at the end of the fighting would govern the subsequent evolution of history—and that these objectives are therefore properly the ones governing nuclear strategy, doctrine, and force structure.

While one can, of course, view with skepticism Soviet statements about doctrine, there can be no doubt that the Soviets continue to invest heavily in programs that are especially relevant to the conduct, and not just the deterrence of nuclear war: e.g., civil defense, air defense, large MIRVed ICBMs and broad qualitative force improvements.

The Soviet leaders make evident through their programs their concerns about failure of deterrence, as well as its maintenance, and their rejection of such concepts as minimum deterrence and assured destruction as all-purpose strategic theories. What troubles us is the heavy emphasis in Soviet military doctrine on the acquisition of war-winning capabilities, or at least the coincidence between their programs and what have been alleged as the requirements of a deliberate war-winning strategy. Even if one does not fully support the idea of the Soviets having embraced a war-winning strategy, it is clearly evident that their military force programs, particularly, nuclear force, far exceed, and are qualitatively different from, those that would be called for by reasonable requirements for deterrence or for an assured retaliatory capability.

The willingness of Soviet political and military leadership to conduct limited nuclear war can only be judged by their day-to-day actions and their statements about the relationship of nuclear weapons to war. Soviet actions are such that they have clearly shown a willingness to use power in whatever way it serves their interests—Afghanistan is only the latest in a long series of examples. This can only lead one to believe that at the very least they have addressed the problem. While it is uncertain what deters the Soviets, the Soviet leadership appear to contemplate at least the possibility of a prolonged exchange if war comes, and in some circles at least, they seem to take seriously the theoretical possibility of victory in such a war.

Question 5. What reaction does the Defense Department expect from the Soviet Union to the targeting policy established in PD-59 and to U.S. strategic force capabilities? What role does the Defense Department believe U.S. nuclear declaratory policy plays in Soviet assessments of U.S. objectives and capabilities? What steps can the Administration take to counter probable Soviet perception that the United States now believes it could use nuclear weapons to achieve U.S. foreign policy, political or military objectives?

Answer. The public reaction of the Soviets to the targeting policy established in PD-59 has been pretty much as expected. Statements have appeared in their media, often quoting ill-informed comments in the United States, and raising the issues of a U.S. "first-strike" philosophy, contending that the United States is seeking to establish nuclear superiority, that the United States is stimulating the arms race, and so forth. This Soviet reaction is not new. There were almost identical reactions to the NSDM-242 policy announcement in 1974. . . .

Question 6. U.S. intentions aside, a credible interpretation can be made that PD-59 and planned U.S. strategic capabilities represent a deliberate move toward a first strike capability against the Soviet Union. If such an interpretation can rationally be made, what are the implications for stability during a period of intense crisis or for Soviet incentive to launch a first strike during United States–Soviet conventional hostilities?

Answer. Nothing in our strategy or our weapons systems justifies such an interpretation. Implementing our strategy requires us to make some changes in our operational planning, such as gradually increasing the scope, variety, and flexibility of options open to us should the Soviets choose aggression. This is not a first strike strategy. Our emphasis on survivability of systems is directly contrary to needs of a first strike strategy. Bombers clearly are not first strike weapons, nor are cruise missiles because of their long flight time. Which is the first strike system, the SS-18, massive and lethal but vulnerable, or the mobile MX, much of whose cost will go to being able to survive a first strike by the other side?

Fixed land-based ICBMs are becoming more vulnerable on both sides. The vulnerability of U.S. ICBMs is increasing at a greater rate than that of the Soviets as a result of aggressive Soviet ICBM modernization programs. But despite United States and Soviet force improvements it is not possible for either the United States or the Soviets confidently to achieve a disarming first strike capability, unless one side or the other allowed the generality of its strategic forces—not only its ICBMs—to become vulnerable. We do not intend to let that happen to us. And there is no reason to expect the Soviets will either. Continued actions in support of the credible survivability of all three legs of the Triad should make the Soviets realize there is little or no incentive for launching a first strike against the United States. Our force modernization programs demonstrate that we have and are determined to retain the capability to absorb a Soviet first strike and still be able to devastate the Soviet Union in retaliation. . . .

Question 10. What effect will PD-59 have on selective employment and general strike planning of SACEUR? What role is envisaged for present or planned U.S. LRTNF [Long-Range Theater Nuclear Force] deployments in Europe in meeting the targeting requirements of PD-59? What role is envisaged for LRTNF deployments outside Europe (e.g., SLCM) in meeting these requirements?

Answer. PD-59 is about options, not about circumstances of use. It is not a prediction of how we would use the weapons, but rather a statement of policy that a full range of options must be available to fit the situation. NATO's strategy of flexible response requires that the alliance have employment options for all levels of the escalatory spectrum for response to aggression. The U.S. countervailing strategy will contribute significantly to the strategy by making available to NATO a larger number of nuclear weapon employment options in defense of the Alliance.

The increased flexibility we are introducing corresponds fully to the concept that the U.S. central strategic forces are available to insure deterrence of attacks on, and defense of, our allies and friends by making clear that we will have responsive, credible and flexible options for use of these forces. Thus, the effect of PD-59 on the selective employment and general strike planning of SACEUR will be that of complementing them because we will have the flexibility for employment of nuclear forces in support of NATO over a broad range of situations. This will enhance the Alliance's defense and make it possible to conduct military operations, should deterrence fail, in a manner favorable to NATO.

PD-59 policy will not affect established procedures for NATO consultations on decisions pertaining to nuclear weapons employment. The policy itself is fully consistent with NATO's flexible response doctrine and reflects our determination to be able to support that doctrine fully.

We will maintain nuclear forces plans to defend our vital national interests wherever they may be imperiled. We will continually seek to enhance deterrence by making it clear to the Soviets that, even short of total destruction from an all-out war, we could by more limited attacks exact an unacceptably high cost from those political, military and industrial/economic targets which Soviet leadership values most.

Question 11. What are the short- and longer-term budgetary implications of the PD-59 requirement for U.S. counterforce and limited nuclear war capability? In weapons delivery systems? Survivable and enduring C³? Improved reconnaissance systems? Civil defense? Protection of civilian and military leadership?

Answer. The increased flexibility for nuclear weapon employment over a range of situations and potentially over a protracted period requires increased endurance for our nuclear capable forces and for the Command, Control, Communications and Intelligence (C³I) systems which support these forces. The acquisition implications of the policy are primarily in the area of C³I needs. Our current forces and those contained in our overall force modernization program will provide the capabilities we need to implement the countervailing strategy. We will need to make some improvements in the overall survivability and endurance of C³I systems. Future force and C³I improvement programs will be evaluated for their consistency with the policy contained in PD-59 to ensure that all such programs fully support this national policy.

Civil defense policy and programs are governed by PD-41. PD-59, as such, will have no budgetary implications. An all-out Soviet attack on U.S. urban-industrial centers would cause many tens of millions of fatalities, even if the United States goes to a major civil defense program to reduce those numbers. The Soviets, who have a major program, can have little confidence of avoiding comparable losses. Evolving U.S. targeting options can further reduce their confidence in this regard. Policy for the protection of leadership and continuity of government is currently under review. Assuring the ability to execute the strategy with availability and continuity of command structure is an important part of a deterrent strategy for a nuclear war of whatever duration.

Question 12. Will PD-59 require that Trident II have counterforce and hard-target kill capabilities? Will it affect the number of MX missiles to be deployed? Will an increase in the number of U.S. strategic weapons be needed beyond projected levels to meet PD-59 targeting requirements? Will these targeting requirements accelerate the trend toward increased numbers of highly accurate warheads?

Answer. PD-59 will not require capabilities beyond that which is currently planned for the current force modernization programs. . . .

14. Can Nuclear War Be Controlled?

Desmond Ball

A strategic nuclear war between the United States and the Soviet Union would involve so many novel technical and emotional variables that predictions about its course—and especially about whether or not it could be controlled—must remain highly speculative.

To the extent that there is a typical lay image of a nuclear war, it is that any substantial use of nuclear weapons by either the United States or the Soviet Union against the other's forces or territory would inevitably and rapidly lead to all-out urban-industrial attacks and consequent mutual destruction. As Carl-Friedrich von Weiszaecker recently wrote, "as soon as we use nuclear weapons, there are no limits."[1]

Among strategic analysts, on the other hand, the ascendant view is that it is possible to conduct limited and quite protracted nuclear exchanges in such a way that escalation can be controlled and the war terminated at some less than all-out level. Some strategists actually visualize an escalation ladder, with a series of discrete and clearly identifiable steps of increasing levels of intensity of nuclear conflict, which the respective adversaries move up—and down—at will. Current U.S. strategic policy, although extensively and carefully qualified, is closer to this second position: it is hoped that escalation could be controlled and that more survivable command-and-control capabilities should ensure dominance in the escalation process. Indeed, reliance on the ability to control escalation is an essential element of U.S. efforts with respect to extended deterrence.

Escalation is neither autonomous and inevitable nor subject completely to the decisions of any one national command authority. Whether or not it can be controlled will depend very much on the circumstances at the time. The use of a few nuclear weapons for some clear demonstrative purposes, for example, could well not lead to further escalation. However, it is most unrealistic to expect that there would be a relatively smooth and controlled progression from limited and selective strikes, through major counterforce exchanges, to termination of the conflict at some level short of urban-industrial attacks. It is likely that beyond some relatively early stage in the conflict the strategic communications systems would suffer interference and disruption, the strikes would become ragged, unco-ordinated, less precise and less discriminating, and the ability to reach an agreed settlement between the adversaries would soon become extremely problematical.

There is of course no immutable point beyond which control is necessarily and irretrievably lost, but clearly the prospects of maintaining control depend to a very great extent on whether or not a decision is taken deliberately to attack strategic command-and-control capabilities.

Command-and-control systems are inherently relatively vulnerable, and concerted attacks on them would very rapidly destroy them, or at least render them inoperable. Despite the increased resources that the U.S. is currently devoting to improving the survivability and endurance of command-and-control systems, the extent of their relative vulnerability remains enormous. The Soviet Union would need to expend thousands of warheads in any comprehensive counterforce attacks against U.S. ICBM silos, bomber bases and FBM submarine facilities, and even then hundreds if not thousands of U.S. warheads would still survive. On the other hand, it would require only about 50–100 warheads to destroy the fixed facilities of the national command system or to effectively impair the communication links between the National Command Authorities and the strategic forces.

This figure would permit attacks on the National Military Command Center, the major underground command posts (including the Alternative

National Military Command Center and the NORAD and SAC Command Posts), the critical satellite ground terminals and early-warning radar facilities, the VLF communication stations, etc., as well as 10 or 20 high altitude detonations designed to disrupt HF communications and generate EMP over millions of square miles. Any airborne command posts and communication links that survived the initial attack could probably not endure for more than a few days. Soviet military doctrine suggests that any comprehensive counterforce attack *would* include strikes of this sort. U.S. strategic targeting plans involve a wide range of Soviet command-and-control facilities, and, while attacks on the Soviet national leadership would probably only be undertaken as part of an all-out exchange, it is likely that attempts would be made to destroy the command posts that control the strategic forces, or at least to sever the communication links between the Soviet NCA and those forces at a much earlier stage in the conflict.

In fact, control of a nuclear exchange would become very difficult to maintain after several tens of strategic nuclear weapons had been used, even where deliberate attacks on command-and-control capabilities were avoided. Many command and control facilities, such as early-warning radars, radio antennae and satellite ground terminals would be destroyed, or at least rendered inoperable, by nuclear detonations designed to destroy nearby military forces and installations, while the widespread disturbance of the ionosphere and equally widespread generation of EMP would disrupt HF communications and impair electronic and electrical systems at great distances from the actual explosions. Hence, as John Steinbruner has argued, "regardless of the flexibility embodied in individual force components, the precariousness of command channels probably means that nuclear war would be uncontrollable, as a practical matter, shortly after the first tens of weapons are launched."[2] Moreover, any attack involving 100 nuclear weapons that was of any military or strategic significance (as opposed to demonstration strikes at isolated sites in northern Siberia) would produce substantial civilian casualties. Even if cities were avoided, 100 nuclear detonations on key military or war-supporting facilities (such as oil refineries) would probably cause prompt fatalities in excess of a million people.

The notion of controlled nuclear war-fighting is essentially astrategic in that it tends to ignore a number of the realities that would necessarily attend any nuclear exchange. The more significant of these include the particular origins of the given conflict and the nature of its progress to the point where the strategic nuclear exchange is initiated; the disparate objectives for which a limited nuclear exchange would be fought; the nature of the decision-making processes within the adversary governments; the political pressures that would be generated by a nuclear exchange; and the problems of terminating the exchange at some less than all-out level. Some of these considerations are so fundamental and so intemperate in their implications as to suggest that there can really be no possibility of controlling a nuclear war.

The origins of a nuclear exchange are relevant because, for example, a strategic nuclear strike by the United States or the Soviet Union against targets in the other's heartland—no matter how limited, precise, or controlled it might be—is most unlikely to be the first move in any conflict between them. Rather, it is likely to follow a period of large-scale military action, probably involving substantial use of tactical nuclear weapons, in an area of vital interest to both adversaries, and during which the dynamics of the escalation process have already been set in motion. Some command-and-control facilities, communications systems and intelligence posts that would be required to control a strategic nuclear exchange would almost certainly be destroyed or damaged in the conventional or tactical nuclear phases of a conflict. And casualties on both sides are already likely to be very high before any strategic nuclear exchange. In the case of a tactical nuclear war in Europe possible fatalities range from 2 to 20 million, assuming extensive use of nuclear weapons with some restraints, up to 100 million if there are no restraints at all.[3] The capabilities of the Warsaw Pact forces (using large and relatively "dirty" warheads) and the Warsaw Pact targeting doctrine make it likely that the actual figure would lie at the higher end of this range.

A war involving such extensive use of nuclear weapons in Europe would almost inevitably involve attacks on targets within the Soviet Union. Indeed, it has long been U.S. policy to use nuclear weapons against the Soviet Union even if the Soviet Union has attacked neither U.S. forces nor U.S. territory. As Secretary Brown expressed it in January 1980, "We could not want the Soviets to make the mistaken judgment, based on their understanding of our targeting practices, that they would be spared retaliatory attacks on their territory as long as they did not employ strategic weapons or attack U.S. territory."[4] The U.S. would attempt to destroy the Soviet theatre nuclear forces, including the MRBMs, IRBMS and bombers based in the western USSR, the reserve forces, and POL and logistic support facilities. Soviet casualties from these attacks could amount to several tens of millions. The prospects for controlling any subsequent strategic exchange would not be auspicious.

In addition to these technical and strategic considerations, the decision-making structures and processes of large national security establishments are quite unsuited to the control of escalatory military operations. The control of escalation requires extreme decisional flexibility: decision-makers must be able to adapt rapidly to changing situations and assessments, and must have the freedom to reverse direction as the unfolding of events dictates; their decisions must be presented clearly and coherently, leaving no room for misinterpretation either by subordinates charged with implementation or by the adversary leadership.

These are not attitudes that are generally found in large national security establishments. In neither the United States nor the Soviet Union are these establishments unitary organizations in which decisions are made and executive commands given on the basis of some rational calculation of the national interest. They are made up of a wide range of civilian and military individuals and groups, each with their own interests, preferences, views and perspectives, and each with their own quasi-autonomous political power bases; the decisions which emerge are a product of bargaining, negotiation and compromise between these groups and individuals, rather than of any more rational processes. The heterogeneous nature of the decision-making process leads, in the first instance, to a multiplicity of motives and objectives, not all of which are entirely compatible, and resolving them generally involves the acceptance of compromise language acceptable to each of the contending participants. The clarity of reception among the adversary leadership is consequently generally poor, and the reactions invariably different from the responses initially sought.

The "fog of war" makes it extremely unlikely that the situation to which NCA believe themselves to be reacting will in fact correspond very closely to the true situation, or that there will be a high degree of shared perception between the respective adversary leaderships. In these circumstances it would be most difficult to terminate a nuclear exchange through mutual agreement between the adversaries at some point short of all-out urban-industrial attacks.

Of course, the pressures to which decision-makers are subject do not come only from within the national security establishment. In the event of a nuclear exchange, the national leadership would also be subject to the pressures of popular feelings and demands. The mood of horror, confusion and hatred that would develop among the population at large as bombs began falling on the Soviet Union and the United States and casualties rose through the millions would inevitably limit the national leaderships' freedom of maneuver. Whether the horror would force them to recoil from large-scale attacks on urban-industrial areas or the hatred would engender rapid escalation must remain an open question—but neither mood would be conducive to measured and considered actions.

The likelihood that effective control of a nuclear exchange would be lost at some relatively early point in a conflict calls into question the strategic utility of any preceding efforts to control the exchange. As Colin Gray has argued, it could be extremely dangerous for the United States "to plan a set of very selective targeting building blocks for prospective rounds one, two and three of strategic force application" while rounds four and five entailed massive urban-industrial strikes.[5] Implementation of such a plan, no matter how controlled the initial rounds, would amount "in practice, to suicide on the instalment plan."[6]

The allocation of further resources to improving the survivability and endurance of the strategic command-and-control capabilities cannot substantially alter this situation. Command-and-control systems are inherently more vulnerable than the strategic forces themselves, and, while basic retaliatory commands would always get to the forces eventually, the capability to exercise strict control and co-ordination would inevitably be lost relatively early in a nuclear exchange.

Furthermore, the technical and strategic uncertainties are such that, regardless of the care and tight control which they attempt to exercise, decision-makers could never be confident that escalation could be controlled. Uncertainties in weapons effects and the accuracy with which weapons can be delivered mean that collateral casualties can never be calculated precisely and that particular strikes could look much less discriminating to the recipient than to the attack planner. The uncertainties are especially great with respect to the operation of particular C^3 systems in a nuclear environment. The effects of EMP and transient radiation on electrical and electronic equipment have been simulated on many components but rarely on large systems (such as airborne command posts). Moreover much of the simulation of nuclear effects derives from extrapolation of data generated in the period before atmospheric nuclear tests were banned in 1963.

Given the impossibility of developing capabilities for controlling a nuclear exchange through to favourable termination, or of removing the residual uncertainties relating to controlling the large-scale use of nuclear weapons, *it is likely that decision-makers would be deterred from initiating nuclear strikes no matter how limited or selective the options available to them.* The use of nuclear weapons for controlled escalation is therefore no less difficult to envisage than the use of nuclear weapons for massive retaliation.

Of course, national security policies and postures are not designed solely for the prosecution of war. In both the United States and the Soviet Union, deterring war remains a primary national objective. It is an axiom in the strategic literature that the criteria for deterrence are different from those for war-fighting, and capabilities which would be deficient for one purpose could well be satisfactory for the other.[7] The large-scale investment of resources in command-and-control capabilities, together with high-level official declarations that the United States would be prepared to conduct limited, selective and tightly controlled strategic nuclear strikes (perhaps in support of extended deterrence), could therefore be valuable because they suggest U.S. determination to act in limited ways—the demonstrable problems of control notwithstanding. However, viable deterrent postures require both capabilities and credibility, and it would seem that neither can be assumed to the extent that would be necessary for the concept of controlled nuclear war-fighting to act as a deterrent. Rather than devoting further resources to pursuing the chimera of controlled nuclear war, relatively more attention might be accorded to another means of satisfying the objectives that limited nuclear options are intended to meet. This is likely, in practice, to mean greater attention to the conditions of conventional deterrence.

NOTES

1. Carl-Friedrich von Weiszaecker, "Can a Third World War Be Prevented?" *International Security* 5, no. 1 (Summer 1980), p. 205.

2. John Steinbruner, "National Security and the Concept of Strategic Stability," *Journal of Conflict Resolution* 22, no. 1 (September 1978), p. 421.

3. Alain C. Enthoven, "U.S. Forces in Europe: How Many? Doing What?" *Foreign Affairs* 53, no. 3 (April 1975), p. 514; Alain Enthoven and K. Wayne Smith,

How Much is Enough? Shaping the Defense Program, 1961–1969 (New York: Harper and Row, 1971), p. 128.

4. Harold Brown, *Department of Defense Annual Report Fiscal Year 1981* (29 January 1980), p. 92.

5. Colin S. Gray, "Targeting Problems for Central War," *Naval War College Review* 33, no. 1 (January-February 1980), p. 9.

6. Ibid., p. 7.

7. See Andre Beaufre, *Deterrence and Strategy* (London: Faber & Faber, 1965), p. 24; and Glenn H. Snyder, *Deterrence & Defense: Toward a Theory of National Security* (Princeton, N.J.: Princeton University Press, 1961), pp. 3–6.

15. Nuclear Strategy: The Case for a Theory of Victory

Colin S. Gray

. . . In 1964–65, the U.S. defense community substantially abandoned the concept of damage limitation. It was believed that strategic stability (the magic concept—far more often advanced and cited than defined), largely by virtue of a logic in technology (a truly American theme), could and should repose in what would amount to a strategic competitive stalemate. Each side could wreak unacceptable damage on the other's society, and neither could limit such prospective damage through counterforce operations or through active or passive defenses. Ballistic missile defense (BMD), in principle if not in contemporary technical realization, did of course pose a potentially fatal threat to this concept. A good part of the anti-ABM fervor of the late 1960s, which extended from Secretary of Defense Robert McNamara to local church and women's groups, can be traced to the strange belief that the goals of peace, security, arms control, stability and reduced resources devoted to defense preparation, could all flow from a context wherein societies were nearly totally vulnerable and strategic weapons were nearly totally invulnerable. It is important to note that U.S. operational planning never reflected any close approximation to the assured destruction concept, and that the legatees of MAD reasoning in the late 1970s have made some adjustments to the doctrine for its better fit with contemporary reality. . . .

To state the central concern of this article, U.S. official thinking and planning does not embrace the idea that it is necessary to try to effect the *defeat* of the Soviet Union. First and foremost, the Soviet leadership fears *defeat*, not the suffering of damage—and *defeat*, as is developed below, has to entail the forcible demise of the Soviet state.

. . . With a healthy strategic (im)balance in favor of the United States on the scale of, say, 1957 or 1962, one can see some logic to strategic flexibility reasoning. However, in the late 1970s and the 1980s, there are many reasons why a Soviet leadership might be less than fully impressed by constrained U.S. strategic execution, and might well respond with a constrained nuclear reply that would (and indeed should) most likely impose a noteworthy measure of escalation discipline upon the United States. Selective nuclear options, even if of a very heavily counter-military character, make sense, and would have full deterrent value, only if the Soviet Union discerned behind them an American ability and will to prosecute a war to the point of Soviet political defeat.

TARGETING THE RECOVERY ECONOMY

Of very recent times, much of the nuclear strategy debate has narrowed down to a dispute over the validity of the thesis that *the real* (and *ultimate*) *deterrent* to Soviet risk-taking/adventure is the threat that our strategic nuclear forces pose to the Soviet recovery economy. . . .

The counter-recovery theory was not a bad one, but in practice several difficulties soon emerged. First, and most prosaically, American understanding of the likely dynamics of the Soviet post-war economy was (and remains) far short of impressive. In the same way that arms controllers have been hindered in their endeavors to control the superpower strategic arms competition by their lack of understanding of how the competition "worked," so our strategic employment planning community has found itself in the position of being required to be able to do that which nobody apparently is competent to advise it how to do. To damage the Soviet recovery economy would be a fairly elementary task, but to damage it in a calculable (even a roughly calculable) way is a different matter. Furthermore, the discovery, year by year through the mid- to late 1970s, of more and more Soviet civil defense preparation, threw into increasing doubt the "damage" expectancy against a very wide range of Soviet economic targets.

Second, it appears that the counter-economic recovery theme is yet another attempt to evade the most important strategic question. Should war occur, would the United States actually be interested in setting the Soviet economy back to 1959, or even 1929? Such an imposed retardation might make sense if it were married to a scheme for ensuring that damage to the American economy were severely limited.

TARGETING THE SOVIET STATE

Nonetheless, the counter-recovery theme of the 1970s has prompted an interesting line of speculation. Namely perhaps the recovery that should be threatened is not economic in character but rather political. . . . Counter-economic targeting should have a place in intelligent U.S. war planning, but only to the extent to which such targeting would impair the functioning of the Soviet state.

The practical difficulties that would attend an endeavor to wage war against the Soviet state, as opposed to Soviet society, have to be judged to be formidable. However, one would at least have established an unambiguous and politically meaningful war aim (the dissolution of the Soviet political system) that could be related to a post-war world that would have some desirable features in Western (and Chinese) perspective. More to the point, perhaps, identification of the demise of the Soviet state as the maximum ambition for our military activity encourages us to attempt to seek out points of high leverage within that system. . . .

As the period of intense debate over SALT II begins, it is fair to note that the United States Government sees merit in strategic flexibility, in *some* counterforce, in *some* degree of direct protection for the American public (though not much), and in the ability, in the last resort, to blow down large sections of the urban Soviet Union. This may be sufficient for deterrence, but a defense community should be capable of providing strategic direction that has more political meaning.

At the present time the United States does not have a strategic posture capable of seeking a military outcome to a war in which Western political authorities could place any confidence. . . .

A non-defense professional might be somewhat puzzled by this discussion. As a general rule, he might observe U.S. war planning surely has *always* been oriented most heavily towards Soviet military targets (strategic forces, projection forces, command and control targets and war-supporting industry and transport networks)—so what is new? The answer lies in the scope of the military targeting, in its ability to cope with a much harder target set than before, and in its design for separation from civil society. . . .

A U.S. SIOP oriented towards different kinds of military targets should be guided by a political logic—what are our war aims? A rewriting and recomputing of the SIOP in an even more heavily counter-military direction than is the case at present could place the United States in a somewhat worse position than that occupied by the (major) allied politicians of World War I—there could be a determination to do *military* damage to the enemy (which is very sensible), but a lack of commitment to the idea of prosecuting the war to the point where the enemy is defeated militarily (unlike World War I). The question of just how the military damage to be wrought is to be translated into political advantage could easily be evaded. . . .

The United States and its allies probably should not aim at achieving the military defeat of the Soviet Union, considered as a unified whole; instead, it should seek to impose such military stalemate and defeat as is needed to persuade disaffected Warsaw Pact allies and ethnic minorities inside the Soviet Union that they can assert their own values in very active political ways. It is possible that a heavily counter-military focused SIOP might have the same insensitivity to Soviet domestic fragilities as may be found in the counter-economic recovery orientation of the 1970s. . . .

With a clear political war aim—to encourage the dissolution of the Soviet state—much of the military war might not need to be fought at all. The apparently resolute determination of the American defense community not to think through its deterrence needs, which would involve addressing the question of war aims, promises to produce yet another marginal improvement in doctrine (after all, U.S. strategic forces have always been targeted against Soviet military power—whatever annual Posture Statements may have said).[1] It may be worth reminding American policy makers in 1979 that the United States had a counter-force doctrine in Vietnam. A focus upon counter-military action, bereft of an overarching political intent, save of the vaguest kind, is unlikely to serve American interests well, except by unmerited luck.

The war-fighting theme which now has limited, though important, official support in Washington, comprises no more than half of the change in thinking that is needed. It is essential for pre-war deterrent effect that Soviet leaders not believe they could wage a successful short war. But, for reasons that none could predict in advance, war might occur regardless of the prewar theories and the postures of the two sides. In that event, it will be essential that the United States has a theory of war responsive to its political interests.[2] Because a counter-military focus in the SIOP is not informed by a clear goal of political victory against the Soviet state, the United States is unlikely to be able to wage an intercontinental nuclear war in a very intelligent fashion. In World War II, American wartime leaders declined to attempt to look beyond the battlefield, so long as the war was still in progress, with results of impressive negative educational value for succeeding generations. How much more intelligent it would be to have explicit war aims that should, in and of themselves, have considerable pre- and intra-war deterrent value.

. . . The U.S. defense community, substantially coerced in its thinking by the adverse trends in the major East-West military balances, has progressed from a counter-economic to a counter-military focus in its nuclear employment reasoning (although the mechanical details of war planning may well have focused more upon Soviet military assets than the U.S. defense community generally understood to be the case), but it has yet to accept a *strategic* focus and advance to a counter-political control thesis. Unlike Soviet defense analysts, Western commentators continue to be bemused by the reality-numbing concept of "war termination." Wars are indeed terminated, but they are also won or lost. Moreover, if the U.S. defense community envisages (as it must, realistically) the sacrifice (presumably unwilling) of tens of millions of Americans in a thermonuclear war, that sacrifice should be undertaken only in a *very* worthwhile cause. If there is no theory of political victory in the U.S. SIOP, then there can be little justification for nuclear planning at all.

The principal intellectual culprit in our pantheon of false strategic gods is the concept of stability. For more than fifteen years, influential members of the U.S. defense and arms control community have believed that it is useful, or even essential, that the Soviet Union have guaranteed unrestricted strategic access to American societal assets. Such unrestricted access *was* believed to have a number of stabilizing consequences. In and of itself it should limit arms competitive activity (such activity as remained would stem from "normal" modernization and from efforts to offset counterforce-relevant developments on the other side),[3] while—more basically—it should promote some relaxation of tension, in that the Soviet Union would, belatedly, be assured of its ability to deter (through punishment) the United States.[4] (This theory has some features in common with the view that the four-fold rise in oil prices in 1973-74 was "good for us"—compelling us to confront the implications of our own profligacy in the energy consumption field.)

Analysts of all (or perhaps most) doctrinal persuasions have come at last to accept the view that the Soviet Union does not relax as a consequence of its achieving a very high quality assured destruction capability: the excellent reason for such continued effort is that the assured destruction of American societal assets plays no known role in Soviet deterrent or wartime planning—save as a threat to deter American counter-economic strikes. In addition, Soviet planners probably see considerable political coercive value for a postwar world in a very large counter-societal threat. Backward though it has seemed to some, the Soviet Union has provided unmistakable evidence of believing that wars, even large nuclear wars, can be won or lost. The mass-murder of Americans makes a great deal of sense in terms of the authority structure of a post-war world (since the Soviet Union cannot consummate a victory properly through the physical occupation of North America), but such a grisly exercise has little or nothing to do with the prosecution of a war (save as a counterdeterrent threat).

American strategic (and arms control) policy, since the mid-1960s, has been misinformed by stability criteria which rested (and rest) upon a near-total misreading of Soviet phenomena. Soviet leaders are opportunists with a war-waging doctrine as their strategic *leitmotiv*. Supposedly sophisticated self-restraint in American arms competitive activity, designed so as not to stimulate "destabilizing" Soviet responses, has simply presented the Soviet Union with an upcoming period of strategic superiority of uncertain duration. The American stability theorem held only for so long as both sides endorsed it. . . .

There is a painful irony of several dimensions in this American intellectual failing. First, among the more pertinent asymmetries that separate the U.S. from the Soviet political system is the acute sensitivity of the former to the *personal* well-being of its human charges. It is little short of bizarre to discover that it is the Soviet Union, and not the United States, that has a serious civil defense program.[5] Second, potentially the strongest element in the overall Western stance vis-à-vis the Soviet Union is its industrial mobilization capacity. Reasonably good American BMD carries healthily terminal implications for Soviet opportunism or adventure. A BMD system that works well enables the United States to wage a long war and to mass produce the military means for eventual victory. So great is American mobilization potential, vis-à-vis the extant strategic posture, that U.S. defense policy, logically, should endorse a defensive emphasis. Such an emphasis is the guarantor of strategic forces in overwhelming numbers *tomorrow*.

Third, if U.S. strategic nuclear forces are to be politically relevant in future crises, the American homeland has to be physically defended. It is unreasonable to ask an American President to wage an acute crisis, or the early stages of a central war, while he is fearful of being responsible for the loss of more than 100 million Americans. If escalation discipline is to be imposed upon the Soviet Union, even in the direst of situations, potential damage to North America has to be limited. . . .

Fourth, even if the arms controllers' argument were correct, that a defensive emphasis would stimulate the Soviet Union into working harder so as to be able to overcome it through offensive force improvement,[6] so what? Generically, the claim that this or that American initiative will catalyze Soviet reactions tends to be accorded far too respectful a hearing. Certainly it is sensible to consider adversary reactions and to take a full systemic look at possibilities, but a country as wealthy (and as responsible for international security) as the United States should not be deterred by the mere prospect of competition from undertaking necessary programs. . . .

Crude though it may sound, the United States would probably achieve more in the field of arms control if it decided to achieve and sustain a politically useful measure of strategic superiority,[7] than if it continues its endorsement of the elusive quality known as essential equivalence.

SUPERIORITY FOR STABILITY

If it is true, or at least probable, that a central war could be won or lost, then it has to follow that the concept of strategic superiority should be revived in popularity in the West. Superiority has a variety of possible meanings, ranging from the ability to dissuade a putative adversary from offering resistance (i.e., deterring a crisis), through the imposition of severe escalation of discipline on opponents, to a context wherein one could prosecute actual armed conflict to a successful conclusion. There is certainly no consensus within the United States defense community today over the issue of whether or not any central war outcome is possible which would warrant description as victory. However, a consensus is emerging to the effect that the Soviet Union appears to believe in the possibility of victory, and that the time is long overdue for a basic overhaul of our intellectual capital in the nuclear deterrence field.

NOTES

1. See Lynn E. Davis, *Limited Nuclear Options: Deterrence and the New American Doctrine*, Adelphi Paper no. 121 (London: IISS, Winter 1975/6); and Desmond Ball, *Deja Vu: The Return to Counterforce in the Nixon Administration* (Los Angeles: California Seminar on Arms Control and Foreign Policy, December, 1974). The Davis characterization, with its focus upon limited nuclear options (LNOs), is very substantially misleading as to the basic thrust of NSDM 242.

In his valuable study of the counterforce debate of the early 1970s, Desmond Ball quotes an Air Force general as claiming (in February 1973) that the SIOP was "never reworked under (President) Johnson. It is still basically the same as 1962" (*Deja Vu: The Return to Counterforce in the Nixon Administration*).

2. Such a war might not be tripped by a military accident that related to political intention on neither side. It might pertain to matters of vital interest to both sides. In short, the U.S. defense community might discover that it did have political goals that far transcended Brodie's prediction that the earliest possible war termination would likely be the superordinate objective.

3. Classic "period-piece" statements of the arms-race stability thesis were George W. Rathjens, *The Future of the Strategic Arms Race: Options for the 1970s* (New York: Carnegie Endowment for International Peace, 1969); and "The Dynamics of the Arms Race," *Scientific American* 220, no. 4 (April 1969), pp. 15–25.

4. Strange to note, the theory of arms race dynamics that featured as its centerpiece the proposition that each side acts and reacts in a fairly mechanistic fashion in pursuit of a secure assured destruction capability has now been discredited pretty well definitively by the historical facts, but the strategic policy premises that flow from that flawed theory have not been over-hauled thoroughly. Since virtually all U.S. commentators agree that the Soviet Union is not attracted to MAD reasoning, the long familiar "instability" case against urban-area BMD and non-marginal civil defense provision is simply wrong. We are still in search of an adequate explanatory model for the strategic arms competition. See Colin S. Gray, *The Soviet-American Arms Race* (Lexington, Mass: Lexington, 1976).

5. To have a serious civil defense program does not mean that a country is preparing for war, any more than equipping a ship with lifeboats means that the shipping line is preparing to operate the ship in a dangerous manner.

6. An argument central to the case against urban area ABM defense was that its banning by treaty would break the action-reaction cycle of the arms race: The Soviet Union would not need to develop and deploy offensive forces to overcome such an American deployment (in order to preserve their assured destruction capability). It is a matter of history that the ABM treaty banned the ABM defense of U.S. cities, but Soviet offensive force improvements have marched steadily onward. The action-reaction thesis was logical and reasonable; it just happened to be wrong (it neglected the local color, the domestic engines of the arms competition). If U.S. MX MPS should induce the Soviet Union to proceed down a similar path, then stability (by anyone's definition) would be promoted. Rubles spent on MPS are rubles not spent on missiles and warheads. It is true that an MPS system in place might attract the Soviet Union to producing large numbers of missiles undetected, to be surge-deployed in a period of acute need. However, the Soviet Government can produce ICBMs secretly now—in the absence of an MPS system they could be fired from presurveyed "soft" sites. The verification argument against MX MPS is not a telling one but—as a hedge—deployment of a fairly thin (preferentially assigned) ballistic missile defense system around the MPS could purchase an extraordinary degree of leverage vis-à-vis any secretly (or suddenly) deployed Soviet missiles. See Colin S. Gray, *The MX ICBM: Multiple Protective Structure (MPS) Basing and Arms Control* HI-2977-P (Croton-on-Hudson, N.Y.: Hudson Institute, February 1979).

7. On the political meaning of strategic power, see Edward N. Luttwak, *Strategic Power: Military Capabilities and Political Utility*, the Washington Papers, vol. 4 (Beverly Hills, Calif.: Sage, 1976).

16. PD-59, NSDD-13, and the Reagan Strategic Modernization Program

Jeffrey Richelson

The Reagan Administration's Strategic Modernization Program calls for the expenditure of $222 billion dollars in the next five years on strategic weapons and command, control, and communications (C³) systems. This includes $78 billion for bombers and cruise missiles, $51 billion for sea-based strategic weapons, $42 billion for ICBMs, $29 billion for nuclear defense, and $22 billion for C³ systems.[1] Additionally, several billion dollars will be invested in the development of new satellite reconnaissance systems.[2]

One possible interpretation of this program is that it is an attempt to acquire the strategic capabilities necessary to implement fully present U.S. nuclear weapons *employment* policy.[3] Such an interpretation is encouraged by the titles of major defense planning documents such as the Defense Resource Board's "Planning Defense Resources to Match Strategy" and the Department of Defense's "Nuclear Weapons Employment and Acquisition Master Plan."[4]

If such an interpretation is valid then the employment policy which serves as the foundation for the Reagan program was not developed by (or under) the Reagan Administration. Rather, it was developed in the Carter Administration and formally codified as Presidential Directive-59 (PD-59), signed by President Carter on 25 July 1980. PD-59 directed an increased emphasis on the targeting of Soviet military and political assets—including hardened ICBM and leadership relocation sites. Additionally, and perhaps most significantly, PD-59 required the U.S. to develop the capability to fight a *prolonged* nuclear war—one lasting months, not just hours or days.[5] Thus, "endurance and flexibility" were added to "survivability and assured retaliation" as essential characteristics for U.S. strategic forces.

In public, Secretary of Defense Caspar Weinberger has echoed the remarks of Carter Administration Defense Secretary Harold Brown concerning U.S. nuclear weapons employment policy.[6] In secret, President Reagan has signed National Security Decision Directive-13 reaffirming PD-59.[7]

This paper focuses on the apparent linkage between U.S. nuclear weapons employment policy and the Reagan strategic modernization program. After examining the evolution of U.S. employment policy through 1976 I consider, in turn, PD-59 and NSDD-13, the strategic capabilities required for their implementation and the manner in which the Reagan program is to provide such capabilities. I then consider the other possible factors in the development of the Reagan program as well as noting the questions that have been raised concerning the feasibility and wisdom of present nuclear weapons employment policy.

U.S. STRATEGIC NUCLEAR WEAPONS EMPLOYMENT POLICY: 1946–1976

It has often been asserted that until PD-59 U.S. war plans, as detailed in the SIOP (Single Integrated Operational Plan), dictated large-scale attacks on Soviet urban-industrial areas in the event of war.[8] In fact for many years prior to the signing of PD-59 such attacks were considered to be an action of last resort, to be undertaken only in response to Soviet attacks on U.S. cities.

The earliest U.S. war plans did call mainly, though by no means exclusively, for attacks on Soviet urban-industrial areas. Thus, the first U.S. war plan, PINCHER (1946), called for an attack on 20 Soviet cities with 50 A-bombs.[9] A subsequent plan, FLEETWOOD (1948), called for attacks on 70 Soviet cities with 133 bombs over a period of 30 days. Moscow was targeted with eight bombs and Leningrad with seven.[10]

This early emphasis on urban-industrial targets was the product of several factors. One factor was the belief that attacking a nation's civilian population could produce a sufficient loss of morale to bring a quick victory. At the same time there were significant technical and economic limitations which seemed to dictate use of the bomb against urban areas—the number of bombs available, their yield and limited accuracy being among the limitations. Further, the U.S. did not have the means of technical intelligence collection it later acquired (for example, U-2, Corona, SAMOS) that allowed for high confidence location of military targets situated away from Soviet cities. Partially as a result of these factors it was felt that the most efficient means of stopping a Soviet military advance into Europe was via the destruction of the workers and factories, located in or near urban areas, that constituted war-supporting industry.

This view was challenged by the Harmon Committee, set up in mid-1949 by Secretary of Defense Forrestal to evaluate the effects of a strategic offensive against the USSR. The committee estimated that the projected attack on 70 cities specified in the FLEETWOOD plan would provide a 30 to 40 percent reduction in Soviet industrial capacity and as many as 2.7 million fatalities. However, the committee also concluded that industrial losses would not be permanent and could either be alleviated by Soviet recuperative action or augmented (depending on the effectiveness of follow-up attacks). Further, the committee concluded that the planned attacks would not necessarily impair the capability of Soviet armed forces to advance rapidly into selected areas of Western Europe, the Middle East and Far East. Nor would such attacks "destroy the roots of Communism or critically weaken the power of the Soviet leadership to dominate the people."[11]

A direct result of the Harmon Committee report was the DROPSHOT war plan, drawn up in 1949 for a war beginning in 1957. DROPSHOT directed attacks on the following target sets:

(i)   stockpiles of weapons of mass destruction, the facilities for their production and means of delivery;
(ii) key governmental and control centers;
(iii) lines of communication, military supply lines, troop concentration and naval targets;
(iv) important elements of Soviet and satellite industrial economy.[12]

The plan required 300 atomic bombs and 20,000 tons of high explosive conventional bombs on about 700 targets.

Hence, DROPSHOT gave substantial attention to targets other than population and industry—although many of the military and political targets were located in or near urban areas. This factor combined with continued targeting of industrial facilities insured a large number of weapons targeted on urban areas. This state of affairs continued throughout the Eisenhower Administration when the first integrated war plan, the SIOP (Single Integrated Operational Plan) was drawn up. The first SIOP called for attacks on all major Soviet and other communist cities in the event of war (as well as attacks on other targets). No alternative options were available and no forces were to be held in reserve. Estimated casualties in the event of war were in the 360–525 million range.[13]

This was the SIOP inherited by the Kennedy Administration and one of the administration's early major tasks was its revision. As a result of the revision a spectrum of attacks was possible, only one of which included direct attacks on cities. The spectrum moved from attacks on Soviet strategic forces to Soviet air defenses away from cities, Soviet air defense near cities, Soviet command and control centers and only finally spasm attacks on cities.[14]

This shift in war planning was first made public in a speech by Defense Secretary McNamara on 17 February 1962. However, it is his now famous commencement address at the University of Michigan which best spelled out the logic of the change in strategy. McNamara stated that:

The U.S. has come to the conclusion that to the extent feasible, basic military strategy in a possible general nuclear war should be approached in much the same way that the more conventional military operations have been regarded in the past. That is to say, the principal military objective, in the event of a nuclear war stemming from a major attack on the Alliance, should be the destruction of the enemy's military forces, not of his civilian population.[15]

A variety of factors compelled McNamara to backtrack quickly and *publicly* from his statement. The most important reason for McNamara's public backtracking was the use of his statements by the Air Force to justify large increases in funding for strategic weapons systems. Thus, in a memo to the President dated 21 November 1962, McNamara stated that "It has become clear to me that the Air Force proposals . . . are based on the objective of achieving a first-strike capability."[16]

Henceforth, McNamara made U.S. ability to destroy specific percentages of Soviet population and industry in a retaliatory strike (the assured destruction capability) the cornerstone of United States *declaratory policy.* Thus, in his 1965 Posture Statement, McNamara spoke of the U.S. need to be able to destroy 25–33 percent of Soviet population and 67 percent of industry to insure deterrence.[17]

Despite this public emphasis on urban-industrial targeting U.S. *employment policy* as evidenced by the SIOP itself remained the same, with attacks on Soviet cities being considered a last resort. Thus, while McNamara talked in public only of the nuclear capability and strategy for the last phase of an all-out nuclear war—"the sufficient reserve striking power to destroy an enemy society if driven to it"—that he mentioned in his Ann Arbor speech, only 7 percent of U.S. nuclear warheads were targeted against Soviet (or other) cities.[18]

This state of affairs remained unchanged throughout the Kennedy and Johnson Administrations. Thus, an Assistant Secretary of Defense in the last years of the Johnson Administration has written that:

The SIOP remains essentially unchanged since McNamara's Ann Arbor speech of 16 June 1962. There have been two developments, however: (1) it has become more difficult to execute the pure-counter-force option, and its value is considered to be diminishing and, (2) all public officials have learned to talk in public only about deterrence and city attacks. No war-fighting, no city-sparing, too many critics can cause too much trouble . . . so public officials have run for cover.[19]

It was in the Nixon Administration that consideration was next given to further revisions in both employment policy (the SIOP) and declaratory policy. Early in the administration a series of studies was undertaken concerning nuclear targeting options.[20] These studies led, eventually, to National Security Study Memorandum (NSSM)-169, approved by President Nixon in late 1973, and in turn to National Security Decision Memorandum (NSDM)-242 signed by the President on 17 January 1974.[21]

It was as a result of NSDM-242 that the first significantly limited nuclear options were developed and announced, although it was several years before these changes were incorporated into the SIOP. At the same time, NSDM-242 implementing documents specified that in the event of a large-scale attack the primary objective of U.S. strategic forces was to delay for as long as possible the Soviet Union's recovery as a major economic and military power. This was to be accomplished by destruction of 70 percent of the Soviet economic recovery base—a requirement which meant destruction of 45–50 percent of the total Soviet economy. A subsequent update of NSDM-242, NSDM-246, was never completed due to the results of the 1976 Presidential election.[22]

PRESIDENTIAL DIRECTIVE-59 AND NATIONAL SECURITY
DECISION DIRECTIVE-13

Shortly after taking office President Carter signed Presidential Directive 18, entitled *U.S. National Strategy*. PD-18 directed the initiation of three studies concerning strategic matters—an ICBM Force Modernization Study, a study on the maintenance of a Secure Reserve Force and a Nuclear Targeting Policy Review.[23] At the same time PD-18 continued NSDM-242 in force until new guidance was issued.

Phase I of the targeting review consisted of a large number of in-house and contractor studies covering a wide range of topics—nuclear war termination, population targeting, Soviet strategic doctrine and perceptions of U.S. strategy, as well as means for destruction of the Soviet state.[24] On the basis of these studies a Phase I report was issued identifying the prime issues of interest. The subsequent Phase II study and report, completed in 1978, focused on the areas singled out for special attention in the Phase I study and made several recommendations which were incorporated into PD-59 and related directives.

PD-59, which bore the title "Nuclear Weapons Employment Policy," altered U.S. strategy for a large-scale nuclear war in two basic ways.[25] First, it mandated a shift in targeting emphasis from the economic recovery targeting mandated by NSDM-242 to the targeting of Soviet political and military assets, strategic military targets, leadership targets, and Other Military Targets (OMT). Hence, the destruction of 70 percent of the Soviet economic recovery base would no longer be the prime objective of U.S. nuclear forces.

According to a report by William Beecher, the target sets which are to be the main object of U.S. targeting are:

(i)   700 underground shelters for key Soviet officials throughout the country;
(ii)   2,000 strategic targets—including 1,400 ICBM silos plus 600 C^2 bunkers, nuclear storage sites and strategic air and naval facilities;
(iii) 3,000 Other Military Targets—including 500 airfields, plus military units, supply depots and critical transportation hubs, power projection forces;
(iv)   200 to 400 key factories.[26]

Reflecting the new concentration on political and military leadership targets, strategic and other military targets the National Strategic Target List was expanded from 25,000 to 40,000 entries.[27] Over 20,000 of the 40,000 entries are military targets and about 15,000 entries represent industrial targets. Two thousand entries constitute leadership and control targets, with the number of such entries likely to grow and include CPSU headquarters, KGB installations and communications facilities.[28]

Second, it required that the U.S. develop the capability to fight a *protracted nuclear conflict*, one which might last months instead of days. The ability to conduct such a conflict requires strategic weapons and C^3 systems which have the characteristics of *endurance* and *flexible response*. That is, such forces must be able not only to survive a first strike and be used in an immediate retaliatory strike but must be able to operate for a significant period of time afterward. Such enduring forces could then be used selectively as the situation required.

The logic behind these aspects of PD-59 has been summarized by Colin Gray:

PD-59 said that deterrent effect is maximized if the Soviet leadership knows that the assets it values most are discretely at nuclear risk; that punishment of Soviet society, while an inevitable by-product of large-scale nuclear war, has little general merit as a deterrent; and that a central homeland-to-homeland war could be protracted, with six months as the consensus guess for the duration of a protracted nuclear war. The assets of highest value in Soviet estimation are believed to be the domestic political control structure and the military power of the state.[29]

PD-59 also continued the trend toward construction of limited nuclear options begun during the Nixon Administration. Thus, in addition to the Major Attack Options (MAOs) continued attention was directed toward the development of Selective Nuclear Options (SNOs) and Regional Nuclear Options (RNOs).[30]

There are at least three important points that should be kept in mind with regard to PD-59. First, it does represent, as Carter Administration Secretary of Defense Harold Brown claimed, an evolutionary rather than revolutionary shift in U.S. targeting policy.[31] Evolutionary rather than revolutionary in that what is being altered is the targeting *emphasis* rather than the target sets per se. At the same time, PD-59 represents a rather significant evolutionary change in strongly increasing the emphasis on targeting political and military assets, an increased emphasis that mandates significant increases in strategic capabilities.

Second, it should be noted that the changes in targeting policy set forth in PD-59 had been under way for approximately two years prior to the signing of PD-59 on 25 July 1980. While the decision to "go public" with PD-59 was, in all likelihood, largely politically motivated, its public exposure (as with the Stealth bomber) represented an attempt to derive political benefits from a program that had been in progress for some time. Thus, the basic elements of PD-59 were laid out as part of Secretary Brown's "countervailing strategy" as explicated in the Department of Defense Annual Reports for FY 80 and FY 81.[32]

Third, two further Presidential Directives, PD-53 and PD-58, which pre-date the signing of PD-59 (but not the targeting changes), were intended to initiate programs considered crucial to PD-59's requirements. PD-53, entitled "National Security Telecommunications Policy" and signed on 15 November 1979, stated that "it is essential to the security of the U.S. to have telecommunications facilities adequate to satisfy the needs of the nation during and after any national emergency" and this required the provision ". . . for connectivity between the National Command authority and strategic and other appropriate forces to support flexible execution of retaliatory strikes during and after an enemy nuclear attack."[33]

PD-58, entitled "Continuity of Government" and signed on 30 June 1980, covered several related topics: evacuation of military and civilian leaders, construction of new hardened shelters for key personnel, data processing and communications systems and improvement of early warning systems. . . . And PD-59 itself acknowledged that "targeting policy imposes requirements in strategic C³ and improvements in our forces must be accompanied by C³ improvements."[34]

As implied above the capabilities required for the implementation of PD-59/NSDD-13 can be divided into two broad categories—those pertaining to strategic weapons and those pertaining to C³I. Each category can be further divided. Thus, strategic weapons capabilities required by PD-59/NSDD-13 can be divided into target coverage and destruction capabilities as well as endurance and flexible response capabilities. C³I capabilities required by PD-59/NSDD-13 can be divided into endurance capabilities and flexible response capabilities.

The target destruction capabilities required of strategic weapons by PD-59/NSDD-13 include not only the ability to destroy soft area and soft point targets (for example, industrial concentrations and airfields) but hardened point targets such as underground leadership shelters, military command and control bunkers, ICBM silos and launch control centers as well as nuclear weapons storage sites. Further, destruction of certain hard-targets is considered to be time-urgent—something that must be accomplished within the approximate 30-minute or less flight time of ICBMs and SLBMs rather than the six or more hours flight time of bombers and cruise missiles. And weapons systems employed must be able to limit collateral damage to allow the Soviet leadership to detect the limited nature of the attack. Additionally, there must be enough warheads so as to allow coverage of the designated target set.

Endurance of a significant portion of U.S. strategic nuclear forces is also mandated by PD-59/NSDD-13. Thus, weapons must not be able simply to survive an initial strike but be available for use over the course of the protracted conflict. Such endurance is obviously a necessity for a strategy which stresses flexible response. Additionally, at least some weapons systems need to be retargetable while others must be able to locate and destroy mobile targets. Retargetable systems are required for re-attack of important targets not destroyed in an initial attack and for use in subsequent stages of the conflict when new targets (for example, rebuilt facilities) appear. Weapons systems capable of locating and destroying mobile targets are required to allow destruction of power projection forces which could significantly alter their location in short periods of time.

C³I capabilities required for endurance include early warning, an enduring warning system, initial leadership survival, continued leadership survival, retaliation decision transmission and continued communications with strategic forces. Three of these capabilities are required for an Assured Destruction Strategy—early warning, initial leadership survival, and retaliation decision transmission. An enduring C³I capability requires continued operation of a warning system to provide advance notice of follow-up attacks, particularly attacks on relocated leadership and military assets. Likewise, endurance requires a maintenance of a functioning leadership and the ability of this leadership to transmit further execution orders to U.S. strategic forces.

A flexible response capability is even more demanding in its implications for C^3I requirements—particularly when compared to the requirements of an Assured Destruction strategy. By its very nature an Assured Destruction Strategy makes no demands on C^3I capabilities with regard to flexible response. PD-59/NSDD-13 make numerous demands. Two-Way Communication with Strategic Forces is required so as to allow Force Status Reporting—the reporting of which forces have survived and are still capable of being employed. Damage Assessment via technical sensors is necessary to determine the exact nature of the appropriate response—that is, which MAO, SAO, or RNO should be employed. Real Time Imagery and Other Intelligence is needed to allow determination of the effectiveness of the attack, to indicate which targets need to be attacked again, and to detect the movement of targets and the creation of new targets. Finally, an ability to communicate with the adversary (that is, the Soviet leadership) is required—without direct communication the "tacit" communication conducted via employment of nuclear warheads may be less than adequate for escalation control. . . .

PD-59, NSDD-13 AND THE REAGAN STRATEGIC MODERNIZATION PROGRAM

The Reagan Strategic Modernization program involves a large number of programs in both the strategic weapons and C^3I areas. The most prominent weapons system is the MX ICBM. As noted above, a major requirement of PD-59/NSDD-13 is the hard-target/time-urgent kill capability—to be used against the 1,400 Soviet ICBM silos, 700 leadership bunkers, and assorted other hard point targets. Attainment of a high damage expectancy against such targets requires approximately 4,200 moderate yield/high accuracy warheads. The requirement of moderate yield and high accuracy results from estimates of Soviet silo hardness in the range of 4,000–6,000 PSI.[35] The requirement of two warheads per target stems from the incorporation of the uncertainties concerning missile reliability into the strategic equation. . . .[36]

As of 1981 the only U.S. ballistic missile warheads with even a moderate hard-target kill capability against Soviet silos were the 900 Minuteman III warheads with the MK-12A re-entry vehicle, consisting of the W-78 335 KT warhead and NS-20 guidance system. The NS-20 gives the MK-12A a CEP of approximately 0.1 nm and hence two MK-12A warheads a probability of .69 of destroying a 4,000 PSI silo and a probability of .60 of destroying a 6,000 PSI silo.[37] Other warheads either had insufficient yield, insufficient accuracy or both or were carried by insufficiently prompt delivery vehicles. Thus, the Titan II and Minuteman II missiles were not sufficiently accurate while the 250 Minuteman II missiles with MK-12 warheads (of 170 kt) had insufficient yield. SLBMs had neither the required yield or accuracy required—a 10 kt Trident I warhead with a .25 nm CEP would have a .14 damage expectancy against a 4,000 PSI silo.

In addition to alleviating the (theoretical) ICBM vulnerability problem, deployment of the MX has been intended to give the U.S. a significant capability to destroy hard-targets such as Soviet ICBMs and leadership bunkers. Use of a 500 kt warhead, as originally intended, and attainment of the MX operational accuracy goal of .075 nm (450 feet) CEP would give two cross-targeted MX warheads a .89 probability of destroying a 4,000 PSI target and a .83 probability of destroying a 6,000 PSI target.[38]

Clearly though, 100 MX missiles with 10 warheads each would not fully provide for coverage of the 2,100 hard-targets mentioned above, nor even for full coverage of Soviet silos. However, two other aspects of the ballistic missile portion of the Reagan strategic program are intended to allow for complete coverage.

First, the Administration may expand the Minuteman III force from 550 to 600 while placing MK 12A re-entry vehicles on all 600 missiles.[39] Hence, in addition to the MX warheads the U.S. would have an additional 1,800 ICBM warheads with moderate capabilities against 4,000 and 6,000 PSI targets. Second, deployment of 264 D-5 SLBMs on 11 Trident submarines (24 per sub) at the end of the decade would give the U.S. the potential of adding another 2,112 warheads with a hard-target destruction capability. D-5s to be used against hard-targets will be equipped with MK-12a re-entry vehicles (with a yield of 335 kt).[40] Combined with the use of stellar inertial guidance to increase accuracy this would give the individual warheads a capability against 4,000 and 6,000 PSI targets similar to that of the MK-12A equipped Minuteman III.[41]

As noted, the ability of U.S. forces to attack the entire Soviet target spectrum—including hard-targets—is only one requirement of PD-59 with respect to strategic weapons systems. A second is that these forces can be used in a flexible manner over a prolonged period of time—that a significant portion of the strategic force must be able to survive not just the immediate attack but for weeks and months.

While the MX may make a significant contribution in terms of hard-target/time-urgent kill capability it may do so only in the early stages of a nuclear war since it is unlikely to be available after several hours. Even if the Closely Spaced basing mode serves to insure initial survivability of a significant portion of MX missiles it will not allow prolonged survival.[42] An enduring ballistic missile capability will have to come from either Deep Underground basing, a system in which missiles are located several thousand feet underground and are transported to the surface via tunnels after an initial attack, or reliance on Trident submarines as part of a reserve force. The higher on-patrol rate for Trident submarines (.66 to .55 for Poseidon submarines) will mean a higher percentage of surviving forces after a first strike.

The enhanced target coverage required by PD-59 can be enumerated as follows: 4,200 warheads required for the strategic hard-targets and one warhead for each of the remaining 600 strategic targets, 3,000 Other Military Targets and 200–400 key factories, hence a total of 8,200 warheads. While U.S. warhead holdings presently exceed this number significantly, fewer than 8,200 would survive a Soviet first strike—even if U.S. forces were on a generated alert.

The required increase in warheads is intended to be achieved by a variety of programs—MX deployment, the deployment of 3,000–4,000 Air Launched Cruise Missiles (ALCMs) on both the B-52 and B-1B bombers (20 on B-52 Js, 12 on B-52 Gs and 22 on B-1Bs), and the total deployment of a Trident fleet of over 20 submarines with both hard and soft area target mission warheads will enhance U.S. warhead holdings by over 5,000.[43]

The flexible response requirement for strategic forces is to be met by the retargeting capability of U.S. ICBMs and SLBMs as well as construction of the B-1B and Advanced Technology Bomber (ATB-Stealth). The latter will allow not only retargeting but an ability to "search and destroy" to locate mobile or new targets and destroy immediately after locating them.

In the area of collateral damage minimization the Administration does not seem to have gone beyond attempts to minimize collateral damage given fixed targeting objectives and available weapons yield. Certainly, even with their projected high accuracy, the planned yield of major systems such as MX, Trident II (D-5) or ALCMs is such as to cast doubt on the feasibility of significant collateral damage minimization—particularly given the predominance of targets located in the densely populated western USSR. . . .

Upon entering office in 1981 Under-Secretary of Defense for Policy, Fred Iklé, initiated a study of U.S. nuclear targeting policy and U.S. weapons acquisition plans. The result of the review apparently supported the doctrinal elements of PD-59 but suggested the need for a revised acquisition strategy—both in terms of increasing acquisition and creating procedures with which strategy and acquisition could be coordinated.

Specific reaffirmation of PD-59 and related Carter directives came in October 1981 with the signing of National Security Decision Directive 13 on nuclear weapons employment policy and National Security Decision Directive 12, entitled "Strategic Communications."[44] The very existence, much less the content of NSDD-13, has not been officially confirmed. However, the classified five-year Defense guidance issued by Secretary of Defense Caspar Weinberger, whose nuclear employment policy portions would echo any Presidential guidance, and leaked to the *New York Times* and *Washington Post* states that:

Should deterrence fail and strategic nuclear war with the USSR occur, the United States must prevail and be able to force the Soviet Union to seek earliest termination of hostilities on terms favorable to the United States.

The United States must have plans that assure U.S. strategic nuclear forces can render ineffective the total Soviet military and political power structure . . . and forces that will maintain, throughout a protracted conflict period and afterward, the capability to inflict very high levels of damage against the industrial/economic base of the Soviet Union . . . so that they have a strong incentive to seek conflict termination short of an all-out attack on our cities and economic assets.

U.S. strategic nuclear forces and their command and communication links should be capable of supporting controlled nuclear counterattacks over a protracted period while maintaining a reserve of nuclear forces sufficient for trans- and post-attack protection and coercion.[45]

Thus, the main difference between PD-59 and NSDD-13 may be that the latter, as a result of the Iklé study, required the Defense Department to produce the "Nuclear Weapons Employment and Acquisition Master Plan."

An interpretation of the Reagan strategic modernization program as an attempt to implement PD-59/NSDD-13 does not imply that there are no other factors in operation. Certainly, bureaucratic and organizational factors may also be influential. It was noted earlier how the Air Force used McNamara's counterforce doctrine as a means of justifying large budget requests. Others may perceive the program as necessary to demonstrate to the Soviet Union a U.S. willingness to match Soviet increases in strategic forces rather than as a requirement of a specific targeting doctrine. None of these motives, however, preclude an explanation which stresses PD-59/ NSDD-13 as the prime catalyst for the Reagan strategic program.

In this regard it is relevant to note that PD-59 and the Reagan Strategic Program are not two sharply divided entities—each belonging exlusively to a particular administration. While the strategic doctrine enunciated in PD-59 came most directly out of the Carter Administration NTPR, it represented the culmination of a re-evaluation process that began during the Nixon/ Ford Administration. Likewise, the Reagan strategic program is far from a Reagan creation. Almost all the initiative involved, and certainly all the major initiatives, can be traced to previous Administrations. The Reagan program involves the approval, acceleration of, and increased spending on many of these programs.

Thus, it is neither implausible nor contradictory that the highest defense policy officials of the Reagan Administration are themselves not reacting to PD-59/NSDD-13 so much as more simplistic views that dictate a large-scale upgrading of U.S. strategic forces—that NSDD-13 is simply approval of a doctrine that justifies the weapons programs desired. Hence, lower-level officials who have been instrumental to the changes of U.S. nuclear weapons employment policy over the past decade would have found allies willing to give them the weapons and C³I capabilities they consider necessary to implement present employment policy.

Of course, the major question is not what the motivations of the highest officials of the Reagan Administration are but whether the capabilities required by PD-59/NSDD-13 are feasible and sensible and if so, whether the Reagan program is the best way to attain those capabilities. Thus, questions can be raised concerning the emphasis on targeting ICBMs and leadership bunkers—with respect to the wisdom of targeting ICBM silos and the feasibility of targeting leadership bunkers.[46] The emphasis in Soviet doctrine on launch-on-warning and launch-under-attack is sufficient to make it doubtful that the Soviet leadership would attempt to ride out an "out-of-the-blue" pre-emptive U.S. strike. In the case where a nuclear conflict had already begun, it is even less likely that they would attempt to ride out an attack. Hence, U.S. ICBMs targeted on Soviet ICBMs would likely detonate on empty silos. With respect to leadership targeting it is not clear that a sufficient portion could even be located. Statements by former Secretary of Defense Harold Brown indicated that the U.S. has had limited success in doing so.[47]

Likewise, the discussion above raises serious questions about whether the stress on ICBM and bomber modernization can actually be expected to produce strategic weapons systems capable of both endurance *and* flexible response. Additionally, the warhead yields on MX, Trident D-5, and the ALCMs are such as to cast doubt on their utility in collateral damage limitation scenarios.

U.S. officials have proclaimed that the United States does not target population per se. But, while there are procedures employed for minimizing collateral damage, this minimization is conducted with respect to a basically fixed target set and weapons arsenal. Such constraints allow little flexibility in attempting to significantly reduce collateral damage. Without significant limitations on collateral damage limited war scenarios have little plausibility.

With respect to C³I systems it is clear that the peacetime C³I system cannot be expected to continue functioning in wartime. Whether the nuclear wartime system being designed—consisting of components embedded in the peacetime system and reconstituted assets—will truly give the U.S. a capability to fight a protracted nuclear war is impossible to judge at this point since the exact nature of the wartime system is not completely designed and the plans for it remain highly classified.

Finally, it should be noted that a prolonged war requires two participants willing to conduct such a conflict. If U.S. targeting policy were to lead to elimination (either intentionally or not) of the Soviet C³I components (human or technical) required for the conduct of such a conflict U.S. acquisition of such a capability would be pointless. . . .

NOTES

1. Center for Defense Information, "Preparing for Nuclear War: President Reagan's Program," *The Defense Monitor* 10, 8, 1982. The $222 billion figure comprises the cost in 1982 dollars ($180 billion) adjusted for inflation.

2. "New Payload Could Boost Shuttle Cost," *Aviation Week and Space Technology*, 14 August 1978, p. 16; "Space Reconnaissance Dwindles," *Aviation Week and Space Technology*, 6 October 1980, pp. 18–20; Clarence A. Robinson Jr., "Defense Research Request Rises 14%," *Aviation Week and Space Technology*, 22 February 1982, pp. 53–55.

3. *Employment policy* refers to the actual policy governing the use of nuclear weapons—what type of targets are to be attacked, under what circumstances, and with what confidence of success. *Declaratory policy* refers to public statements by authorized government officials about how nuclear weapons would be used—these statements may be misleading because of oversimplification, or a reluctance to discuss

politically sensitive—either domestically or internationally—aspects of employment policy.

4. George C. Wilson, "Planners Say Defense Budget Is Insufficient," *Washington Post*, 8 March 1982, pp. 1, 4; "Why C³I Is the Pentagon's Top Priority," *Government Executive*, January 1982, pp. 14 ff.

5. See Senate Committee on Foreign Relations, *Nuclear War Strategy* (Washington: Government Printing Office, 1981); William Beecher, "U.S. Drafts New N-War Strategy vs. Soviets," *Boston Globe*, 27 July 1980, pp. 2, 12.

6. See Harold Brown, *Department of Defense Annual Report Fiscal Year 1981* (Washington: Government Printing Office, 1980), p. 656; Testimony of Caspar Weinberger in Senate Committee on Foreign Relations, *Strategic Weapons Proposals*, *Part 1* (Washington: Government Printing Office, 1981), p. 193.

7. Robert Scheer, "Pentagon Plan Aims at Victory in Nuclear War," *Los Angeles Times*, 15 August 1982, pp. 1, 23; Jack Nelson, "Weinberger Expands on Nuclear Plans," *Los Angeles Times*, 26 August 1982, pp. 1, 7.

8. For a view of the reasons for misinterpretations of PD-59 found in the press see Fred Kaplan, "Going Native Without a Field Map: The Press Plunges into Limited Nuclear War," *Columbia Journalism Reviews*, January/February 1981, pp. 23–29.

9. Gregg Herken, *The Winning Weapon: The Atomic Bomb in the Cold War, 1945–1950* (New York: Vintage, 1981), p. 286.

10. Ibid., p. 271.

11. Aaron Freidberg, "A History of U.S. Strategic Doctrine, 1945–1980," *Journal of Strategic Studies*, 3 December 1980, pp. 37–71.

12. Anthony Cave Brown (ed.), DROPSHOT: *The American Plan for World War III Against Russia in 1957* (New York: Dial Press/James Wade, 1978), p. 23.

13. Desmond Ball, *Deja Vu: The Return to Counterforce in the Nixon Administration* (Santa Monica: California Seminar, 1974), p. 11.

14. Ibid., p. 12.

15. Cited in William W. Kaufman, *The McNamara Strategy* (New York: Harper and Row, 1974), pp. 114–20.

16. Robert Scheer, "Fear of a U.S. First Strike Seen as Cause of Arms Race," *Los Angeles Times*, 8 April 1982, pp. 1, 13.

17. C. Johnston Conover, U.S. *Strategic Nuclear Weapons and Doctrine* (Santa Monica: Rand Corporation, 1977), p. 13; Benjamin T. Plymale, "The Evolution of U.S. Declaratory Strategic Policy," *Journal of International Relations*, Fall 1979.

18. R. Jeffrey Smith, "Another in a Series of Counterforce Weapons," *Science*, 7 May 1982, p. 598.

19. Ball, op. cit., p. 16.; Desmond Ball, *Politics and Force Levels: The Strategic Missile Program of the Kennedy Administration* (Berkeley: University of California Press, 1980), Ch. 1.

20. John Edwards, *Superweapon: The Making of MX* (New York: W. W. Norton, 1982), pp. 67–70.

21. Desmond Ball, "Counterforce Targeting: How New? How Viable?" *Arms Control Today*, 11, 2, 1981, pp. 1 ff.

22. "U.S. War Plan Calls for Striking Russian Industry," *Los Angeles Times*, 2 February 1977, p. 1.

23. Desmond Ball, "Developments in U.S. Strategic Nuclear Policy Under the Carter Administration," *ACIS Working Paper No. 21*, 1980.

24. Ibid., p. 173.

25. Colin Gray, *Strategic Studies and Public Policy: The American Experience* (Lexington: University Press of Kentucky, 1982), p. 157.

26. William Beecher, "U.S. Drafts New N-War Strategy vs. Soviets," *Boston Globe*, 27 July 1980, pp. 1, 12.

27. The Defense Intelligence Agency maintains a computer tape, known as the Target Data Inventory (TDI), which lists all possible targets in the Soviet Union—approximately 500,000. From the TDI the NSTL is drawn and consists of those facilities which are considered the most significant potential targets. From the NSTL the actual war plans—the SIOP—are developed. See Major General J. F. O'Malley, "JSTPS: The Link Between Strategy and Execution," *Air University Review* 28, 4, 1977, pp. 38-41.

28. Ball, "Counterforce Targeting . . . ," op. cit.

29. Colin Gray, *Strategic Studies and Public Policy: The American Experience* (Louisville: University Press of Kentucky, 1982), p. 158.

30. Ball, "Developments . . . ," op. cit.

31. Senate Committee on Foreign Relations, op. cit., p. 9.

32. Harold Brown, *Department of Defense Annual Report for FY 1980* (Washington: Government Printing Office, 1980), pp. 65-68.

33. Senate Committee on Armed Services, *Department of Defense Authorization for Appropriations for FY 1982 Part 7* (Washington: Government Printing Office, 1981), p. 4210; Joseph Albright, "The 'Message Gap' in Our Crisis Network," *Washington Post*, 19 October 1980, p. C-1.

34. Senate Committee on Armed Services, op. cit., p. 4210.

35. See Robinson, op. cit.; Colin Gray, *The MX ICBM and National Security* (New York: Praeger, 1981), p. 100, on silo hardness. PSI refers to pounds per square inch of overpressure—the pressure above the normal atmospheric pressure of 14.7 pounds per square inch. Since blast overpressure decreases rapidly from the point of detonation it is necessary that the warhead have reasonably high accuracy for there to be a significant probability of silo destruction. The kill probability of a missile warhead against a hardened silo is a function of the yield in megatons, accuracy, and reliability of the warhead as well as PSI. For a statement of the formula used in computing kill probabilities, see Thomas Brown, "Missile Accuracy and Strategic Lethality," *Survival* 18, 1976, pp. 52-59.

36. It is generally assumed that there is only a .8-.85 chance that all factors concerned with missile launch, warhead separation and detonation will function properly. Hence, the maximum kill probability will be in the .8-.85 range when a single warhead is used. Fratricide constraints limit warheads per target to two. A figure of .8 reliability is employed in all calculations.

37. CEP refers to Circle of Equal Probability—the radius of the smallest circle around the target in which 50 percent of the warheads would be expected to land. CEP is the measure of accuracy used in formulas used to determine kill probabilities.

38. MX accuracy goal is given in Senate Committee on Appropriations, *Department of Defense Appropriations FY 1982 Part 5* (Washington: Government Printing Office, 1981), p. 445. On the MX warhead see Walter Pincus, "Pentagon Plans to Build MX Warhead that Air Force and OMB Don't Want," *Washington Post*, 24 February 1982, p. 16. Recent testimony indicates a lower yield warhead in order to conserve special nuclear material.

39. Clarence A. Robinson, Jr., "U.S. Upgrading Its Strategic Arsenal," *Aviation Week and Space Technology*, 8 March 1982, pp. 27 ff.

40. Ibid.

41. Ibid.; House Committee on Armed Services, *Hearings on Military Posture and HR 2970 Part 3* (Washington: Government Printing Office, 1981), p. 158.

42. "MX Closely Spaced Basing" (mimeo: Department of the Air Force, Washington, 28 June 1982).

43. "Statement of Caspar W. Weinberger," in Senate Committee on Appropriations, op. cit., pp. 409–11; Arms Control and Disarmament Agency, *Fiscal Year 1983 Arms Control Impact Statements* (Washington: Government Printing Office, 1981), p. 67.

44. Senate Armed Services Committee, *Strategic Force Modernization Programs* (Washington: Government Printing Office, 1982), p. 232.

45. Michael Getler, "Adminstration's Nuclear War Policy Stance Still Murky," *Washington Post*, 10 November 1982, p. A24.

46. See Ball, "Counterforce Targeting . . . ," and Louis Rene Beres, "Tilting Toward Thanatos: America's 'Countervailing' Strategy Nuclear Strategy," *World Politics* 34, 1, 1981, pp. 25–46.

47. Harold Brown, *Department of Defense Annual Report Fiscal Year 1981* (Washington: Government Printing Office, 1980), p. 78.

Bibliography: Part One

Allison, Graham T. *Essence of Decision: Explaining the Cuban Missile Crisis.* Boston: Little, Brown and Co., 1971.

Ball, Desmond. *Targeting for Strategic Deterrence.* Adelphi Paper no. 185. London: International Institute for Strategic Studies, 1983.

Bundy, McGeorge. "To Cap the Volcano." *Foreign Affairs* 48 (October 1969):1–20.

Freeman, Lawrence. *The Evolution of Nuclear Strategy.* Boston: St. Martin's, 1983.

Howard, Michael. "On Fighting a Nuclear War." *International Security* 5 (Spring 1980):3–17.

Iklé, Fred. "Can Nuclear Deterrence Last Out the Century?" *Foreign Affairs* 51 (January 1973):267–285.

Johnson, Robert H. "Periods of Peril: The Window of Vulnerability." *Foreign Affairs* 61 (Spring 1983):950–970.

Kahan, Jerome. *Security in the Nuclear Age: Developing U.S. Strategic Arms Policy.* Washington, D.C.: Brookings, 1975.

Kahn, Herman. "Escalation as a Strategy." In Henry Kissinger, ed., *Problems of Nuclear Strategy.* New York: Praeger, 1965.

McNamara, Robert S. "The Role of Nuclear Weapons: Perceptions and Misperceptions." *Foreign Affairs* 62 (Fall 1983):59–80.

Panofsky, Wolfgang. "The Mutual Hostage Relationship Between America and Russia." *Foreign Affairs* 52 (October 1973):109–118.

Steinbruner, John D. "Nuclear Decapitation." *Foreign Policy* 45 (Winter 1981/82):16–28.

PART TWO

THE TWO FACES OF SOVIET NUCLEAR STRATEGY

CHAPTER 4

Victory and Parity

It is extraordinarily difficult for anyone of any nationality, outside a small Soviet ruling circle, to be at all confident of Soviet motives and intentions. Secretive, lacking an open press, Soviet society repels intimacy and understanding. Nowhere is this more evident than in the area of Soviet nuclear strategy.

From what is known of Soviet strategy in the West, it is clear that Soviet leaders approach nuclear war and arms control in a manner fundamentally different from that of their counterparts in the West. Soviet military science is based on the works of Clausewitz and Lenin, and this is well delineated in the excerpts from Sokolovsky's *Soviet Military Strategy.* But there are many disciples of Clausewitz in the West, so this is not the major difference. Rather, it is a persistent and publicly stated commitment to the achievement of victory in nuclear war in recent years that sets Soviet strategy apart. Military spokesmen repeatedly identify victory as the goal of Soviet nuclear strategy. Selections in this chapter include a number of their statements; especially notable is the "Military Strategy" entry in the *Soviet Military Encyclopedia* by the former chief of staff, Marshal N. V. Ogarkov.

In the later years of the Brezhnev regime, Soviet political leaders preferred to speak of parity and flatly denied that they sought nuclear superiority or would initiate an aggressive nuclear war. The address by Leonid Brezhnev at Tula in January 1977 is a good example of a Soviet political formulation of this kind. This political stance has not relieved concern in the West, for the commitment to victory has not changed, and the difference may represent disagreements within the Soviet leadership. However, the politicians' denials of a desire for superiority have induced an ambiguity about Soviet intentions that is more troubling than either the military or political positions might have been alone, for together they suggest an effort at deception.

The Soviet commitment to victory in nuclear war together with the relative thoroughness of Soviet preparations for the conduct and endurance of nuclear war, from military decontamination units to extensive civil defense measures, also provokes deep concern in the West, as Fritz Ermath observed in "Contrasts in American and Soviet Strategic Thought." But that concern, and the close attention to Soviet strategy it fosters, have not led to unanimity. Western experts agree that the Soviets approach nuclear strategy in a fundamentally different way than it is approached in the United States. Personifying the difference, Freeman Dyson suggested that U.S. strategists say, "Whatever else happens, if you drive us to war, you shall not survive." Soviet strategists say, "Whatever else happens, if you drive us to war, we shall survive."[1] Western experts agree that the Soviets do not make a distinction between deterrence and war-fighting; the Soviet position is that being prepared to fight is the best deterrent. Some Western experts even agree on the ideological and prudential origins of the Soviet military stance. It would be an admission of the falsity of Marxism-Leninism if Soviet leaders were to state publicly that in a clash with capitalism a communist system would not prevail. Further, to accept the logic and force requirements of Assured Destruction would be to put the survival of the Soviet regime totally in capitalist hands, an impossibility for Soviet leaders.

However, Western experts disagree about the implications of the differences between Soviet and U.S. strategy. Moreover, their different conclusions are important, for they have often become the rationale for U.S. responses to the Soviet Union. In the debate with Richard Pipes, Raymond Garthoff contended that the significance of the differences had been exaggerated. Soviet strategy, Garthoff maintained, neither inclined Soviet leaders to launch a nuclear war nor prevented the conclusion of mutually beneficial arms control agreements. To others, such as Richard Pipes, the Soviet preference for war-fighting and the studied ambiguity over victory and parity demonstrated the readiness of the Soviet leadership to resort to force when the military balance favored it, not to shrink even from nuclear war, a war they had prepared for and believed (and said) they would win.[2]

NOTES

1. Freeman Dyson, *Weapons and Hope* (New York: Harper & Row, 1984), pp. 231–232, chaps. 19 and 20.

2. Richard Pipes, "Why the Soviet Union Thinks It Could Fight and Win a Nuclear War," *Commentary* 64, no. 1 (July 1977):21–34.

17. The Nature of Modern War

Marshal V. D. Sokolovsky

One of the primary tasks of the theory of military strategy is to study and define the nature of wars: their characteristic military strategy and technical features. The correct scientific solution of this problem is attainable primarily through the Marxist-Leninist analysis of the actual conditions of historical and social development (which makes it possible to establish the essential socio-political nature of various wars), why and how wars break out, and the material base upon which they are waged.

The scientific forecast of the nature of a future war is important because only thus can the political and military leaders correctly and optimally direct the development of the armed forces and successfully and forcefully execute the mission of preparing the whole country for war with an aggressor.

At present, accurate anticipation of the nature of the initial period of a war is exceptionally important for the theoretical and practical problems of military strategy. In this initial period, the effect of the armed conflict upon the course and outcome of modern war will be fundamentally different from that of the past. Consequently, grave new demands already confront the armed forces, the state, and the people.

THE MARXIST-LENINIST CONCEPT OF THE ESSENTIAL MEANING OF WAR IN THE PRESENT EPOCH

The question: What is the essential meaning of war? is decisive for the solution of all basic theoretical and practical problems in military strategy. It is also crucial for clarifying the nature of any particular war. The tenets of historical materialism; the Marxist-Leninist study of war; and the more important documents of the program for the Communist and Workers' Parties, defining their theoretical, political, and practical activity, at present provide an exhaustive answer to this question. The military events of our epoch are convincing proof of the correctness of the Marxist-Leninist concept of the essential meaning of war and the causes and conditions which give rise to it. . . .

Marxism-Leninism teaches that war is a socio-historical phenomenon arising at a definite stage in the course of social development. It is an extremely complex social phenomenon, whose essential meaning can be revealed solely by using the only scientific method: Marxist-Leninist dialectics. When discussing the use of Marxist epistemology in the study of war, V. I. Lenin stated that "dialectics require a comprehensive study of a given social phenomenon as it develops, as well as a study of information which is seemingly extraneous to basic motive forces, to the development of productive forces and to the class struggle."

Historical experience shows that even the greatest world war, no matter how all-encompassing it may seem, represents only one aspect of social development and completely depends upon the course of that development and upon the political interactions between classes and states.

V. I. Lenin insisted that *war is part of a whole and that whole is politics.* He also stated that war is a continuation of politics and that politics also "continues" during war. Lenin's proposition is extremely important and basic; it takes note of bourgeois theories of the comprehensive, all-encompassing nature of war, and of "class peace" during war. It explains that even during war, politics continues; that is, class relations do not cease, and the class conflict continues in every way and by every means (ideological, political, economic, etc.). . . .

On this Marxist-Leninist basis, war can be defined as armed violence: organized armed conflict between different social classes, states, groups of states and nations, in order to achieve definite political goals.

In peacetime, classes, states, and nations have always attempted to attain their goals by the most diverse means and forms of struggle: ideological, political, economic, and others. When antagonisms were sharply aggravated, they resorted to forms of armed conflict—to war.

All of this demonstrates that war is only one form of political and class conflict. V. I. Lenin showed, in particular, that "civil war is the most acute form of class conflict, when a series of economic and political clashes occurring repeatedly, gather momentum, spread, become more intense, and are transformed into acute and armed conflict. . . ." Another of Lenin's ideas is that: "In certain periods of acute economic and political crises, the class struggle ripens into a direct civil war, i.e., armed conflict. . . ."

The meaning of Lenin's explanation (that war is the continuation of politics by other—violent—means) is that war is not synonymous with politics in general but comprises only part of it, and that politics, in addition to war, commands a large arsenal of various nonviolent means which it can enlist to attain its goals, without resort to war. This now closely guides the Communist Party of the Soviet Union and the Soviet government in its summons to the Western powers to solve all international disputes by negotiation and not by war.

The theory of Soviet military strategy also takes into account another aspect of the problem, namely, that war has a special and concrete character in contrast to other political means. A special system, a military organization, is created to wage war; military equipment is produced; and methods of combat are developed. The conduct of war itself has always been a special form of human activity, where each opposing side has directed its efforts toward the defeat of the other, toward the capture of enemy territory, or to holding its own territory, striving thus to attain its political goals.

In this epoch, society's power to produce has grown enormously, thus generating new, superpowerful and superlong-range instruments of destruction. The formation of a world socialist system has also caused radical changes in the terms of political struggle. Hence in a future world war, the participants will achieve their political goals not only by the defeat of the armed forces but also by the complete disruption of the enemy's economy and the demoralization of his population. *For this reason, the essential meaning of war, as a continuation of politics with the instruments of armed violence, and its concrete character now stand out even more prominently than in the past, and the modern instruments of violence become increasingly important.*

Thus, armed conflict has become an even more characteristic form of human activity. The causes are the following: first, enormous masses of people are drawn into modern war as armed forces increase and the civilian population is enlisted on a large scale to cope with a number of the military and semimilitary problems of defending the rear. Second, the complexity of the widely available modern military equipment requires special military knowledge and skills. And finally, modern war, as never before, demands the utmost use of the economy to provide for the needs of war and, simultaneously, requires the existence of a large defense industry as well as a special material and technological base, which are specially created to satisfy the needs of the armed conflict.

In modern warfare, nuclear weapons can be employed for various missions: strategic, operational, and tactical. From a purely military point of view, a nuclear weapon is incomparably more effective than a conventional weapon. It permits the execution of military missions in a considerably shorter time than was possible in past wars. For this reason, experts believe that the nuclear weapon is the most powerful and effective instrument by which to destroy an opponent in any type of operation, or in war as a whole. The introduction of this weapon into the Soviet Armed Forces markedly increased their combat capabilities and put a powerful instrument in the hands of Soviet military strategy for curbing an aggressor, protecting the achievements of socialism, and assuring peace.

The aggressors also equip their armed forces with nuclear weapons on a large scale. The United States is the main nuclear power in the West. England has a certain supply of nuclear weapons, and France is beginning to create them. Revanchist West Germany is especially feverish in its efforts to obtain nuclear weapons from the United States and to organize its own production.

It is not impossible that in time some other states of both military groupings will have nuclear weapons.

The nuclear industries in the Soviet Union and the United States have achieved such a level that each country has enormous stockpiles of nuclear weapons. . . .

From this, we conclude that the Soviet Union's Armed Forces and those of the other socialist countries must be prepared above all to wage a war where both antagonists make massive use of nuclear weapons. Therefore, the key task of strategic leadership and theory is to determine the correct, completely scientific solution to all the theoretical and practical questions related to the preparation and conduct of just such a war.

In addition to nuclear weapons, military missiles of various types and purposes have been rapidly developed during the last decade, especially missiles for destroying targets on the ground and in the air. At the end of the fifties, massive introduction of missiles into the Soviet Armed Forces began. . . .

The intensive development and enormous military effectiveness of strategic missiles have led to the creation of a new branch of the Soviet Armed Forces: Strategic Missile Forces. These forces can, if necessary, be used to carry out the main missions of war: the destruction of the aggressor's means of nuclear attack—the basis of his military power—and the defeat of the main formations of his armed forces, as well as the destruction of the basic, vitally important enemy targets. . . .

Execution of these missions by the Missile Forces will create the conditions for the successful operations by the other branches of the Armed Forces, for defending the rear of the country against the enemy's nuclear blows, and for the rapid achievement of the military, political, and strategic war aims and final victory.

The Strategic Missile Forces now have so many launchers, missiles, and nuclear warheads, including multimegaton warheads, that they can completely carry out their missions. . . .

Thus missiles are the most effective and promising instruments of armed combat. Their massive employment with nuclear warheads radically alters the methods of conducting war and its nature, making it decisive and destructive to the nth degree.

One of the important tenets in Soviet military doctrine is that a world war, if the imperialists initiate it, will inevitably assume the character of a nuclear war with missiles, i.e., a war in which the nuclear weapon will be the chief instrument of destruction, and missiles the basic vehicle for their delivery to target.

The massive employment of atomic and thermonuclear weapons and the unlimited capabilities of missiles to deliver them to any target within minutes will make possible the most decisive military results at any distance, over an enormous area, and within an extremely brief period.

One must emphasize that the present international system and the present state of military technology will cause any armed conflict to develop, inevitably, into a general war if the nuclear powers are drawn into it.

The logic of war is such that if American aggressive circles initiate a war, it will be transferred immediately to the territory of the United States of America. All weapons—intercontinental ballistic missiles, submarine-launched missiles, and other strategic weapons—will be used in this military conflict.

Countries on whose territory NATO and American military bases are located and countries which build these military bases for aggressive purposes would suffer crushing blows in such a war. A nuclear war would, in an instant, spread over the entire globe. . . .

The development and introduction of missiles, nuclear weapons, and electronic equipment have led to fundamental changes in almost all other weapons. As a result, the relative importance and strategic purpose of the various branches of the armed forces and their military employment have so changed that a war of a wholly new nature is foreordained.

Regardless of the future wartime role of such instruments of strategy as the Strategic Missile Forces, victory over the aggressor can be achieved only by the combined exertions of all the war-waging forces, namely: the Ground Forces, the National PVO, the Air Force, and the Navy, with the active participation of the people.

In a future war, the socialist coalition will aim at conclusive political and military goals. To attain these goals, it will not be enough just to destroy the enemy's means of nuclear attack, to defeat his main forces by missile blows and to disorganize his rear. For final victory in what would clearly be a class war, it will be absolutely necessary to smash the enemy's armed forces completely, deprive him of strategic areas of deployment, liquidate his military bases, and occupy his strategically important regions. In addition, the enemy's air and naval forces must not be permitted to make landings, and his ground forces must be prevented from invading the territories of the socialist countries. These territories must be held, and the internal security of the socialist states must be protected against enemy subversion. All these and a number of other problems can be solved only by the Ground Forces. . . .

In all previous wars, the belligerents sought principally to defeat or weaken the enemy's armed forces and, consequently, to capture and occupy vitally important regions or administrative and political centers. Once such goals were achieved, the political goals of the war were usually also realized.

The adversaries therefore conducted offensive and defensive operations, or a combination of both, in accordance with their political and strategic goals and the capacities of their armed forces. The main events took place in military theaters of operation (ground or naval), both sides being in direct contact, since no long-range destructive instruments of strategy then existed. . . .

What will be the characteristic strategic aims of a future war and how will it be conducted?

Assuming that the belligerents of both camps strive for the goals described above, they will employ the most decisive instruments of war—above all, many nuclear weapons—in order to annihilate the opponent or force him to surrender in the shortest possible time.

The question arises: Given such conditions, what is the main strategic goal of war? Is it, as in the past, the defeat of the opponent's armed forces or is it the annihilation and devastation of targets deep within a country in order to break up the organization of the country?

Soviet military strategy answers as follows: The attainment of both these goals must be simultaneous. The annihilation of the opponent's armed forces, the destruction of targets deep in his territory, and the disorganization of the country will be a single, continuous process of the war. The two chief reasons are, first, the need to defeat the aggressor as thoroughly and as quickly as possible, which requires that he be deprived of the military, political, and economic capacity to wage war; and, second, the use of military instruments [now] at hand might well accomplish these [military, political, and economic] aims simultaneously.

The probable enemy's targets, which comprise his economic, moral, and political potential and his military might, are located over an enormous area, often deep within his territory and on other continents. To destroy them, long-range strategic weapons above all will be required, together with appropriate methods of armed combat. The proportion of such [long-range] military operations in the whole struggle will greatly increase. At the same time, military operations on a relatively shallow front, where the opponent's ground forces are concentrated, will be much less important in a future war, and chiefly ground forces in contact with the opponent's troops will conduct these operations.

All this shows that the relationship between the role and importance of armed combat, on the one hand, waged by forces in direct contact with the enemy in the zone of military operations and employing tactical and operational weapons of destruction, and of armed combat, on the other hand, waged beyond this zone by strategic weapons of destruction has changed sharply in the direction of increasing the role and importance of the latter.

Thus, the mission of strategic weapons, whose action is beyond the range of tactical and operational weapons, has become much more important than that of troops in direct contact with the opponent.

Thus, the instruments acting on the opponent, the techniques of their use, and the way in which the war of the future will be fought will differ radically from previous wars, including also the past world wars.

Massive missile-nuclear blows will be of decisive importance in attaining the goals of a future world war. Such blows will be the principal and decisive means of conducting the war, and their delivery cannot be made directly dependent upon the course of the battle between adversaries in direct contact on the ground front. . . .

To achieve the most decisive results in the shortest time in a future world war, the Soviet Armed Forces and those of the entire socialist camp will have to employ their main military forces from the very outset of the war, literally during the very first hours and minutes. This is a strategic requirement because the enemy's first massive nuclear assaults could cause such civilian and military losses that the people and the country would be put in an extremely difficult situation. Therefore, not only is a high degree of combat readiness required of the Armed Forces but the entire country must also be specially prepared for war against the aggressor.

War conducted in such a way can fundamentally alter former concepts of how fighting develops during various stages of a war. At the same time, it attests to the extraordinary increase in the importance of the initial period of the war.

In the very first minutes of the war, the belligerents may use up their carriers, missiles and aircraft, accumulated in peacetime along with their stockpiles of nuclear weapons in order to destroy and devastate the enemy's most important targets throughout his entire territory, and to achieve their main political and military-strategic aims within a brief period of time and at the very outset of the war. Therefore the initial period of modern missile war will obviously be the main and decisive period and will predetermine the development and outcome of the entire war. The fighting in this period obviously will be most fierce and destructive. . . .

In the present epoch (in spite of the absence of a fatal inevitability of war and in spite of the unrelenting struggle for peace by the Soviet Union, the entire socialist camp, and all men of good will), the outbreak of war is not excluded. This conclusion derives from the insoluble economic and political contradictions of imperialism, the bitter international class struggle, the aggressive policy of world reaction, particularly of the American monopolists, and the intensified imperialist war preparations.

If the imperialist bloc initiates a war against the USSR or any other socialist state, it will inevitably become a *world war*, with the majority of the world's countries participating in it.

In its political and social essentials, *a new world war will be the decisive armed collision of two opposing world social systems. This war will inevitably end with the victory of the progressive, communist social and economic system over the reactionary, capitalist social and economic system, which is historically doomed to destruction.* The true balance of political, economic, and military forces of the two systems, which has changed in favor of the socialist camp, guarantees such an outcome of war. However, victory in a future war will not come of itself. It must be thoroughly prepared for and secured in advance. . . .

Given the acute class nature of a future world war, in which each side will aim for conclusive political and military results, the peoples' relation to the war will be of very great importance. In spite of the fact that large quantities of new kinds of military equipment will be used, *massive armed forces* will engage in combat. Many millions of people will be needed to meet war needs and to work in the economy. Therefore the attitude of the masses of the people toward the war will inevitably exert a decisive influence on the outcome of a future war.

From the point of view of weapons, a third world war will be a *missile and nuclear war.* The massive use of nuclear weapons, particularly thermonuclear, will make the war unprecedentedly destructive and devastating. Entire states will be wiped off the face of the earth. Missiles carrying nuclear warheads will be the main instruments for attaining the war's aims and for accomplishing the most important strategic and operational missions. Consequently, the leading branch of the armed forces will be the Strategic Missile Forces, and the role and mission of the other branches of the armed forces will be essentially changed. However, final victory will be attained only as a result of the combined efforts of all the branches of the armed forces. . . .

Since modern weapons permit exceptionally important strategic results to be achieved in the briefest time, both the *initial period of the war* and the methods of breaking up the opponent's aggressive plans by dealing him in good time a crushing blow will be of *decisive significance for the outcome of the entire war.* Hence, the main task of Soviet military strategy is working out means for reliably *repelling a surprise nuclear attack by an aggressor.* . . .

The enormous capacities of missiles and nuclear weapons make it possible to achieve the purposes of war within relatively short periods of time. Therefore, in the interest of our Motherland, it is necessary to develop and perfect the instruments and methods of combat with a view to *attaining victory over the aggressor, above all in the shortest possible time,* with minimum losses; but at the same time, it is necessary to make serious preparations for a protracted war.

The ability of the country's economy to mass produce military equipment, especially missiles, and to establish superiority over the enemy in modern weapons are the material prerequisites of victory. *The ability of the economy to assure the maximum power to the Armed Forces for dealing an annihilatory blow to the aggressor in the initial period of the war will be decisive for the outcome of a future war.*

Victory in war is determined not only by superiority in the military and technical sense, which, in turn, depends on the superiority of a nation's social, economic, and political system, but also by the ability to organize the defeat of the enemy and make effective use of available weapons. For this purpose, the country must be thoroughly and scientifically prepared for war against the aggressor; and a high level of military skill must be required of the command staff and troops. Success in a future war will also depend on how far the level of development of military strategy meets the demands of modern war. . . .

18. The Nature of the Offensive Under Conditions Where Nuclear Weapons Are Employed

Colonel A. A. Sidorenko

One of the most important features of offensive combat in nuclear war will be its greater *resoluteness* than formerly. This character of the offensive is determined by the political content of a future war and the sharply increased combat capabilities of the troops.

The clearly expressed class character of a future war will cause the unusual heating of the conflict and resoluteness of political and military goals which, naturally, will also affect the resoluteness of the battle.

The raising of the resoluteness of the goals of offensive battle is now based on the capabilities for the rapid smashing of the enemy with nuclear weapons, the use of the increased mobility and maneuverability of the troops, and the high morale and combat qualities of the personnel of the Armed Forces of the USSR.

Having nuclear weapons and long-range means for delivering them to the targets as well as highly-mobile troops, the attacker can now destroy the defensive simultaneously throughout its entire depth, dependably and in a short time with their skillful use, and can smash the enemy force without even having numerical superiority over him.

The presence of missiles possessing great range of fire provides the opportunity to concentrate the main efforts quickly on the selected direction by the maneuver of nuclear strikes with the dispersion of the disposition of the launchers themselves frontally and in depth. This makes it difficult for the enemy to determine the direction of the main blow, assures the attainment of surprise, and increases the survivability of the nuclear missile weapons. At the same time, the complete motorization of the troops and the high mobility of the *podrazdeleniye* and *chast'* permit exploiting the results of nuclear strikes with the dispersed disposition of the *podrazdeleniye* over a relatively large area.[1]

Consequently, the principle of mass in the contemporary offensive is expressed in the concentration of the main fire efforts on the main direction and, primarily, a large portion of the nuclear power, and in the rapid exploitation of the results of the nuclear strikes by the troops. The massing of forces and means now is closely linked with their disposition. In this, the dispersion of the troops on the field of battle should not lead to a reduction in the combat capabilities of the *podrazdeleniye* and *chast'*. Therefore, it is expedient only within limits which assure the successful accomplishment of the combat mission.

An important feature of the modern offensive is its highly *maneuverable*, dynamic character. This is caused by a number of reasons and, first, by the mass employment of nuclear weapons and the complete motorization and high degree of mechanization of the ground forces. If, in the last war, conditions for maneuver actions in the offensive arose only after breakthrough of the tactical depth of the enemy defense, now it is created from the very start of the offensive.

Nuclear weapons permit dependably neutralizing the enemy defense in the shortest time, forming breaches and gaps in it, and making maximum use of the extremely high mechanization of the ground forces. . . .

With the employment of nuclear weapons, offensive actions will be conducted primarily on tanks, armored personnel carriers, and helicopters. Battles in dismounted combat formations are possible only where the enemy offers strong resistance and where the terrain hinders the actions of the *podrazdeleniye* on vehicles. In many cases, displacement and maneuver will comprise the basic content of the combat actions of the attacking troops.

From this it follows that the *podrazdeleniye* and *chast'* will achieve success if they execute the maneuver skillfully, exploiting favorable conditions in the situation, deploy from march formations into approach march and combat formations in short times (on signal or command), organize and launch strikes with their own weapons, attack the enemy swiftly and, after completing his defeat, immediately close into columns and advance forward with the maximum utilization of march capabilities.

A characteristic feature of the offensive under contemporary conditions which is caused by the employment of nuclear weapons and the absence of a continuous front is the *conduct of the offensive over directions* of attack. This feature began to manifest itself back during the years of the Great

Patriotic War, primarily with the actions of tank and mechanized *soyedineniye* in the operational depth. Now, the offensive on directions of attack is also characteristic for overcoming the enemy's tactical defensive depth. This is caused, on the one hand, by the fact that with the great destructive capabilities of nuclear weapons and the high mobility of the attacking troops, there is no necessity as formerly to concentrate a tremendous mass of troops and combat equipment on narrow sectors and for actions on a continuous front and, on the other, by the necessity to disperse the troops in view of the danger of their destruction by nuclear weapons. . . .

A new characteristic feature of the offensive in nuclear war is the *conduct of combat actions under conditions of the presence of vast zones of contamination, destruction, fires, and floods.*

As a result of the mass employment of nuclear weapons by the warring sides, tremendous areas will be subjected to radioactive contamination; populated places, bridges, and other structures will be destroyed; and big centers of conflagration and inundation will be formed. The *podrazdeleniye* will not only be forced to fight on contaminated terrain, but also to overcome destruction, rubble, and other obstacles which may also be contaminated with radioactive substances. All this will have a great influence on the nature and methods of operation by the attacking troops.

The employment of large quantities of combat equipment, the conduct of the offensive to a great depth and at high rates, and the strain of the battles will require *a great expenditure of material*, and the enemy's influence with nuclear and chemical weapons will lead to *mass losses of troops and equipment* which will cause the necessity to eliminate the aftereffects of nuclear and chemical attack. . . .

The employment of various methods for the conduct of the offensive and the defeat of the enemy will be characteristic of modern war. In two world wars, with the establishment of a continuous positional front the only way of overcoming the enemy defense was the breakthrough. It consisted of making a breach in the enemy defense for its entire depth by launching a blow with a powerful troop formation on a narrow sector of the front. Second echelons, reserves, and mobile groups which consisted primarily of armored troops were committed to the breakthrough which had been made to develop the success into the depth and toward the flanks. . . .

In case of their employment, nuclear weapons will become the main means for destroying the enemy in a battle and operation. The success of the actions of the *soyedineniye* and *chast'* and the attainment of the goals of the offensive in short times will be directly dependent on the results of their employment. This obliges commanders to know the combat properties of nuclear weapons and the basic principles of their employment and to be able to analyze the situation quickly, perform the necessary calculations, and exploit the results of nuclear strikes. . . .

It is envisioned that surface nuclear bursts will be employed in those cases where, in accordance with the situation, it will not complicate the actions of the attacking troops by radiation contamination of the terrain with high radiation levels. In particular, nuclear strikes with a surface burst may be inflicted under favorable meteorological conditions against troops under cover in strong shelters as well as against sturdy objects for their destruction.

It is believed that nuclear weapons, as the main means of destruction, should be employed in all cases for the destruction of the most important targets and objectives. These include, first of all, enemy means of nuclear attack, large concentrations of his troops and especially armored troops, reserves, in particular tank reserves, artillery in firing positions, bridges, crossings, control posts and communications centers, objects in the troop rear area, defensive structures, and others. . . .

Depending on the nature of the objectives, the assigned degree of destruction, the availability of nuclear means, the terrain configuration, meteorological conditions, and other data on the situation, nuclear weapons can be employed in the form of single, group, and massed nuclear strikes.

A *single nuclear strike* is launched against an objective with one nuclear weapon. It is employed in those cases where the yield of the nuclear weapon assures inflicting the assigned degree of destruction on the objective.

A *group nuclear strike* is launched simultaneously by several nuclear weapons. It is employed when one nuclear weapon cannot attain the assigned degree of destruction of an important objective or when the employment of one more powerful nuclear weapon is impossible because of the situation.

A *massed nuclear strike* is a strike inflicted by a large number of nuclear weapons simultaneously or with the shortest possible time. Its goal is the destruction of enemy means of nuclear attack which have been discovered, the inflicting of destruction on the main formations of his troops, and disorganization of the rear, economy, and troop control.

Single and group nuclear strikes are usually employed in offensive combat. . . .

The principle of the employment of nuclear weapons in combination with other means of destruction follows from the fact that it is impossible to destroy all varied objectives on the battlefield with nuclear weapons alone. It is believed that nuclear weapons, as the main means of destruction, will be employed only for the destruction of the most important objectives; all other targets are neutralized and destroyed by the artillery, aviation, and the fire of tanks and other weapons. In other words, nuclear weapons are employed in combination with other means in accordance with the concept of the battle. . . .

The simultaneous launching of nuclear strikes is recognized as especially advantageous at the start of an offensive because, in this case, the troops will be able to exploit their results successfully. It is considered inexpedient to launch separate nuclear strikes during this period with large time intervals and at a considerable distance by the foreign military specialists. In their opinion, the maximum utilization of all their specific capabilities should be envisioned to obtain the greatest effect from the employment of this means of destruction.

The simultaneous launching of nuclear strikes throughout the entire depth of the enemy disposition is possible only with the presence of the corresponding quantity of nuclear ammunition and means for its delivery to the target as well as complete and reliable data on the enemy objectives throughout the entire depth of his defense. But the attacker will not always have such capabilities and data. It is considered difficult to determine the location of all objectives reliably and accurately for the launching of a simultaneous nuclear strike against them even with the presence of modern means of reconnaissance. And before destroying such objectives, final reconnaissance is required to refine their location. A portion of the objectives may be in motion or appear anew. Hence, the conclusion is drawn that one can hardly count on the fact that the attacker will succeed in destroying all important objectives with one simultaneous nuclear strike. In the course of the offensive it will often be necessary to launch nuclear strikes as the attacking troops advance and targets are disclosed for destruction by nuclear weapons.

The successive launching of nuclear strikes as the offensive is developed deprives the enemy of the opportunity to eliminate the aftereffects of nuclear bursts and restore the defense prior to the approach of the attacking troops. However, even with such a method of employing nuclear weapons the first nuclear strike, according to the assertion of the foreign military specialists, in principle should be the most powerful. If the attacker succeeds in destroying the most important targets and objectives of the defender with the very first nuclear volley, he will no longer be able to offer resistance to the attacker with either nuclear or conventional weapons. . . .

The employment of nuclear weapons creates the opportunity for the reduction in the duration of the fire preparation in comparison with the last war since the volume of missions accomplished by conventional means is reduced. In addition, it is necessary to also consider the increased effectiveness of the weapons themselves as well as the ammunition which they employ. In turn, the reduction in the fire preparation in connection with the employment of nuclear weapons permits increasing the use coefficient of technical capabilities of artillery. . . .

NOTES

1. The century-old military history, including the history of the Soviet Armed Forces, is convincing evidence that in an armed conflict of any scale—be it an engagement of *podrazdeleniye, chast'*, or *soyedineniye* or a battle of operational *ob"yedineniye*—only the offensive leads to the attainment of victory over the enemy. The offensive is the only type of combat actions of the troops, the employment of which attains the complete rout of the enemy and the seizure of important objectives and areas.

The essence of the offensive consists of having the troops that are conducting it destroy the enemy with all available means and, exploiting the results obtained, advance swiftly into the depth of his disposition, destroy and capture personnel, armament, and combat equipment belonging to the enemy, and seize specific territory.

[Editors' note] *Ob"yedineniye* is a Soviet term which refers to a major field force, such as a *front* or an army.

Soyedineniye is used by the Soviets to refer to a corps, a division, or a brigade. The components may be from a single arm or from various arms and services. The term also is used loosely for an army.

Chast' is a Soviet term which designates any unit of regimental or smaller size that is administratively self-contained and separately numbered. Examples of this are a rifle regiment, an engineer battalion of a rifle division, and a corps signal battalion.

Podrazdeleniye is the Russian term for "subdivision." It is used to refer to a subordinate unit of a *chast'*; it is any unit which cannot be fully identified numerically except by reference to the larger unit of which it is an integral part: battalions, companies, and platoons of a rifle regiment; the battalions and batteries of an artillery regiment; the companies of an engineer or signal battalion.

19. Modern Means of Waging War and Operational-Level Strategy

Col. General N. A. Lomov

An important feature of a modern offensive is the high maneuverability and dynamism of warfare as a result of equipping the troops with nuclear weapons and their complete motorization.

During the years of the last war, the advancing side first penetrated the tactical depth of the enemy defenses and only after this was able to conduct maneuvering operations in strategic depth. Nuclear weapons make it possible in the shortest period of time to cause great losses to the defending side, and to create breaches in its battle formations, while the high motorization of troops allows the use of these conditions for rapidly shifting efforts in depth. In other words, a possibility for conducting active maneuvering actions under the conditions of using nuclear weapons can be created from the very outset of the offensive.

The high maneuverability and dynamic nature of warfare cannot help but lead to sharp changes in the situation. Nuclear weapons create an opportunity to quickly alter the composition of the troop movings and the balance of forces of the sides, as well as make it necessary to change not only the methods but also the types of warfare. The increased role of the time factor is also caused by this. One of the decisive conditions for success in an operation is the anticipating of the enemy in making nuclear strikes, particularly against the enemy's nuclear missile weapons.

The radiation situation, as well as the presence of vast zones of destruction, fire and flooding will be a new factor having a major effect on the character and methods of an offensive in a nuclear war.

It is quite obvious that as a result of the massed use of nuclear weapons by the warring parties, inevitably enormous areas of radioactive contamination will occur, population points will be destroyed, and major areas of fires and flooding will be formed. This will lead to a change in the appearance and character of the terrain. In the course of the offensive the troops must cross not only individual areas of contamination, but also conduct warfare for a long time on contaminated terrain. This will complicate the conditions of the offensive, and will make it necessary to strengthen the protection of personnel. It will also necessitate the carrying out of additional measures for engineering support and the solution of a whole series of other tasks. All of this cannot help but tell substantially on the character and methods of actions of the advancing troops.

The most characteristic features of an offensive occurring under the effect of this factor are the following: 1) an offensive along the directions with the least radiation levels, bypassing areas of fires, and extensive ruins; 2) the consecutive crossing of zones of contamination depending upon the degree of protection for personnel; 3) in the course of the offensive a combination of different types of combat operations, the rapid transition from one to another, and the conducting of them simultaneously. These features inevitably will cause changes in the organization of battle formations of the troops, as well as in the depth and content of their missions. . . .

Combat under the conditions of prepared enemy defenses, with the impossibility of skirting them or capturing them on the move, will begin with a breakthrough which consists in breaking the defenses by nuclear weapons and air strikes, by artillery and tank fire, and an offensive of the subunits with the subsequent development of actions in depth.

The offensive starts by making nuclear strikes with tactical missiles and aviation for the purpose of destroying the means of mass destruction and defeating the basic enemy grouping on the axis of the main strike of advancing troops.

After the nuclear strikes, for neutralizing and destroying the enemy which has not been destroyed by the nuclear weapons on the front line and in the tactical depth, preparatory firing and fire support of the advancing subunits will be carried out.

The experience of wars teaches that the concentration of basic efforts on the main axis is always a most important condition for success in combat. It is impossible to be superior over the enemy in all areas, but there must be superiority on the main, decisive axis in order to make a crushing strike and achieve victory.

The concentrating of basic efforts on the main axis maintains its significance under present-day conditions; however, its content has changed. In contrast to previous wars, when on selected axes, for example, in conducting combat by the subunits of the land forces, a multifold superiority was created over the enemy in infantry, tanks, and artillery, under present-day conditions, superiority in forces and means is achieved primarily by making nuclear strikes at the main enemy groupings with the means of the superior chiefs. By making nuclear strikes primarily against the nuclear means and tank groupings it is possible to quickly and sharply alter the balance of forces and achieve the necessary superiority over the enemy.

In modern combat, superiority over the enemy is achieved primarily by concentrating the fire efforts of the forces and mainly the nuclear strikes. The concentration of the troops on the main axis should be carried out on the basis of strict and detailed calculations so that their grouping and number make it possible primarily to utilize the results of the nuclear strikes, to successfully pierce the enemy defenses, and rapidly complete the destruction of the forces which have remained after the nuclear strikes. It is extremely important to concentrate the necessary forces and means on the direction of the main strike in a rapid and covert manner, from different directions and only for the time necessary for making the strike. As soon as such necessity is passed, the troops must be immediately dispersed. This is caused by the constant threat of the enemy's use of nuclear weapons. The dispersion of the troops on the battlefield at the same time should provide for carrying out the mission; it should not obstruct a new concentration of the units and subunits for making a strong strike against the enemy, and at the same time should exclude the possibility of the simultaneous destruction of several subunits by a medium-powered nuclear explosion. . . .

Surprise in combat makes it possible to catch the enemy unaware, to spread panic in his ranks, to paralyze the will to resist and to sharply reduce the combat capability of troops, to disorganize command, and to create favorable conditions for achieving victory even over superior forces. The presence of nuclear weapons, the increased mobility and maneuverability of tanks and other means of motorization make it possible to achieve surprise strikes and attacks. Surprise is achieved by confusing the enemy of one's intentions, by keeping secret the overall purpose of the forthcoming actions and preparations for them, by rapid and concealed concentration and deployment of forces in the region of making the strikes, by the unexpected use of weapons, and particularly nuclear ones, as well as by the use of tactical procedures and new weapons unknown to the enemy. In other words, surprise is achieved by making strikes against the enemy at a place and at a time where he does not expect them. . . .

Under conditions of using nuclear weapons, when rapid and abrupt changes occur in the situation, while the subunits and ships are forced to fight in open combat and approach march formations (formations and orders) to avoid mass destruction, with the presence of spaces and exposed flanks, the significance of maneuvering forces and means in the course of combat has immeasurably risen.

In any type of combat and under any conditions a maneuver can achieve its goal in the instance that it is carried out quickly. The slightest delay under present conditions can nullify a correctly conceived maneuver. This is why the maneuver should be simple in terms of goal and not require a great deal of time for implementing it.

20. Military Strategy

Marshal N. V. Ogarkov

In the development of Soviet military strategy after the Second World War, two periods can be distinguished. The first covers the years from 1945 to 1953, when Soviet military strategy was being developed on the basis of the rich experience of the Great Fatherland War, but when at the same time the presence of nuclear weapons in the United States was taken into consideration. The Armed Forces of the USSR were equipped with more improved military equipment. Methods of conducting strategic operations were worked out in the light of increased striking and firing power, mobility and troop maneuverability.

The second period of development of Soviet military strategy (post-1954) is linked to the introduction of nuclear weapons and missiles into the Armed Forces and the emergence of new armed services and troop branches. It is characterized by major changes in both the theory and the practice of military affairs. Soviet military strategy was confronted with a number of fundamentally new tasks, the most important of which were: analysis of the nature and methods of conducting nuclear war, structuring the armed forces in accordance with the possibility of repulsing the aggressor's massive nuclear strikes, maintaining the Armed Forces in a state of ever high combat readiness and ensuring their orderly entry into the war under whatever conditions it may begin, etc. The creation of the Warsaw Pact Organization in 1955 confronted military strategy with a new task: that of elaborating the general bases of military strategy of the Socialist Community in which are organically combined the international and national interests of the allied countries. . . .

Like Soviet military doctrine in general, Soviet military strategy has a particularly defensive orientation and does not envisage any forestalling strikes or premeditated attacks. Its main tasks are to work out methods of repulsing an aggressor's attack and destroying him through the conduct of decisive operations. In contrast to the military strategy of the imperialist states, which openly proclaim an arms race and the establishment of military-technical superiority, Soviet military strategy proceeds from the necessity of securing for the Soviet Armed Forces all that is required for the country's defense and the aggressor's destruction and the maintenance of means at that level which will guarantee the homeland's security. Soviet military strategy takes into account the possibility that the USSR and other Socialist countries may not permit a probable enemy to achieve military-technical superiority, but at the same time does not make it as its aim to achieve military-technical superiority over other countries. . . . Soviet military strategy proceeds from the possibility of averting war in the modern era, but at the same time takes into account the military preparations of imperialist states and the danger of a war beginning as a result of their aggressive policies.

Soviet military strategy considers a future world war, should the imperialists succeed in unleashing it, as a decisive clash between two opposing world socio-economic systems—socialism and capitalism. It is assumed that the majority of the world's states may be drawn into such a war, simultaneously or successively. It will be a global fight of multi-million-men coalitions, unprecedented in scale and violence. The armed forces will also pursue the most decisive political and strategic goals without any compromise. The entire military, economic and spiritual might of the belligerent states, coalitions and social systems will be fully utilized in the course of such a war.

Soviet military strategy takes into account the possibility that a world war can begin, and be waged for a certain time, with the use of only conventional weapons. However, the expansion of military operations can lead to its escalation into a general nuclear war, the chief means of which will be nuclear weapons, primarily strategic ones. At the foundation of Soviet military strategy lies the thesis that the Soviet Union, proceeding from its principled policy, will not be the first to use these weapons. In principle, it is against the use of *mass destruction weapons*. However, any possible aggressor must distinctly realize that in the event of a nuclear missile attack on the Soviet Union or on the other Socialist countries, he will receive an annihilating retaliatory strike.

It is considered that with modern mass destruction, a world war will be of comparatively short duration. However, given the enormous military and economic potentials of the belligerent states' coalitions, the possibility cannot be excluded that the war could also be protracted. Soviet military strategy proceeds from the fact that if a nuclear war is foisted upon the Soviet Union, then the Soviet people and their armed forces must be ready for the most severe and prolonged trials. In this case, the Soviet Union and the fraternal Socialist states will, compared with the imperialist states, possess certain advantages resulting from their just war aims and the progressive nature of their social and state order. This provides them with objective possibilities for achieving victory. However, in order for these possibilities to be realized, the timely and all-around preparation of the country and the armed forces is necessary.

Soviet military strategy also takes into consideration the possible occurrence of *local wars* whose political character is determined on the basis of class positions and the Leninist thesis on just and unjust wars. While supporting national liberation wars, the Soviet Union resolutely comes out against local wars unleashed by the imperialists, considering not only their reactionary essence, but also the great danger linked with the possibility of their expansion into a world war.

Assessing the strategic content of wars, Soviet military strategy considers that war consists of a complex system of interconnected, large-scale, simultaneous and successive strategic operations, including operations in continental theaters of combat operations (TVD). The general objective of every such operation is an element of the overall military-political aims of the war, connected with ensuring the defense and holding of important regions of one's territory and, when possible, with the destruction of specific strategic groupings of the opponent. . . .

Within the framework of the strategy, operations in continental TVD can be conducted: by initial and subsequent operations on fronts and in coastal areas; by initial and subsequent operations of fleets, air, anti-air, airborne, naval landing, combined landing and other operations, as well as by inflicting nuclear missile and aviation strikes. Other types of strategic operations can also be conducted. Contemporary operations will be characterized by their enlarged scope, the bitter struggle to seize and retain the strategic initiative, high maneuverability of operations by armed forces groups in separate directions and under conditions of an absence of continuous fronts, deep and mutual penetrations of the sides, and swift and abrupt changes in the operational-strategic situation. The aims of all these operations—just as victory in war in general—can be attained only by the continued efforts of all armed forces services and troop branches. In this respect, one of the most important principles of Soviet military strategy is organization and support of close and uninterrupted interaction of strategic operations.

Soviet military strategy considers that waging modern war requires the presence of multi-million-men mass armies. Since their upkeep in peacetime is practically impossible and uncalled for in terms of the country's defense requirements, a suitable mobilization of the Armed Forces is envisaged. In connection with the possibility of an aggressor's surprise attack, a special place in Soviet military strategy is assigned to ensuring the combat readiness of the Armed Forces. . . .

While considering the offensive as the main type of strategic operation, Soviet military strategy nevertheless acknowledges an important role for defense in war, the necessity and possibility of organizing and conducting it on a strategic scale for the purpose of frustrating or repulsing the opponent's offensive, holding (defending) certain territory, gaining time to concentrate needed forces, and economizing forces in one direction while creating superiority over the opponent in another direction. Along with this, there is the consideration that defense on any scale must be active in order to create the conditions for shifting to the offensive (counter-offensive) for the purpose of achieving the total destruction of the enemy. . . .

On the one hand, troop control in modern conditions is becoming ever more complicated and is continuously increasing the volume of work to be executed by the organs of strategic leadership; on the other hand, the time for executing [such work] is decreasing. This increases the need for firmness, flexibility, operability, and secrecy of control under conditions of electronic counter-measures on the part of the opponent.

Soviet military strategy serves peace, the people's security and the interests of defending the achievements of socialism against the encroachments of any aggressor. It is developed and perfected in accord with changes in the world's military-political situation and the continually growing economic and morale-political potential of the Soviet Union and the Socialist community.

21. The Morale Factor in Modern War

Major General S. K. Il'in

A great deal of attention in the Soviet Union has been, and continues to be, devoted to the study and analysis of the character and role of the morale factor in modern war, and especially in a nuclear war, and to its significance for the conduct of the war and the attainment of victory. Such studies and analyses are said to be part of Soviet military science and strategy, of assessments of the military balance with the West and of the search for ways and methods of achieving Soviet superiority in this respect over all possible enemies.

The new means of armed struggle and their methods of use in combat introduce very profound changes in the understanding of the role and significance of the morale factor in the attainment of victory over the enemy. . . . In modern war the demands on the political, combat and psychological qualities of the warring masses increase immeasurably.

In a war in which nuclear-missile weapons are employed, in a certain sense changes will occur in the interdependence of the components, the two sides of the morale factor: the strength of the morale of the people and the morale of the army. . . .

In a modern war the rear (i.e., homeland) will be engulfed in its entirety by the flame of war. The tasks performed by those working in the rear will at times entail no less risks to their lives than to those of the participants in combat actions, and will also require enormous courage, steadfastness and dedication on the part of everyone at his post. . . .

The morale of the population must be very high in order to stand up to such tests and in order not to lose self-control, the will to fight and faith in victory over the enemy under conditions of mass fatalities and destruction. Despite this, the urban inhabitants, while suffering incredible privations and dangers, will have to restore destroyed installations and continue the production of weapons, ammunition and everything that is required by the front.

It is obvious that victory in such a war will depend to an enormous degree on the ability of the population to maintain essential steadfastness and courage, not to succumb during moments of great trials to feelings of hopelessness and despair, and to preserve its strength and will to struggle with the enemy. In this connection, civil defense acquires exceptional importance because its effectiveness in large measure depends on the viability of the organs of administration, industrial enterprises and other installations in the rear. Work in military-patriotic indoctrination and the morale-psychological tempering of the population must be conducted still more actively.

As the aggressive, imperialist forces prepare for war against the USSR and other countries of the Socialist fraternity, they cannot fail to be concerned about the stability of the morale factor of their own populations in the event of nuclear missile strikes on their rears. . . .

Naturally, a new war would also be a great misfortune for the Soviet people; it would cause them innumerable calamities and suffering. Even so, our people, united and monolithic in the social-political sense, clearly perceiving the justness of the character of the struggle against imperialism, have infinitely more "reserve of firmness" of their morale forces than the populations of capitalist countries. Their higher morale spirit will assure the fuller utilization of the country's economic capabilities for the purpose of achieving victory, and will provide a firm foundation for high morale-political and combat qualities of the fighting troops. . . .

Nuclear weapons will exercise an enormous influence on the mass of fighting troops. Resort to nuclear weapons will alter the entire character of the battle, incredibly complicate the character of combat actions. . . . Now, even if the people survive nuclear detonation, they cannot be certain that the threat to their lives is over. This threat may continue to exist in the form of radioactive contamination, and this state of uncertainty, the continuing fear for one's life, may have no less a psychological effect than the nuclear explosion itself.

One of the factors reinforcing the effects of nuclear weapons on the people's psychology is the surprise and newness of its use. . . .

Soldiers and officers will be exposed to enormous psychological tension while crossing the zones of radioactive contamination, when there will be a real, yet invisible threat to people's lives. A difficult problem will arise in the psychological preparation of soldiers in connection with the liquidation of the effects of enemy nuclear strikes. . . .

The problem of overcoming fear in combat is becoming greater. One can sometimes hear it alleged, as it applies to our army which is based on a high level of political class consciousness, that it is incorrect to speak of such phenomena. Such arguments appear superficial, unscientific. Fear is a person's psychological reaction to a threat. . . . It serves the instinct of self-preservation. . . . It is another matter that a person strong in spirit can suppress his fear. . . .

Panic can arise in subunits which lack firm leadership, a reliable, well-informed combat activist group and the necessary information about the course of the battle. . . .

Because lack of information is an important cause of negative phenomena in the actions of troops in dangerous situations, all necessary measures must be taken to maintain it [information] at the necessary level. The more complex the situation, the greater is the need to provide the troops with information on the state of their units and subunits and enemy actions. This is one of the essential conditions to prevent the occurrence of panic.

An important measure in the prevention of panic is the determined struggle against all types of rumors which may stir up the troops. . . .

Strategy takes into account the correlation of material and morale forces of the sides on the eve of war and from its beginning. . . .

In solving specific questions in the conduct of war, strategy takes into account the correlation of the morale of the warring sides. It determines the type of strategic operations, the influencing of one or another link in the enemy's coalition, the sequence and tempo in carrying out operations, etc. In a coalition war it is required that the morale of the troops and population of each warring state be taken into account and plans for strategic operations be developed accordingly.

Should a new war occur, the use of new means of combat will demand that the commanders take into account to an even greater degree the morale factor in the implementation of strategic plans and intentions. This is explained, first of all, by the fact that military strategy now has at its disposal decisive means of armed combat. It must take into account the profound morale-psychological effects of nuclear missile strikes on the enemy's important installations and military-strategic targets. At the same time, the high command will take into account to a maximum extent the possibility of nuclear strikes on the country's [USSR's] rear installations and their effect on the population's morale. It is obvious that military strategy will have to measure in a largely different manner the morale of the armies of the two sides, the attitudes of the troops to the policies of the warring states and their war aims. It will also be necessary to take account of the degree of steadfastness in the face of nuclear missile strikes exhibited by the armies of individual countries that are members of the aggressive imperialist grouping.

22. Contrasts in American and Soviet Strategic Thought

Fritz W. Ermath

We are having trouble with Soviet strategic doctrine. Soviet thinking about strategy and nuclear war differs in significant ways from our own. To the extent one should care about this—and that extent is a matter of debate— we do not like the way the Soviets seem to think. Before 1972, appreciation of differences between Soviet and American strategic thinking was limited to a small number of specialists. Those who held it a matter of high concern for policy were fewer still. Since that time, concern about the nature, origins, and consequences of these differences is considerably more widespread, in large measure as a result of worry about the Soviet strategic arms buildup and the continued frustrations of achieving a real breakthrough in SALT. . . .

THE NEED TO UNDERSTAND STRATEGIC DOCTRINE

Let us define "strategic doctrine" as a set of operative beliefs, values, and assertions that in a significant way guide official behavior with respect to strategic research and development (R&D), weapons choice, forces, operational plans, arms control, etc. The essence of U.S. "doctrine" is to deter central nuclear war at relatively low levels of arms effort ("arms race stability") and strategic anxiety ("crisis stability") through the credible threat of catastrophic damage to the enemy should deterrence fail. In that event, this doctrine says it should be the aim and ability of U.S. power to inflict maximum misery on the enemy in his homeland. Making the world following the outbreak of nuclear war more tolerable for the United States is, at best, a lesser concern. Soviet strategic doctrine stipulates that Soviet strategic forces and plans should strive in all available ways to enhance the prospect that the Soviet Union could survive as a nation and, in some politically and militarily meaningful way, defeat the main enemy should deterrence fail—and by this striving help deter or prevent nuclear war, along with the attainment of other strategic and foreign policy goals.

These characterizations of U.S. and Soviet strategic doctrine and the differences between them are valid and important. Had U.S. strategic policy been more sensitive over the last ten years to the asymmetry they express, we might not find ourselves in so awkward a present situation. We would have been less sanguine than we were about prospects that the Soviets would settle for an easily defined, non-threatening form of strategic parity. We would not have believed as uncritically as we did that the SALT process was progressing toward a common explication of already tacitly accepted norms of strategic stability. . . .

The most influential factor that has inhibited lucid comparisons of U.S. and Soviet strategic thinking has been the uncritically held assumption that they had to be very similar, or at least converging with time. Many of us have been quite insensitive to the possibility that two very different political systems could deal very differently with what is, in some respects, a common problem. We understood the problem of keeping the strategic peace on equitable and economic terms—or so we thought. As reasonable men the Soviets, too, would come to understand it our way. . . .

For many years the prevailing U.S. concept of nuclear war's consequences has been such as to preclude belief in any military or politically meaningful form of victory. Serious effort on the part of the state to enhance the prospect for national survival seemed quixotic, even dangerous. Hence stems our relative disinterest in air defenses and civil defenses over the last fifteen years, and our genuine fear that ballistic missile defenses would be severely destabilizing. Growth of Soviet nuclear power has certainly clinched this view of nuclear conflict among critical elements of the U.S. elite. But even when the United States enjoyed massive superiority, when the Soviet Union could inflict much less societal damage on the United States, and then only in a first strike (through the early 1960s), the awesome destructiveness of nuclear weapons had deprived actual war with these weapons of much of its strategic meaning for the United States.

The Soviet system has, however, in the worst of times, clung tenaciously to the belief that nuclear war cannot—indeed, must not—be deprived of strategic meaning, i.e., some rational relationship to the interests of the state. It has insisted that, however awful, nuclear war must be survivable and some kind of meaningful victory attainable. . . .

In recent years the changing strategic balance has had the effect of strengthening rather than weakening the asymmetry of the two sides' convictions on this matter. Dubious when the United States enjoyed relative advantage, strategic victory and survival in nuclear conflict have become the more incredible to the United States as the strategic power of the Russians has grown. For the Soviets, however, the progress of arms and war-survival programs has transformed what was in large measure an ideological imperative into a more plausible strategic potential. For reasons to be examined below, Soviet leaders possibly believe that, under favorable operational conditions, the Soviet Union could win a central strategic war today. Notwithstanding strategic parity or essential equivalence of force, they may also believe they could lose such a conflict under some conditions.

The concept of deterrence early became a central element of both U.S. and Soviet strategic belief systems. For both sides the concept had extended or regional dimensions, and a good deal of political content. There has, in short, been some functional symmetry between the deterrence thinking of the two sides: restraint of hostile action across a spectrum of violence by the threat of punishing consequences in war. Over time and with shifts in the overall military balance, latent asymmetries of thinking have become more pronounced. For the United States, strategic deterrence has tended to become the only meaningful objective of strategic policy, and it has become progressively decoupled from regional security. For the Soviets, deterrence—or war prevention—was the first, but not the only and not the last objective of strategy. Deterrence also meant the protection of a foreign policy that had both offensive and defensive goals. And it was never counterposed against the ultimate objective of being able to manage a nuclear war successfully should deterrence fail. The Soviet concept of deterrence has evolved as the strategic balance has improved for the Soviet Union from primary emphasis on defensive themes of war prevention and protection of prior political gains to more emphasis on themes that include the protection of dynamic processes favoring Soviet international interests. Repetition of the refrain that detente is a product of Soviet strategic power, among other things, displays this evolution.

Strategic stability is a concept that is very difficult to treat in a comparative manner because it is so vital to U.S. strategic thinking, but hardly identifiable in Soviet strategic writings. In U.S. thinking, strategic stability has meant a condition in which incentives inherent in the arms balance to initiate the use of strategic nuclear forces and, closely related, to acquire new or additional forces are weak or absent. In an environment dominated by powerful offensive capabilities and comparatively vulnerable ultimate values, i.e., societies, stability was thought to be achievable on the basis of a contract of mutually vulnerable societies and survivable offensive forces. Emphasis on force survivability followed, as did relative uninterest in counterforce, active, and passive defenses.

Soviet failure to embrace these notions is sufficiently evident not to require much elaboration. One may argue about Soviet ability to overturn stability in U.S. terms, but not about Soviet disinclination to accept the idea as a governing principle of strategic behavior. Soviet acceptance of the ABM agreement in 1972 is still frequently cited as testimony to some acceptance of this principle. It is much more probable, however, that the agreement was attractive to Moscow because superior U.S. ABM technology plus superior U.S. ABM penetrating technology would have given the United States a major advantage during the mid- to late 1970s. In a unilateral sense, the Soviets saw the ABM agreement as stabilizing a process of strategic catch-up against a serious risk of reversal. But it did not mean acceptance of the U.S. stability principle.

The United States has always been relatively sensitive to the potential of technology to jeopardize specific formulae for achieving stability, although it has been relatively slow to perceive the pace and extent to which comparative advantage has shifted from passive survivability to counterforce technologies. The Soviets have also been sensitive to destabilizing technologies. But they have tended to accept the destabilizing dynamism of technology as an intrinsic aspect of the strategic dialectic, the underlying engine of which is a political competition not susceptible to stabilization. For the Soviets, arms control negotiations are part of this competitive process. Such negotiation can help keep risks within bounds and also, by working on the U.S. political process, restrain U.S. competitiveness.

Soviet failure to embrace U.S. strategic stability notions as strategic norms does not mean, as a practical matter, that the Soviets fail to see certain constellations of weapons technology and forces as having an intrinsic stability, in that they make the acquisition of major advantages very difficult. What they reject is the notion that, in the political and technical world as they see it, those constellations can be frozen and the strategic competition dimension thereby factored out of the East-West struggle permanently or for long periods. . . .

It has long been U.S. policy to assure that U.S. strategic nuclear forces are seen by the Soviets and our NATO allies as tightly coupled to European security. Along with conventional and theater nuclear forces, U.S. strategic nuclear forces constitute an element of the NATO "triad." The good health of the alliance politically and the viability of deterrence in Europe have been seen to require a very credible threat to engage U.S. strategic nuclear forces once nuclear weapons come into play above the level of quite limited use. For more than twenty years NATO's official policy has had to struggle against doubts that this coupling could be credible in the absence of clear U.S. strategic superiority. Yet the vocabulary we commonly employ itself tends to strain this linkage in that theater nuclear forces are distinguished from strategic. Ironically, the struggle to keep so-called Forward Based Systems out of SALT, because we could not find a good way to bring in comparable Soviet systems, tended to underline the distinction. In our thinking about the actual prosecution of a strategic conflict, once conflict at that level begins we tend to forget about what might be the local outcome of the regional conflict that probably precipitated the strategic exchange.

The Soviets, on the other hand, appear to take a more comprehensive view of strategy and the strategic balance. Both in peacetime political competition and in the ultimate test of a central conflict, they tend to see all force elements as contributing to a unified strategic purpose, national survival and the elimination or containment of enemies on their periphery. The U.S.S.R. tends to see intercontinental forces, and strategic forces more generally, as a means to help it win an all-out conflict in its most crucial theater, Europe. Both institutionally and operationally, Soviet intercontinental strike forces are an outgrowth and extension of forces initially developed to cover peripheral targets. Land combat forces, including conventional forces, are carefully trained and equipped to fight in nuclear conditions. In the last decade, the emergence of a hostile and potentially powerful China has more firmly riveted the "rimland" of Eurasia into the Soviet strategic perspective.

Whatever the consequence of a central U.S.-Soviet nuclear conflict for their respective homelands, it could well have the effect of eliminating U.S. power and influence on the Eurasian landmass for a long time. If, by virtue of its active and passive damage-limitation measures, the Soviet Union suffered measurably less damage than did the United States, and it managed to intimidate China or destroy Chinese military power, the resultant Soviet domination of Eurasia could represent a crucial element of "strategic victory" in Soviet eyes. In any case, regional conflict outcomes seem not to lose their significance in Soviet strategy once strategic nuclear conflict begins.

Comparative study of U.S. and Soviet strategic doctrine should give attention to a closely related matter: how we perceive and measure force balances. . . .

The important point, however, is a conceptual one: Unlike the typical U.S. planner, the Soviet planner does not appear to see the *force* balance prior to conflict as a kind of physical reification of the war outcome and therefore as a measure of strategic strength by itself. Rather he seems to see the force balance, the "correlation of military forces," as one input to a complex combat process in which other factors of great significance will play, and the chief aim of which is a new, more favorable balance of forces. The sum of these factors is strategy, and strategy is a significant variable to the Soviet planner.

As a generalization, then, the Soviet planner is very sensitive to operational details and uncertainties. Because these factors can swing widely, even wildly, in different directions, a second generalization about Soviet force analysis emerges: a given force balance in peacetime can yield widely varying outcomes to war depending on the details and uncertainties of combat. Some of those outcomes could be relatively good for the Soviet Union, others relatively bad. The planner's task is to improve the going-in force balance, to be sure. But it is also to develop and pursue ways of waging war that tend to push the outcome in favorable directions.

This kind of thinking occasions two very unpleasant features in Soviet military doctrine: a strong tendency to preempt and a determination to suppress the enemy's command and control system at all costs. The Soviets tend to see any decision to go to nuclear war as being imposed on them by a course of events that tells them "war is coming," a situation they bungled memorably in June 1941. It makes no difference whose misbehavior started events on that course. Should they find themselves on it, their operational perspective on the factors that drive war outcomes places a high premium on seizing the initiative and imposing the maximum disruptive effects on the enemy's forces *and* war plans. By going first, and especially disrupting command and control, the highest likelihood of limiting damage and coming out of the war with intact forces and a surviving nation is achieved, virtually independent of the force balance.

This leads to a final generalization. We tend rather casually to assume that, when we talk about parity and "essential equivalence" and the Soviets about "equal security," we are talking about the same thing: functional strategic stability. We are not. The Soviets are talking about a going-in force balance in which they have an equal or better chance of winning a central war, if they can orchestrate the right scenario and take advantage of lucky breaks. It is the job of the high command to see that they can. If it fails to do so, the Soviet Union could possibly lose the war. This is not stability in our terms.

Again, this is not to argue that the Soviets do not foresee appalling destruction as the result of any strategic exchange under the best of conditions. In a crisis, Soviet leaders would probably take any tolerable and even some not-very-tolerable exits from the risk of such a war. But their image of strategic crisis is one in which these exits are closing up, and the "war is coming." They see the ultimate task of strategy to be the provision of forces and options for preempting that situation. This then leads them to choose strategies that, from a U.S. point of view, seem not particularly helpful in keeping the exits open, and even likely to close them off. . . .

In sum, there are fundamental differences between U.S. and Soviet strategic thinking, both at the level of value and at the level of method. The existence of these differences and, even more, our failure to recognize them have had dangerous consequences for the U.S.-Soviet strategic relationship.

One such might be called the "hawk's lament." Failing to appreciate the character of Soviet strategic thinking in relation to our own views, we have underestimated the competitiveness of Soviet strategic policy and the need for competitive responsiveness on our part. This is evident in both our SALT and our strategic force modernization behavior.

A second negative effect might be termed the "dove's lament." By projecting our views onto the Soviets, and failing to appreciate their real motives and perceptions, we have underestimated the difficulties of achieving genuine strategic stability through SALT and over-sold the value of what we have achieved. This has, in turn, set us up for profound, perhaps even hysterical, disillusionment in the years ahead, in which the very idea of negotiated arms control could be politically discredited. If present strategic trends continue, it is not hard to imagine a future political environment in which it would be difficult to argue for arms control negotiations even of a very hard-nosed sort.

The third and most dangerous consequence of our misunderstanding of Soviet strategy involves excessive confidence in strategic stability. U.S. strategic behavior, in its broadest sense, has helped to ease the Soviet Union onto a course of more assertive international action. This has, in turn, increased the probability of a major East-West confrontation, arising not necessarily by Soviet design, in which the United States must forcefully resist a Soviet advance or face collapse of its global position, while the Soviet Union cannot easily retreat or compromise because it has newly acquired global power status to defend and the matter at issue could be vital. In such conditions, it is all too easy to imagine a "war is coming" situation in which the abstract technical factors on which we rest our confidence in stability, such as expected force survival levels and "unacceptable damage," could crumble away. The strategic case for "waiting to see what happens," for conceding the operational initiative to the other side—which is what crisis stability is all about—could look very weak. Each side could see the great operational virtues of preemption, be convinced that the other side sees them too, and be hourly more determined that the other side not have them. This, in any case, could be the Soviet way of perceiving things. Given the relative translucence of U.S. versus Soviet strategic decision processes, however, our actual ability to preempt is likely to be less than the Soviets', quite apart from the character of the force balance. Add to that the problem of a vulnerable Minuteman ICBM force and you have a potentially very nasty situation.

What we know about the nature of our own strategic thinking and that of the Soviet Union is not at all comforting at this juncture. The Soviets approach the problem of managing strategic nuclear power with highly competitive and combative instincts. Some have argued that these instincts are largely fearful and defensive, others that they are avaricious and confident. My own reading of Russian and Soviet history is that they are both, and, for that, the more difficult to handle.

23. Remarks on the Objectives of Soviet Nuclear Strategy: 1

Leonid Brezhnev

SPEECH IN TULA DURING
PRESENTATION OF A GOLD
STAR MEDAL TO THE CITY

Comrades! The lessons of the last war summon us to vigilance. Yes, fascism was overthrown. But there are still fascists and profascist regimes. Some people still dream of revenge. There are aggressive forces that are by no means inactive. This must not be forgotten. . . .

Soviet people warmly approve of the Party's foreign policy. They know that this policy safeguards their homeland from war, is in the interests of all peoples, opens up scope for friendship and cooperation among them and serves the cause of social progresss throughout the planet. . . .

No country has ever put before mankind a program aimed at lessening and then fully eliminating the danger of a new war that was as broad, concrete and realistic as the Soviet Union's. This program includes such a global measure as a World Treaty on the Nonuse of Force in International Relations. It encompasses all the main problems connected with the arms race and outlines effective steps for curbing that race and for disarmament. It is aimed at preventing new types and new systems of weapons of mass destruction from appearing and at completely banning nuclear tests. The Soviet Union has proposed to the U.S. that the two countries mutually refrain from the creation of new types of submarines and strategic bombers. . . .

The important proposals set forth at the recent meeting of the Warsaw Treaty's Political Consultative Committee are convincing new proof of the inherently peace-loving nature of the GDR [German Democratic Republic, or East Germany], Poland, Rumania, the USSR and Czechoslovakia. I mean the proposals that all participants in the all-European conference pledge not to be the first to use nuclear weapons against any other participant and that the number of members of the Warsaw Treaty and NATO not be expanded. . . .

Of course, comrades, we are improving our defenses. It cannot be otherwise. We have never neglected the security of our country and the security of our allies, and we shall never neglect it. (*Prolonged applause.*)

But the allegations that the Soviet Union is going beyond what is sufficient for defense, that it is striving for superiority in armaments with the aim of delivering a "first strike," are absurd and utterly unfounded. At a meeting with prominent representatives of the American business world not so long ago I said, and I want to reemphasize, that the Soviet Union has always been and continues to be a staunch opponent of such concepts. (*Applause.*)

Our efforts are aimed at preventing both first and second strikes and at preventing nuclear war altogether. (*Applause.*) Our approach to these questions can be formulated thusly: The Soviet Union's defense potential should be sufficient to deter anyone from disturbing our peaceful life. (*Prolonged applause.*) Not a course aimed at superiority in armaments but a course aimed at their reduction, at lessening nuclear confrontation—that is our policy. (*Applause.*)

On behalf of the Party and the entire people, I declare that our country will never embark on the path of aggression and will never lift its sword against other peoples. (*Stormy applause.*)

It is not we but certain forces in the West that are jacking up the arms race ever higher, above all, the race in nuclear arms. It is not we but those forces that are throwing hundreds of billions into the bottomless pit of military preparations that are the initiators of swollen military budgets. It is these forces, acting under the false pretext of a "Soviet menace," that constitute the aggressive line in the international politics of our time.

As we all know, the easing of international tension is being achieved at the cost of enormous efforts. Moreover, it is not easy to preserve the accumulated political capital of detente. But no difficulties and obstacles will make us retreat. There is no more urgent and vitally important task than to make peace lasting and inviolable. (*Prolonged applause.*)

Statesmen who understand their responsibility to millions of people, their responsibility for the fate of nations, are obliged to consider the peoples' will for peace. As far as the Soviet Union is concerned, we shall not be found wanting. (*Applause.*)

We are prepared, in conjunction with the new administration in the United States, to carry out a major new advance in relations between our countries.

First of all, we are convinced that the work of preparing a strategic arms limitation agreement on the basis of which an accord was reached in Vladivostok in late 1974 ought to be completed as early as possible. Some politicians in Washington are now voicing their regret that this agreement still remains unsigned. But regrets are only regrets—lost time cannot be regained, and it is important that practical conclusions be drawn from this.

In the U.S., too, people are asking what will happen if such conclusions are not drawn. A few days ago an influential American newspaper wrote that in such an event "the Soviet Union and the United States will start the creation of a new generation of nuclear arms that in practical terms will be impossible to control."

Such a prospect does not suit us. I repeat: Time will not wait, and the conclusion of an agreement must not be postponed.

Needless to say, the Soviet Union is prepared to go further in questions of strategic arms limitation. But first it is necessary to consolidate what has already been achieved and to implement what was agreed upon in Vladivostok, especially because the interim agreement expires this October. Then we could at once move on to talks on more far-reaching measures. Otherwise, what might happen is that, by adding new questions to those now being discussed, we would only complicate and delay still more the solution of the problem as a whole.

It is urgently necessary to do a more reliable job of presenting the proliferation of nuclear weapons and to increase the effectiveness of the nonproliferation regime established by the well-known treaty. We are prepared to hold businesslike talks on this score.

We would like to reach an agreement as soon as possible on the reduction of armed forces and armaments in Central Europe. We have no objections to discussing questions related to the topic at any level and at any site: Vienna, Bonn, Washington, Moscow—wherever it is convenient.

24. Remarks on the Objectives of Soviet Nuclear Strategy: 2

Leonid Brezhnev

RESPONSE TO A QUESTION BY A PRAVDA CORRESPONDENT

Question: President Ronald Reagan of the United States said recently that the Soviet Union, supposedly judging by conversations of its leaders "among themselves," considers that victory in a nuclear war is possible. Thereby he tried to justify his course of the accelerated build-up of the U.S. nuclear arsenal.

What could you say, Leonid Ilich, concerning this statement of the American President?

Answer: While leaving on Mr. Reagan's conscience his remark that he supposedly knows what Soviet leaders are talking about among themselves, I will say the following concerning the substance of the question.

The thoughts and efforts of the Soviet leadership, just as of the Soviet people as a whole, are directed at preventing nuclear war altogether, at eliminating the very danger of its outbreak.

Among ourselves we are saying the same that was stated by me publicly from the rostrum of the 26th Congress of the CPSU, namely that it is a dangerous madness to try to defeat each other in the arms race and to count on victory in nuclear war.

I shall add that only he who had decided to commit suicide can start a nuclear war in the hope of emerging a victor from it. No matter what might the attacker possesses, no matter what method of unleashing nuclear war he chooses, he will not attain his aims. Retribution will ensue ineluctably.

Such is our principled viewpoint. It would be good if the President of the United States too, would make a clear and unambiguous statement rejecting the very data of nuclear attack as a criminal one.

Why should not the United States, may I ask, support the proposal made by the Soviet Union at the current session of the United Nations General Assembly concerning one's not being the first to use nuclear arms?

For if there is no first nuclear strike then, consequently, there will be no second or third nuclear strikes. Thereby all talk about the possibility or impossibility of victory in nuclear war will become pointless—the question of nuclear war as such will become pointless—the question of nuclear war as such will be removed from the agenda of the day.

And this is exactly what all peaceloving people on earth strive for, what the Soviet Union and its leadership are consistently working for. So now it is up to the United States and its leadership.

25. A Debate Between Raymond Garthoff and Richard Pipes on Soviet Nuclear Strategy

Richard Pipes and Raymond Garthoff

RICHARD PIPES: SOVIET STRATEGIC DOCTRINE

In his 1978 article,* Ambassador Garthoff argued that there exists a fundamental identity of view as well as interest between the United States and the Soviet Union concerning the nature and use of strategic nuclear weapons. He maintained that the Soviet regime looks upon military power as a defensive or deterrent instrumentality *par excellence*—a means of protecting itself from "imperialist" designs—and that it expects to expand its global influence not by military means but through the "rise of the working class." As concerns specifically strategic nuclear weapons, he told us that the Soviet leadership does not believe nuclear war can be won, that it shares our belief in the strategy of mutual deterrence, and that, consequently, it genuinely seeks strategic parity and equitable strategic arms limitation agreements.

In order to adduce a convincing case about Soviet strategic doctrine and intentions, one must draw on three types of evidence. One is authoritative material bearing on the overall political and military objectives of the Soviet Government. Another consists of the literature on its strategic doctrine. A third is data on actual Soviet strategic programs and deployments. Only from these disparate kinds of evidence is it possible to arrive at a convincing notion of how the leaders of the Soviet Union view strategic nuclear

---

*Raymond Garthoff, "Mutual Deterrence and Strategic Arms Limitation in Soviet Policy," *International Security* 3 (Summer 1978):112-147. An updated version of the article appeared in Derek Leebaert, ed., *Soviet Military Thinking* (London: George Allen and Unwin, 1981), pp. 92-124.

weapons, and to decide whether or not there exists a fundamental similarity between their and our perceptions of this subject.

The Evidence of Soviet Hardware

Let me state at the outset that in his article Ambassador Garthoff entirely ignored the third kind of evidence, that bearing on Soviet hardware. Admittedly, in a brief essay he could not have covered adequately both Soviet theories and deployments, but to omit completely any reference to the immense data we in the United States have at our disposal on Soviet offensive and defensive strategic systems seriously weakens his case.

If Ambassador Garthoff's view is to prevail, he must explain the reasons for the steady Soviet emphasis on heavy, land-based ballistic missiles, culminating in the decision taken in the 1960s to proceed with the production of the SS-18 and SS-19, two major counterforce systems of great throw-weight and high accuracies with MIRV payloads two to three times those of Minuteman III. He would further need to explain the attention lavished in the USSR on strategic defenses, including air defenses and civil defense—programs that do not fit mutual deterrence doctrines and have no adequate counterparts in the United States. . . .

Ideological Mainsprings of Soviet Military Doctrine

. . . Ambassador Garthoff would have us believe that Soviet leaders, as good Marxist-Leninists, look at military power mainly as a means of protecting their domain from the "capitalist military threat" and not as "the decisive element in advancing the historical process, which will progress when conditions are ripe through indigenous action by the rising working class."

Contrary to Ambassador Garthoff, Marxism, and especially Marxism in its Leninist interpretation, is permeated with faith in the utility of military power as a means of both internal coercion and external expansion—i.e., as an offensive instrumentality. Nor is there any inclination in Marxism-Leninism to await fatalistically for "conditions ripe" for revolution to emerge. . . .

Lenin's contribution to political practice may best be characterized by the term "militarization of politics." Lenin removed from Marxist socialism its democratic component and put in its place coercion, of which armed force is the centerpiece. He was the first dictator to have grasped the need to pursue revolutionary objectives in a manner consistent with the tenets of modern military science, and, conversely, to make all military activity subordinate to revolutionary objectives. His admiration for Clausewitz knew no bounds, nor did his contempt for pacifists. Time and again he upbraided his followers for advocating disarmament and placing their hope in the global spread of the revolution on the kind of processes that Ambassador Garthoff proposes, namely the "indigenous action of the rising working class."

Interpreting Soviet Writings and Pronouncements

. . . The bulk of Ambassador Garthoff's citations and references comes from open sources: Soviet newspapers and magazines, and monitored radio broadcasts. These consist of public statements by Soviet public figures to the effect that the Soviet Union is a peaceful country which regards modern war as unmitigated disaster for all humanity, the Soviet Union included, and seeks nothing but parity and stability in arms. I am at a loss how to persuade him that such evidence must not be taken at face value. If we did take it at face value, we would have to believe that Soviet troops invaded Afghanistan in order to expel "foreign interventionists," that Solidarity is a product of the CIA, or that the Soviet submarine which wandered into Swedish waters did so because of a navigational error.

Soviet public pronouncements have a certain utility in telling us (as well as Soviet citizens) what the Politburo wishes the public to believe: they have little value as indicators of what the Soviet leadership is actually thinking. Sometimes the pronouncements are floated deliberately to provide ammunition to those who argue as does Ambassador Garthoff. Thus, of late there has been a spate of high-level Soviet pronouncements unequivocally denying belief in either limited nuclear war or victory in a general nuclear conflict. These pronouncements are so obviously designed to counter and, if possible, neutralize the strategic doctrines adopted by the Reagan Administration that one surely cannot take them seriously, especially given the Soviet strategic buildup that points in the opposite direction.

Quite rightly, Ambassador Garthoff singles out for particular attention another, much more reliable source: the classified Soviet publication *Military Thought*, translations of which, for most of the decade 1963–1973, have been released by the CIA. This journal, written by the military for the military and not in general circulation, has been billed by a high Soviet officer as "the main military theoretical organ of the Ministry of Defense of the USSR." As such, it is a genuinely authoritative source. I am happy to meet Ambassador Garthoff on this common ground.

Joseph Douglass and Amoretta Hoeber, the authors of *Soviet Strategy for Nuclear War*, which was published one year after the appearance of Ambassador Garthoff's article, list in their bibliography 86 articles from *Military Thought* devoted directly or inferentially to nuclear strategy. Out of this corpus of 86 articles, Ambassador Garthoff relies for his repertoire of citations on half a dozen, a tiny proportion and statistically a very flawed sample. I suppose this is because the other 80 or so items offer no comfort whatever to his point of view, even with the most generous interpretation.

Specific Soviet Military Writers

Even with that small sample he has culled, Ambassador Garthoff takes all kinds of liberties, as can be seen from a closer look at the context in which the citations he provides appear. . . .

Item 1 is the article by Marshal Krylov, the then (1967) Commander-in-Chief of Soviet Strategic Rocket Forces.[1] Ambassador Garthoff cites the

Soviet Marshal twice. The first of these passages states that a preemptive nuclear attack will not save the aggressor "from great destruction and human losses." The second citation asserts that a surprise attack will not prevent a punishing retaliatory counterblow, possibly in the form of a "launch-on-warning" response, which merely reinforces the previous point. Ambassador Garthoff interprets these statements in the spirit of mutual assured destruction. The main body of Krylov's article, however, flatly contradicts such an interpretation. The point Marshal Krylov wants to get across is that, given the fact that the victim of a first strike can always retaliate, it is essential to consider strong *defensive measures*. This is what he says in a passage which Ambassador Garthoff does not cite: "Under any conditions, victory in nuclear war, from the economic and military standpoints, must be completely prepared in peacetime." This is hardly a recipe for MAD. In another part of his essay, General Krylov asserts that "the mass use of nuclear weapons in the first moments and hours of the war . . . will ensure the achievement of the main strategic goals of the war in the very beginning and determine the entire subsequent military-political situation."

The same holds true of Item 2, an article by General Vasendin and Colonel Kuznetsov.[2] The passage Ambassador Garthoff cites from this article again insists on the unavoidability of a devastating retaliatory strike on the part of the victim of a first strike. The passages he fails to cite, however, are the ones which throw light on the purpose of these warnings, to wit:

The increase in the technical capabilities of achieving surprise and the danger of the first nuclear strike of the enemy which has colossal destructive power in modern conditions compel us to examine again the problems of the *defense* capability of the states. In conditions of the unlimited use of nuclear-rocket weapons *countries which were unable to prepare for repulsing an aggressor or to disrupt or even weaken the nuclear attack can suffer in a short time destruction the extent of which will play a decisive role in the further conduct of war.*

And they continue, outlining the scenario for a first strike by a putative "aggressor":

The preparation by an aggressor for a sudden nuclear attack involves an increase in the protection of the means of attack and objectives of the deep rear from retaliatory nuclear strikes, as well as the creation of highly mobile strategic nuclear means capable of changing the launching point in time after the launch.

It should be clear that Vasendin and Kuznetsov emphasize the awesome destructive power of a retaliatory strike not in order to conclude that nuclear war can produce no winners, but, on the contrary, to press for effective defensive preparations as a means of winning a nuclear war, should it break out. Like Krylov, in another passage also uncited by Ambassador Garthoff, they assert that countries which fail to have made such peacetime preparations of a defensive nature "can suffer in a short time destruction the extent of which will play a *decisive* role in the further conduct of war."

U.S.-Soviet Doctrinal Equivalence?

These statements, indeed, reflect a general pattern. The only occasion when Soviet sources speak of an "annihilating" retaliatory strike is when they refer to a Soviet response to an American preemptive attack on their country. There is never, to the best of my knowledge, any admission that the reverse proposition also holds true, nor any attempt to arrive at a general theory of mutual assured destruction. This evidence confirms my assessment of Soviet strategic theory, as stated in my July 1977 *Commentary* article: namely, that the Russians believe in the fact of deterrence, but "whereas American theorists of mutual deterrence regard this condition as mutually desirable and permanent, Soviet strategists regard it as undesirable and transient: they are entirely disinclined to allow us the capability of deterring them."[3] Or, to put it in other words, Soviet strategists regard "mutual deterrence" to be a reality of the balance of nuclear forces as presently constituted, but they mean to alter this balance in their favor and in this manner secure a monopoly on deterrence.

Perhaps the most puzzling of all is Ambassador Garthoff's rendering of the Bochkarev article. He cites from it at great length and concludes that the Soviet general is so overawed by the destructiveness of nuclear weapons that his primary concern is to prevent nuclear war by means of deterrence. Now, it is true that General Bochkarev is very conscious of the horrors of nuclear war and warns that it must be avoided. Yet, in citing a long quotation from page 16 of Bochkarev's article, Ambassador Garthoff omits—I must assume, deliberately—the key passage:

The fundamental line of our Party is expressed [in the resolution of the Central Committee] with utmost precision and clarity: if imperialism commits a crime and plunges mankind into the abyss of nuclear war, imperialism will perish, and not "both sides," not socialism, although the socialist countries, too, will face supreme tests and suffer immense losses.

It is a pity Ambassador Garthoff deprived his readers of this critical citation, because from all I have read of Soviet writings on the subject and from all I have learned of Soviet strategic deployments, it faithfully reflects the dominant Soviet view. This view posits immense losses in any nuclear exchange for all the combatants and hence demands avoidance of such a war, but it also assumes that should nuclear war nevertheless break out, Russia's superiority in offensive and defensive strategic forces would result in a Soviet victory. This outlook may, indeed, provide a basis for a "stable" strategic relationship, but only if it is not misunderstood or misrepresented as an equivalent of U.S. doctrine.

Strategic Advantage Versus Stability

What does it all add up to? The overall conclusions simply do not support Ambassador Garthoff's case.

The Soviet High Command can be depicted as adhering to mutual deterrence only if the "warwinning" strategy is arbitrarily defined to mean total immunity from any retaliatory punishment for the power launching a first strike. Ambassador Garthoff interprets any evidence that the Russians acknowledge the unavoidability of retaliation—and this evidence is, indeed, incontrovertible—to mean that they subscribe to the doctrine of mutual deterrence. He does this despite the fact that in the Soviet military literature such statements are almost invariably coupled with claims that, heavy costs notwithstanding, victory in nuclear war is entirely attainable. As one previous critic of Ambassador Garthoff has pointed out, his manner of arguing "is a perfect example of the invalid common attempt to mirror-image Soviet doctrine in terms of U.S. doctrine. The straw man is the demand for *retaliation-proof* first-strike capability of the Soviet Union as an absolute precondition for preemption" (emphasis added). The author of these lines, Mr. Lewis A. Frank, went on to say that the Russian military do not think in terms of "absolute safety": "Political-military doctrine always involves matters of *relative* not absolute advantage."[4]

Soviet military writings lay such heavy emphasis on the costs to be paid by the initiator of a first strike for reasons entirely different from those which Ambassador Garthoff adduces. One of these reasons is to refute irresponsible talk in some quarters in the Soviet Union, and the United States, in the 1950s and 1960s to the effect that a preemptive strike would bring a quick, decisive and painless victory. The other is to get across a point which has become central to Soviet nuclear strategy: namely, that *defensive* preparations for nuclear war are as important as offensive ones. It is only a combination of offensive and defensive measures that, in the Soviet view, can assure victory in nuclear war. This manner of thinking helps explain the massive Soviet defensive preparations for the contingency of nuclear war—preparations which, of course, run quite contrary to the spirit of mutual deterrence.

It is important to keep in mind the strategic balance at the time the articles in *Military Thought* were written. The United States held a substantial edge in all categories of strategic forces. It would have been to the Soviets' advantage to embrace a MAD (Mutual Assured Destruction) doctrine, since their opponents were far more capable of winning a nuclear war than they. Significantly, their warwinning doctrine preceded by a decade their ability to implement it.

RAYMOND GARTHOFF: A REBUTTAL

Fundamental to Dr. Pipes' thesis has been the idea that to the Soviet leaders war remains "permanently valid" despite military-technological changes, because they accept the Clausewitzian principle that "war as a continuation of politics does not change with changing technology and armament." (This is a quotation from Marshal Sokolovsky in 1962, which Dr. Pipes alleges "was certainly hammered out with all the care that in the United States is lavished on an amendment to the Constitution" [*Commentary*, July 1977, p. 30]. That incredible assertion tells us something not only about Dr. Pipes' style, but also about his regard for accuracy.)

My article had cited authoritative Soviet sources which made clear that there was no contradiction between recognition of the catastrophic consequences of a nuclear war and need to avoid it, and reaffirmation of the long-standing Clausewitzian and Leninist conceptions that war is a political phenomenon and, wisely or unwisely, successfully or unsuccessfully, is always a continuation of politics (or policy, as it may alternatively be translated, both from the German and the Russian). Of course it is! Although some political and military writers have chosen at various times to use this as a debating point (as discussed and illustrated in my article, and updated in my 1981 revision, although not in Pipes'), it is very clear that authoritative Soviet political and military leaders have rejected the idea that this truism carries any implications as to the acceptability of war under existing conditions of vast nuclear arsenals on both sides.

Dr. Pipes does not now take issue with my discussion of this point in the article (or more extensively in the 1981 revision). Soviet political and military writers who were sparring over the question earlier had agreed by the end of the 1970s, in the words of one authoritative spokesman, that "no statesman, no military strategist—if he is in his right mind—can regard nuclear missile war as a means of achieving any political aim." Rising Politburo member Konstantin Chernenko stated in the Lenin Day address on behalf of the leadership (April 22, 1981) that nuclear war poses such a threat "to the whole of civilization, or even to life in this world" that it would be "criminal to look upon nuclear war as a rational, almost 'legitimate' continuation of policy." So much for Dr. Pipes' concluding argument in his 1977 article that "as long as the Russians persist in adhering to the Clausewitzian maxim on the function of war, mutual deterrence does not really exist." (That statement was, in any case, also contradicted by Dr. Pipes' own acknowledgment in that same article that "the Russians certainly accept the *fact* of deterrence" [pp. 34 and 29].)

In his 1977 article, Dr. Pipes still claimed that "Since the mid-1960s, the proposition that thermonuclear war would be suicidal for both parties has been used by the Russians largely as a commodity for export," and within the Soviet Union "such talk is generally denounced as 'bourgeois pacifism' " (p. 30). This statement was demonstrably incorrect, and a number of statements by Brezhnev and other Soviet political and military leaders were cited in my 1978 article (and more in the 1981 revision) refuting it. Dr. Pipes is silent. What can he say, when Brezhnev authoritatively addresses the XXVI Party Congress and states: "to count on victory in a nuclear war is dangerous madness" (*Pravda*, February 24, 1981)?

Dr. Pipes also attempts no rebuttal to my conclusions on Soviet acceptance of parity and disavowal of superiority as an aim, now frequently stated by Soviet political and military leaders.

Now to his criticisms. Dr. Pipes and I are fully in accord that in analyzing Soviet strategic thinking one must take into account authoritative material bearing on Soviet objectives, on strategic doctrine, and data on Soviet strategic programs and deployments. He finds lacking in my article explicit discussion of Soviet military programs, but he fails to show how it would have changed my conclusions. (In other writings I have done so explicitly, as for example in an article in *Problems of Communism* in 1975 cited in a footnote to my article.) Yet Dr. Pipes himself fails to cite anything in this connection apart from noting Soviet emphasis on large land-based missiles and strategic defenses and civil defense. These and other Soviet military programs do need to be taken into account in addressing the overall questions of Soviet military doctrine and aims. They certainly need to be much more seriously addressed than Dr. Pipes did in his *Commentary* article, which discussed such aims; they were not addressed in my article because of its more limited focus. Dr. Pipes is wrong in suggesting I would have difficulty reconciling them with my interpretations of Soviet policy and doctrine, and he adduces no evidence to support that assertion.

Dr. Pipes does not need to remind me of the military writings of Engels and Lenin, writings which I analyzed in several books published beginning in 1953. His comments on Soviet "militarization of politics" are hardly a novel contribution. Thirty years ago, I had written: "Soviet military (and political) doctrine is based on a military model of political relations derived from the fundamental Bolshevik conflict-image of the world. . . . This 'military' conception . . . pervades all Soviet politics" (Garthoff, *Soviet Military Doctrine*, 1953, p. 9). . . .

I do not accept Dr. Pipes' "summary" characterization of my views. For example, I do not argue that "the Soviet regime" expects to expand *its* global influence through "the rise of the working class"—I had clearly stated that I was characterizing the *ideological* element in Soviet thinking. Soviet diplomacy is active, and military power is one instrumentality of Soviet policy. Dr. Pipes prefers evidently to set up views he wishes to refute. Such "Pipes-dreams" require no reply.

I stand by my interpretation of Marxism-Leninism and the role of military power. Marxist-Leninist rejection of "spontaneity" does nothing to answer the real question, both in theory and practice: what actions, and courses of action, are considered expedient? Nowhere have I said that the Soviet leaders counsel "waiting fatalistically." The questions—key questions—of what it is prudent and possible to do are matters of policy, not of ideology. And that is why Soviet views of the nuclear relationship are critically important. . . .

The major part of Dr. Pipes' critique is leveled against a few of the documentary sources I cite. He exhibits a very peculiar double-standard: first he agrees as to the importance of such sources, then he deprecates them. He says "citations rarely prove anything" (after having introduced several to support his own argument), and then claims that I "select only a small portion of the citations available." That is ludicrous. Anyone, including Dr. Pipes in his writings, selects only a small portion of the almost infinite number available. On the other hand, I object very strongly to his unjustified charge that I ignore quotations "that are just as authoritative and far more voluminous, but indisputably lead to different conclusions." That is simply untrue. Moreover, if he believes it to be true, and wishes to persuade others, why does he not cite any of these voluminous quotations that "indisputably lead to other conclusions"? Yet he does not, in sixteen pages of typescript, introduce even *one* new contradictory reference. (I shall turn in a moment to those of the references I had cited which he does address.)

Dr. Pipes objects that "the bulk" of my references are to "open sources"— as, of course, are *all* the Soviet sources he cited in *Commentary*. He does credit me with singling out the classified Soviet publication *Military Thought*. But even there he is unhappy because of the limited *quantitative* selection I have made! He states that the Douglass and Hoeber book cites 86 articles from *Military Thought*, while I cite only "half a dozen, a tiny proportion and statistically a very flawed sample." What does that line of argument do to Dr. Pipes' citation of three references from Lenin's more than forty volumes of collected works? His comment is typical: acid, darkly suggestive, but irrelevant. "I suppose," he says, "this is because the other 80 or so items [sic; articles] offer no comfort whatever to his point of view." He is wrong; in fact, "the others" number about 800 and give me no discomfort. (To date, 81 issues—typically with about ten articles in each—from the period 1963 through 1973 have been released. At the time of my article, 54 through 1969 had been made available.) But the Douglass-Hoeber book— in addition, of course, to being much larger—had a much broader subject. Dr. Pipes is wrong in implying that I had chosen to ignore references, which he wishes the reader to assume are many, that did not fit my thesis. I was not able to find any. And he is evidently unable to find even one, among the 86 articles quoted by Douglass and Hoeber or presumably in the original 800-plus articles. . . .

The heart of Dr. Pipes' critique is found in his attempts to challenge my use of source references from *Military Thought*.

First, Marshal Krylov's article. Dr. Pipes tries to help Krylov along by interpreting for us the point "he [Krylov] wants to get across" as the essentiality of defensive measures in support of an offensive war-waging capability. He quotes Krylov to the effect that victory requires preparation in peacetime, which Dr. Pipes asserts supports an argument for strategic *defensive* preparation. In fact, that sentence is immediately followed by another indicating then SRF Commander Krylov's thought: "The Strategic Rocket Forces must always be in constant combat readiness for the immediate delivery of an annihilating retaliatory nuclear-missile strike against the aggressor in case he risks unleashing a nuclear war." Contrary to Dr. Pipes' interpretation, that *is* consistent with mutual deterrence and even MAD, assuming both sides take such a stance, as they have.

Dr. Pipes then quotes a statement by two Soviet military writers whom I had cited, General Vasendin and Col. Kuznetsov, again to argue his theme of a stress on "defensive preparations." The statements he quotes in no way contradict those that I had cited from the same article, as he implies; they also do not make the point he is trying so hard to demonstrate. He quotes the Soviet writers as stating: "The preparation by an aggressor for a sudden nuclear attack involves an increase in the protection of the means of attack and objectives deep in the rear. . . ." In fact, that sentence, as it appears in the original reads: "Therefore, the preparation by an aggressor for a sudden nuclear attack *directly* involves *not only an increase in the combat readiness of strategic nuclear means and the achievement of effectiveness of a first strike, but also* an increase in the degree of protection. . . ." Why did he omit the italicized passage, and why did he not at least indicate the omission by ellipses, in accordance with scholarly convention? Again, all of this digression by Dr. Pipes in no way affects the quotation I had cited by these writers, which is clearly written to apply both to the United States and the Soviet Union: "Everyone knows that in contemporary conditions in an armed conflict of adversaries comparatively equal in power (in number and especially in quality of weapons) an immediate retaliatory strike of enormous destructive power is inevitable."

Dr. Pipes chooses to generalize in his own words in rather quickly passing over the article by Generals Ivanov and Zemskov. He states that while they refer to retaliation, "their claims apply *only to the ability of the Soviet Union* [his italics] to inflict what we would call 'unacceptable damage' on the United States." He claims "in neither of these articles is there an indication that the authors regard 'annihilating' retaliatory blows" as applicable both to the United States and USSR. He is flatly wrong in both cases. He then generalizes to claim that "the only occasion when Soviet sources speak of an 'annihilating' retaliatory strike is when they refer to a Soviet response to an American preemptive attack. . . . There is never, to the best of my knowledge, any admission that the reverse proposition also holds true. . . . This evidence [!] confirms my assessment of Soviet strategic theory, as stated in my July 1977 *Commentary* article. . . ."

Incredible! Dr. Pipes ignores many Soviet statements I cite that directly contravene his contention, and then blithely says that the absence "to the best of his knowledge" of such statements is "evidence" supporting his own theories. . . .

Dr. Pipes purports to find puzzling my interpretation of another article by General Bochkarev. He cites a passage I "omitted" ("I must assume deliberately," he darkly adds) and finds it "a pity Ambassador Garthoff deprived his readers of this critical citation." It means, he tells us, that "Russia's superiority in offensive and defensive strategic forces would result in a Soviet victory."

On the contrary, I am delighted to have the opportunity to look with Dr. Pipes at the Bochkarev statement he cites. What is the context of the statement? It is a general political peroration at the conclusion of the article. The passage immediately *preceding* the one Dr. Pipes quotes is this: "'In case the imperialist aggressors nevertheless dare to unleash a new world war,' it states in the Program of the CPSU, 'the peoples will no longer tolerate a system which plunges them into devastating wars. They will sweep away and bury imperialism.'" So it is not, as Dr. Pipes would have it, "Russia's superiority in offensive and defensive strategic *forces*," but *the peoples* who emerge from the radioactive ruins enraged at imperialism which started the war who will prevail, and they will provide not "Soviet victory" in the sense Dr. Pipes implies, but rather will "sweep away and bury imperialism." There may or may not be many Russians or Americans left. Whatever confidence General Bochkarev has in the future after a nuclear war rests not on SS-18s, but on the "throw-weight" of History.

General Bochkarev does, in fact, also speak of "the *possibility* of the victory of the socialist states in a global nuclear war," in a passage cited in my article, but only in order to state: "The Communist Party of the Soviet Union and the other Marxist-Leninist parties decisively reject . . . the desirability of world nuclear war from the point of view of the interests of the world proletarian revolution. . . ."

In noting the need to recall the military balance in 1969 when the *Military Thought* articles cited in my article appeared, Dr. Pipes stresses the essential need for strategic defensive as well as offensive capabilities in preparation for nuclear war. Apart from the fact that these articles in 1969 did, indeed (as I cited them in my article), stress the particular importance of ballistic missile defense to the nuclear balance, and that they preceded SALT I, Dr. Pipes—who offers counsel on looking also at Soviet military programs—ignores totally the Soviet acceptance both in fact and by the ABM treaty of a virtual ban on ballistic missile defenses. Today the Soviet Union has 32 operational ABM launchers, compared to 64 in 1969, and a treaty limit of 100 at one location. Does that fact support his view of an all-out Soviet drive for a comprehensive strategic defensive-offensive posture? It is, however—despite Dr. Pipes' contortions in wrestling with the doctrinal evidence—compatible with Soviet acceptance of a state of mutual deterrence.

NOTES

1. Marshal N. Krylov, "The Nuclear Missile Shield of the Soviet State," *Voyennaya mysl'* [Military Thought], No. 11, November 1967, pp. 13–21.

2. Major General N. Vasendin and Colonel N. Kuznetsov, "Contemporary War and Surprise," *Voyennaya mysl'*, No. 6, June 1968, pp. 42–48.

3. Richard Pipes, "Why the Soviet Union Thinks It Could Fight and Win a Nuclear War," *Commentary*, July 1977, p. 29.

4. Letter to the Editor, *Problems of Communism*, September–October 1975, p. 88.

——————— **Bibliography: Part Two** ———————

Berman, Robert P., and Baker, John C. *Soviet Strategic Forces.* Washington, D.C.: Brookings, 1982.

Douglas, John D., Jr., and Hoeber, Amoretta. *Soviet Strategy for Nuclear War.* Stanford: Hoover Institution Press, 1979.

Frank, Lewis Allen. *Soviet Nuclear Planning.* Washington, D.C.: American Enterprise Institute, 1977.

Hanson, Donald W. "Is Soviet Strategic Doctrine Superior?" *International Security* 7 (Winter 1982-83):61–83.

Hyland, William. "The USSR and Nuclear War." In Barry M. Blechman, ed., *Rethinking the U.S. Strategic Posture.* Cambridge, Mass.: Ballinger, 1982.

Nerlich, Uwe, ed. *Soviet Power and Western Negotiating Policies.* Vols. 1 and 2. Cambridge, Mass.: Ballinger, 1983.

Pipes, Richard. "Why the Soviet Union Thinks It Could Fight and Win a Nuclear War." *Commentary* 64 (July 1977):21–34.

PART THREE

INTERMEDIATE-RANGE NUCLEAR FORCE CONTROVERSIES

CHAPTER 5

Europe and Asia

Most of the headlines about U.S. and Soviet nuclear weapons in Europe have addressed the controversy as an element in the larger picture of Soviet-American relations. As indicated by President Reagan's "Zero Option" address and Yuri Andropov's "162 Systems" response, the superpower dispute is a major part of the story. But Helmut Schmidt's "Dual Track" proposal and the Japanese White Paper on defense make clear that the countries of Western Europe and Asia are keenly aware of the issues and deeply affected by them. The Soviet deployment of the potent, MIRVed SS-20 missile has continued, and this has brought even the most reluctant European NATO members, Belgium and the Netherlands, closer to acceptance of the new U.S. regional nuclear weapons, the Pershing II and ground-launched cruise missiles.

In retrospect, the "walk in the woods" idea canvassed during several strolls alone, away from their delegations (and any listening devices), by the U.S. arms control negotiator Paul Nitze and his Soviet counterpart Yuli Kvitsinsky has a strong appeal. The "walk in the woods" plan called for the Soviets to reduce the numbers of its SS-20s from more than 300 to 75 (three MIRV warheads each) and, in return, for the United States and its allies to deploy no Pershing IIs and 75 cruise missile launchers (four missiles each). In the spring and summer of 1983, when the idea originally surfaced, one suspects that neither the Soviet Union nor the United States would have been pleased to adopt it. The Soviets were reluctant because they could not be certain until after the West German elections that they would not succeed in splitting the West Europeans from the Americans. The United States preferred not to have to accept the deal because it smelled victory, and as each day passed the Reagan administration's conviction grew stronger that the United States would succeed in turning the tables on the Soviets. To underline their position the Soviet Union withdrew from both strategic and theater nuclear negotiations with the United States. When the Christian Democrats won the election in West Germany in 1983, the Soviets had no illusions—they had lost a round. Following the reelection of President Reagan in 1984 and the easing of the Soviet succession problems with Mikhail Gorbachev's emergence as party chairman, the Soviets moved quickly to reopen arms control discussions with the United States. The two sides met in Geneva in January 1985 and subsequently began regular negotiations on intermediate-range nuclear forces starting in March, alongside negotiations aimed at controlling strategic nuclear and space weapons.

26. A "Dual Track" Proposal

Helmut Schmidt

THE 1977 ALASTAIR BUCHAN MEMORIAL LECTURE

THE NECESSITY OF ARMS CONTROL

Most of us will agree that political and military balance is the prerequisite of our security, and I would warn against the illusion that there may be grounds for neglecting that balance. Indeed, it is not only the prerequisite for our security but also for fruitful progress in East-West detente.

In the first place we should recognize that—paradoxical as it may sound— there is a closer proximity between a hazardous arms race, on the one hand, and a successful control of arms, on the other, than ever before. There is only a narrow divide between the hope for peace and the danger of war.

Second, changed strategic conditions confront us with new problems. SALT codifies the nuclear strategic balance between the Soviet Union and the United States. To put it another way: SALT neutralizes their strategic nuclear capabilities. In Europe this magnifies the significance of the disparities between East and West in nuclear tactical and conventional weapons.

Third, because of this we must press ahead with the Vienna negotiations on mutual balanced force reductions (MBFR) as an important step towards a better balance of military power in Europe.

No one can deny that the principle of parity is a sensible one. However, its fulfillment must be the aim of all arms-limitation and arms-control negotiations and it must apply to all categories of weapons. Neither side can agree to diminish its security unilaterally.

It is of vital interest to us all that the negotiations between the two super-powers on the limitation and reduction of nuclear strategic weapons should continue and lead to a lasting agreement. The nuclear powers have a special, an overwhelming responsibility in this field. On the other hand, we in Europe must be particularly careful to ensure that these negotiations do not neglect the components of NATO's deterrence strategy.

We are all faced with the dilemma of having to meet the moral and political demand for arms limitation while at the same time maintaining a fully effective deterrent to war. We are not unaware that both the United

States and the Soviet Union must be anxious to remove threatening strategic developments from their relationship. But strategic arms limitations confined to the United States and the Soviet Union will inevitably impair the security of the West European members of the Alliance vis-à-vis Soviet military superiority in Europe if we do not succeed in removing the disparities of military power in Europe parallel to the SALT negotiations. So long as this is not the case we must maintain the balance of the full range of deterrence strategy. The Alliance must, therefore, be ready to make available the means to support its present strategy, which is still the right one, and to prevent any developments that could undermine the basis of this strategy.

At the meeting of Western heads of State and Government in London last May I said that the more we stabilize strategic nuclear parity between East and West, which my Government has always advocated, the greater will be the necessity to achieve a conventional equilibrium as well.

Today, again in London, let me add that when the SALT negotiations opened we Europeans did not have a clear enough view of the close connection between parity of strategic nuclear weapons, on the one hand, and tactical nuclear and conventional weapons on the other, or if we did, we did not articulate it clearly enough. Today we need to recognize clearly the connection between SALT and MBFR and to draw the necessary practical conclusions. . . .

At the same meeting in May I said that there were, in theory, two possible ways of establishing a conventional balance with the Warsaw Pact states. One would be for the Western Alliance to undertake a massive build-up of forces and weapons systems; the other for both NATO and the Warsaw Pact to reduce their force strength and achieve an overall balance at a lower level. I prefer the latter.

The Vienna negotiations have still not produced any concrete agreement. Since they began, the Warsaw Pact has, if anything, increased the disparities in both conventional and tactical nuclear forces. Up to now the Soviet Union has given no clear indication that she is willing to accept the principle of parity for Europe, as she did for SALT, and thus make the principle of renunciation of force an element of the military balance as well.

Until we see real progress on MBFR, we shall have to rely on the effectiveness of deterrence. It is in this context and no other that the public discussion in all member states of the Western Alliance about the "neutron weapon" has to be seen. We have to consider whether the "neutron weapon" is of value to the Alliance as an additional element of the deterrence strategy, as a means of preventing war. But we should not limit ourselves to that examination. We should also examine what relevance and weight this weapon has in our efforts to achieve arms control.

27. Zero Option

Ronald Reagan

. . . Today, I wish to reaffirm America's commitment to the Atlantic Alliance and our resolve to sustain the peace. And from my conversations with allied leaders, I know that they also remain true to this tried and proven course.

NATO's policy of peace is based on restraint and balance. No NATO weapons, conventional or nuclear, will ever be used in Europe except in response to attack. NATO's defense plans have been responsible and restrained. The allies remain strong, united and resolute. But the momentum of the continuing Soviet military build-up threatens both the conventional and the nuclear balance. Consider the facts. Over the past decade:

The United States reduced the size of its armed forces and decreased its military spending. The Soviets steadily increased the number of men under arms. They now number more than double those of the United States. Over the same period the Soviets expanded their real military spending by about one-third.

The Soviet Union increased its inventory of tanks to some 50,000 compared to our 11,000. Historically a land power, they transformed their navy from a coastal defense force to an open ocean fleet, while the United States, a sea-power with transoceanic alliances, cut its fleet in half.

During a period when NATO deployed no new intermediate-range nuclear missiles, and actually withdrew 1,000 nuclear warheads, the Soviet Union deployed more than 750 nuclear warheads on the new SS-20 missiles alone.

Our response to this relentless build-up of Soviet military power has been restrained but firm. We have made decisions to strengthen all three legs of the strategic triad—sea-, land-, and air-based. We have proposed a defense program in the United States for the next 5 years which will remedy the neglect of the past decade and restore the eroding balance on which our security depends.

I would like to discuss more specifically the growing threat to Western Europe which is posed by the continuing deployment of certain Soviet intermediate-range nuclear missiles.

The Soviet Union has three different missile systems—the SS-20, the SS-4, and the SS-5—all with a range capable of reaching virtually all of Western Europe. There are other Soviet weapons systems which also represent a major threat. The only answer to these systems is a comparable threat to Soviet targets. In other words, a deterrent preventing the use of these Soviet weapons by the counter-threat of a like response against their own territory. At present, however, there is no equivalent deterrent to these Soviet intermediate missiles. And the Soviets continue to add one new SS-

20 a week. To counter this the allies agreed in 1979, as part of a two-track decision, to deploy as a deterrent land-based cruise missiles and *Pershing* II missiles capable of reaching targets in the Soviet Union. These missiles are to be deployed in several countries of Western Europe. This relatively limited force in no way serves as a substitute for the much larger strategic umbrella spread over our NATO allies. Rather, it provides a vital link between conventional, shorter-range nuclear forces in Europe and inter-continental forces in the United States. Deployment of these systems will demonstrate to the Soviet Union that this link cannot be broken.

Deterring war depends on the perceived ability of our forces to perform effectively. The more effective our forces are, the less likely it is that we will have to use them. So, we and our allies are proceeding to modernize NATO's nuclear forces of intermediate range to meet increased Soviet deployments of nuclear systems threatening Western Europe.

Let me turn now to our hopes for arms-control negotiations. There is a tendency to make this entire subject overly complex. I want to be clear and concise.

I told you of the letter I wrote to President Brezhnev last April. Well, I have just sent another message to the Soviet leadership. It's a simple, straight-forward, yet historic message: the United States proposes the mutual reduction of conventional, intermediate-range nuclear, and strategic forces.

Specifically, I have proposed a four-point agenda to achieve this objective in my letter to President Brezhnev. The first, and most important, point concerns the Geneva negotiations. As part of the 1979 two-track decision, NATO made a commitment to seek arms-control negotiations with the Soviet Union on intermediate-range nuclear forces. The United States has been preparing for these negotiations through close consultation with our NATO partners. We are now ready to set forth our proposal. I have informed President Brezhnev that when our delegation travels to the negotiations on intermediate-range land-based nuclear missiles in Geneva on the 30th of this month, my representatives will present the following proposal: The United States is prepared to cancel its deployment of *Pershing* II and ground-launched cruise missiles if the Soviets will dismantle their SS-20, SS-4, and SS-5 missiles. This would be an historic step. With Soviet agreement, we could together substantially reduce the dread threat of nuclear war which hangs over the people of Europe. This, like the first footstep on the moon, would be a giant step for mankind.

We intend to negotiate in good faith and go to Geneva willing to listen to and consider the proposals of our Soviet counterparts. But let me call to your attention the background against which our proposal is made. During the past six years, while the United States deployed no new intermediate-range missiles and withdrew 1,000 nuclear warheads from Europe, the Soviet Union deployed 750 warheads on mobile, accurate ballistic missiles . . . they now have 1,100 warheads on the SS-20, SS-4 and SS-5 missiles and the United States has no comparable missiles. Indeed, the United States dismantled the last such missile in Europe over 15 years ago.

As we look to the future of the negotiations, it is also important to address certain Soviet claims which, left unrefuted, could become critical barriers to real progress in arms control. The Soviets assert that a balance of intermediate-range nuclear forces already exists. That assertion is wrong. By any objective measure . . . the Soviet Union has an overwhelming advantage, on the order of six-to-one.

Soviet spokesmen have suggested that moving their SS-20s beyond the Ural mountains will remove the threat to Europe. . . . The SS-20s, even if deployed behind the Urals, will have a range that places almost all of Western Europe, the great cities—Rome, Athens, Paris, London, Brussels, Amsterdam, Berlin and so many more—all within range of these missiles, which incidentally are mobile and can be moved on short notice.

The second proposal I have made to President Brezhnev concerns strategic weapons. The United States proposes to open negotiations on strategic arms as soon as possible next year. I have instructed Secretary Haig to discuss the timing of such meetings with Soviet representatives. Substance, however, is far more important than timing. As our proposal for the Geneva talks this month illustrates, we can make proposals for genuinely serious reductions but only if we can take the time to prepare carefully. The United States has been preparing carefully for resumption of strategic arms negotiations because we do not want a repetition of past disappointments. We do not want an arms-control process that sends hopes soaring only to end in dashed expectations.

I have informed President Brezhnev that we will seek to negotiate substantial reductions in nuclear arms which would result in levels that are equal and verifiable. Our approach to verification will be to emphasize openness and creativity—rather than the secrecy and suspicion which have undermined confidence in arms control in the past.

While we can hope to benefit from work done over the past decade in strategic arms negotiations, let us agree to do more than simply begin where these efforts previously left off. We can and should attempt major qualitative and quantitative progress. Only such progress can fulfill the hopes of our own people and the rest of the world. Let us see how far we can go in achieving truly substantial reductions in our strategic arsenals. To symbolize this fundamental change in direction, we will call these negotiations START—strategic arms reduction talks.

The third proposal I have made to the Soviet Union is that we act to achieve equality at lower levels of conventional forces in Europe. The defense needs of the Soviet Union hardly call for maintaining more combat divisions in East Germany today than were in the whole Allied Invasion Force that landed in Normandy on D-Day. The Soviet Union could make no more convincing contribution to peace in Europe—and in the world—than by agreeing to reduce its conventional forces significantly and constrain the potential for sudden aggression.

Finally, I have pointed out to President Brezhnev that to maintain peace, we must reduce the risks of surprise attack, and the chance of war arising out of uncertainty or miscalculation. I am renewing our proposal for a conference to develop effective measures that would reduce these dangers. At the current Madrid meeting of the Conference on Security and Co-operation in Europe, we are laying the foundation for a western-proposed conference on disarmament in Europe. This conference would discuss new measures to enhance stability and security in Europe. Agreement on this conference is within reach. I urge the Soviet Union to join us and the many other nations who are ready to launch this important enterprise. All of these proposals are based on the same fair-minded principles: substantial, militarily significant reductions in forces; equal ceilings for similar types of forces; and adequate provisions for verification.

My administration, my country and I are committed to achieving arms reductions agreements based on these principles. Today I have outlined the kinds of bold, equitable proposals which the world expects of us. But we cannot reduce arms unilaterally. Success can only come if the Soviet Union will share our commitment; if it will demonstrate that its often-repeated professions of concern for peace will be matched by positive action.

Preservation of peace in Europe and the pursuit of arms reductions talks are of fundamental importance. But we must also help to bring peace and security to regions now torn by conflict, external intervention and war.

The American concept of peace goes well beyond the absence of war. We foresee a flowering of economic growth and individual liberty in a world at peace. . . .

28. 162 Systems

Yuri Andropov

Question: There is clearly no observable progress at the Geneva talks on medium-range weapons in Europe. The American proposal, which contains the "zero option" and, more recently, the "interim version," seems incompatible with the Soviet proposal to reduce the number of its missiles to the number of West European medium-range missiles. What, in your opinion, is the essence of the differences, and do you consider a compromise possible?

Answer: We understand the concern that questions of the limitation of medium-range nuclear arms in Europe, which are being discussed at the Geneva talks, are not being resolved. These talks, to put it bluntly, are deadlocked.

Why were they started? The Soviet side proposed the talks and began them with the resolve to work for a reduction in the USSR's and the NATO countries' medium-range nuclear weapons in that region and for a radical and reciprocal lowering of the level of nuclear confrontation. It has been found that the U.S.'s goal at the Geneva talks is, at all costs, to add powerful new arms to NATO's already extensive nuclear arsenal. The only thing the U.S. wants to reduce is Soviet missiles.

As you see, two opposing lines—two fundamentally different approaches, I would say—face each other in Geneva. We ran into this line from the Americans even before Geneva. As is known, the U.S. scuttled the SALT-II Treaty and withdrew from a whole series of talks that were making progress or nearing a successful conclusion. I might remind you that the U.S. broke off, and is still avoiding the resumption of, the talks on the general and complete prohibition of nuclear weapons tests, on antisatellite systems, on the limitation of deliveries and sales of conventional arms and on the limitation of military activity in the Indian Ocean. I will add to this the treaties with the Soviet Union on the limitation of the underground testing of nuclear weapons and on nuclear explosions for peaceful purposes, which the United States has not ratified to this day. All this speaks for itself. As for the Geneva talks, the present U.S. administration entered them, as is known, with great reluctance. . . .

Recently, talk has been heard in the U.S. and in some other NATO countries about the Soviet side's "unyieldingness." But what, in point of fact, should we yield? It's demanded of us that we enter into an agreement on missiles alone—on the existing Soviet medium-range missiles and the American missiles that plans call for bringing into Europe. The Americans simply refuse to come to an agreement on the other components of medium-range nuclear weapons. They declare that they don't want to hold talks on this, and that's all there is to it.

They want us to pretend that we don't notice the more than 400 warheads on British and French sea- and land-based missiles, which are aimed at the Soviet Union and at other socialist countries. The Americans, and in their wake representatives of other NATO countries, call the British and French missiles a "deterrent" force. I am ready to assume that this is the case. Then the question arises: Why, while recognizing France's and Britain's right to deterrence, do they deny us the right to have our own corresponding deterrent force—in exactly the same amount as the French and the British have?

It is also said that the nuclear arms of France and Britain shouldn't be counted, since they have some kind of "independent status." But have these countries ceased being members of the North Atlantic Alliance? Do they conceal the fact that their nuclear weapons are directed against the Soviet Union? And isn't it symptomatic that the French government not only has not dissociated itself from the plans for the deployment of American nuclear missiles in Western Europe but, on the contrary, is avidly supporting them? What kind of "independent status" is that! On the one hand, the well-known NATO decision is presented as a kind of mandate to the United States from all the members of this alliance, while on the other hand, when they start to count NATO's armaments it turns out that the British and French missiles have nothing to do with them, that they are "independent," supposedly. There's no logic in that.

Try to look at the situation from the standpoint of the Soviet Union and its legitimate interests: On what grounds, by what right, do they want to leave us disarmed in the face of these British and French nuclear missiles targeted on our country?

It's clear that we cannot agree to this and never will agree to it. The Soviet people have the same right to security as the peoples of America, Britain, France and other countries.

We are also asked to close our eyes to the fact that, if we were to accept the Americans' proposals, they would retain intact in the European zone their aircraft stationed at airfields and on aircraft carriers—i.e., forward-based systems for delivering nuclear weapons capable of reaching the territory of the Soviet Union. And there are hundreds of delivery vehicles and thousands of nuclear warheads.

Q: The West and the East reprove each other for seeking military superiority. The Americans even claim that the Soviet representatives at the Geneva talks are seeking to perpetuate the USSR's superiority in medium-range missiles on the European continent. What should an approximate balance look like?

A: We know about these claims. American military and political figures are especially zealous on this account. However, some other Western politicians aren't far behind them. There is not a grain of truth in these claims. One must stand on the facts. In the recent period alone the Soviet Union, wishing to move the talks off dead center, has submitted a whole series of proposals of an obviously constructive nature for consideration by the American side and the other NATO countries.

Suffice it to recall our proposals at the same Geneva talks on the limitation of medium-range nuclear arms in Europe. At present, each side has in Europe approximately 1,000 delivery vehicles for medium-range nuclear weapons, plus several thousand tactical nuclear warheads. If our most far-reaching proposal had been accepted, a proposal which, incidentally, the West bashfully prefers to ignore, on the European continent there would remain no types of nuclear weapons designed to strike targets in Europe—neither medium-range nor tactical weapons. Will Europe and European security win or lose if this proposal is implemented? The answer is clear and unequivocal.

The same applies to another proposal of ours. If it had been implemented, there would have been a radical reduction: Each side would have reduced the number of its medium-range arms by two-thirds. In other words, we would have covered two-thirds of the road toward the complete freeing of Europe from these weapons.

Finally, if the West had accepted the variant that we proposed in December 1982, the USSR and the NATO countries would have retained 162 missiles each—i.e., exactly as many as Britain and France now have on the NATO side. Each side would also have retained 138 medium-range planes. I would like to emphasize that even these figures are not an absolute. They could be reduced on a reciprocal basis if NATO agreed to it.

All these proposals by the Soviet Union remain in force. They fully ensure genuine equality and a real balance of forces.

Where can one find in all this a striving by the USSR for military superiority?

The complexity and danger of the present situation consists in the fact that the arms race imposed by the West is outdistancing the talks. To avoid this, to create favorable conditions for holding talks, common sense suggests the need to freeze the two sides' nuclear arsenals. Doing this would be the most reasonable thing until other solutions are found. We have proposed that medium-range weapons be frozen, as well as strategic arms. In our view, this would be only the first step. One might object: Why a freeze, when we should be talking about reductions? This does sound nice, but the trouble is that while this is being declared the arms buildup is continuing. So what we have is no freeze and no reduction.

The contrast is no less striking when one compares the approaches of the USSR and the U.S. to other problems of arms limitation and reduction.

I have already mentioned our commitment not to be the first to use nuclear weapons. If the U.S. and other NATO members that possess nuclear weapons were to make an analogous statement, people all over the world would breathe a sigh of relief. Why isn't this done?—one wonders. But our initiative is encountering a wall of silence.

Here's another example. What justifies the West's failure to this day to respond to the honest proposal, dictated by the best of intentions, that the Soviet Union and the other socialist countries made for the conclusion of a Treaty on the Nonuse of Military Force and the Maintenance of Relations of Peace between the Warsaw Treaty countries and the NATO countries? It's hard to find a sensible explanation of this.

Or let us look at how the two sides are behaving at the strategic arms talks. Compare the two positions. We propose substantial—by more than one-fourth—reductions in the total number of strategic delivery vehicles, with no exceptions. The number of nuclear warheads on those vehicles would also be reduced to equal levels. But what is the U.S. insisting on? It is only talking about reductions, while in fact it's working for a substantial— a really "radical"—buildup in its strategic arms. There are plans for the additional deployment of more than 12,000 long-range cruise missiles alone. Let me cite the latest news from Washington: Yet another plan has been placed on the President's desk—to deploy in the next few years a substantial additional number of intercontinental ballistic missiles of a new type.

This is an unbridled arms race. There's nothing else you can call it.

29. France and the Euromissiles: The Limits of Immunity

Pierre Lellouche

. . . It is a fact that France is reacting quite differently from the rest of the Western world to the nuclear debate provoked by the INF affair. With the exception of a few large demonstrations—mostly orchestrated and dominated by the French Communist Party—there is simply no such thing as a genuine and powerful French "peace movement" analogous to those existing in West Germany, Holland or the United States. In fact, French public opinion reveals a high degree of passivity and indifference vis-à-vis Pershings and the nuclear debate as a whole. Government officials readily interpret this as a "consensus" in support of French security policy, and there again, their assessment is correct—at least in part. . . .

. . . For France, . . . the issue at stake is of fundamental importance: it is whether, in East-West relations today, the country can continue to exercise a margin of absolute independence in the conduct of her security policy—particularly when this involves French nuclear weapons; or, to phrase it differently, whether France must clarify in one way or another the basic ambiguity of her security policy with respect to her precise contribution to NATO and European security.

When asked why their country is apparently immune to the peace campaign that has overwhelmed all of Western Europe—and even Reagan's America—the standard answer given by French officials (and generally by most of the French defense establishment) is that France has had her own nuclear weapons for some 25 years, that she left NATO's integrated military command in 1966, and that, because they are used to relying on themselves for their own defense, the French—unlike their European neighbors—are less vulnerable to the neutralist or pacifist temptation. . . .[1]

Interestingly enough, while France clearly enjoys what Dominique Moïsi has called an "agreement on fundamentals" on security policy, this consensus rests much less on the actual military value of French weapons in case of war than on a set of rather abstract and highly ambiguous principles which the French have been taught to see as deriving from the possession of a national deterrent. Very much like the Americans or the British, the French do not like to think of the possible use of their nuclear weapons. Opinion polls reveal some serious doubt as to the usability of these weapons even as a last resort, and even if France herself is invaded (let alone to defend Germany).[2] Yet, a vast majority of French citizens (including, since 1977-78, members of the PCF [French Communist party] and the PS [Socialist party]) find it "normal" for France to have *independent* nuclear weapons[3] and no one—even in the most Atlanticist circles—would ever envisage a return to the NATO fold. In truth, then, the French "consensus" centers primarily on the symbolic value of nuclear weapons—on the notion of "independence," and on France's partial membership in NATO (which most French citizens understand as no membership at all, since that is what they have been told for years by their media).

As a result, as soon as one starts digging a little deeper into that "consensus," and starts asking questions about what French weapons are really for in case of war, deep differences begin to appear. And here, interestingly enough, the differences do not follow France's right-left ideological polarization. Instead one finds a fundamental cleavage in each political family (except for the Communist Party which is, as usual, monolithic) between those who interpret "independence" and "deterrence" as, in effect, a status of de facto armed neutrality, and those who remain profoundly committed to the rest of Europe and the Alliance.

Even more fascinating perhaps is the composition of each of these two camps. In the first group, which I call the "nationalist-neutralist" group, one finds of course the Communists, but also the left-wing of the Socialist Party (the CERES) as well as some parts of the right-wing establishment who interpret de Gaulle's foreign policy as one of strict neutrality and equidistance between the "blocs." Here, the *force de frappe* is acclaimed to be both the symbol and key instrument of national independence, and France's main adversary remains the United States and/or Germany, not the Soviet Union.

In the other, "European-Atlanticist," group, one finds the major part of the PS (behind Mitterrand himself), the centrist and center-right parties, and more recently, the Gaullist Rally for the Republic (RPR) under Jacques Chirac. For this group, current political and strategic realities exclude any neutralist temptations on the part of France: the adversary *is* the Soviet Union, and France's forces, though independent, are also to be committed to the security of her neighbors and of Germany in particular. This commitment, it is felt, should in the long run take the form of a genuine European defense cooperation linked (in varying degrees) to a continued U.S. involvement in Europe.

Until now, no major confrontation has taken place between these two rather strange coalitions. The reason for this is twofold: First, sensing the divisions in their own parties, French politicians have carefully avoided any serious debate on defense each time the opportunity for such a discussion has arisen.[4] Second, the strategic doctrine defined more than 20 years ago by de Gaulle (and put on paper only once, in the 1972 Defense White Paper) is a masterpiece of ambiguity. Officially, the ambiguity was aimed at keeping the Soviets guessing as to what France would do in case of war; in practice, however, it satisfied everybody in France herself. In his theory of the "three circles," one of the 1972 White Paper's chief authors, General Poirier, defined nuclear weapons as protecting the "national sanctuary" (namely France herself—the "first" circle), but also stated that France's security was directly linked to developments on her immediate periphery (namely Germany, the "second" circle).[5]

It is on this comfortable ambiguity that the "consensus" was built, and through it that France was able to capitalize politically on her nuclear forces in the conduct of her foreign policy throughout the 1960s and the first half of the 1970s. If it was best for French interests, however, such a doctrine—and therefore, the consensus itself—could only last so long as three conditions were met:

- that the U.S. nuclear guarantee (which the French had declared worthless for themselves since 1959) still worked to protect all of non-nuclear Europe (which meant that the overall and regional East-West balances of forces were to be maintained);
- that Germany remained secure and stable within NATO, thereby securing a European geostrategic status quo which, for the first time in their history, put the French in the "second line" from their likely adversary; and
- that France was able to keep modernizing her deterrent, unconstrained by arms control or diplomatic pressure.

The central argument of this paper is that, as a result of the evolution of the overall correlation of forces between East and West since the late 1960s, and of the parallel transformation of both East-West and West-West relations during that period, each of these conditions has been—and is still— profoundly called into question. That fact, in turn, is bound to alter in a fundamental way the basic structure of French security policy. In this environment, the INF affair, which in reality cuts across each one of the three elements just listed, may well act as the catalyst for a major period of reassessment and redefinition of French security policy.

In a sense, that process of reassessment has already begun, since the first two conditions (the credibility of the extended U.S. deterrent over Western Europe, and the stability of Germany) have been perceived in France as dangerously slipping away ever since the mid-1970s. It is this new situation which has prompted a major rapprochement with NATO since 1974 (with the signing of the Ottawa Declaration),[6] and a new debate in the French defense community (since 1975-76) as to what concrete contribution France can provide to the Alliance and to Germany in particular.[7] The objective, then, as is still the case under the present Socialist government, was to compensate at least in part for the declining credibility of the U.S. security guarantee in order to help reassure and stabilize Germany. This produced, in 1976, a new concept called "enlarged sanctuarization" and "forward battle," which in turn triggered the first serious political clash over defense policy since 1958. The vehemence of Gaullist and Communist opposition to the new concept (which they saw as a betrayal of national independence and strict nuclear deterrence) led President Valéry Giscard d'Estaing to quickly retreat back into ambiguity . . . and silence. Thereafter, Giscard carefully avoided any departure from the standard "Gaullist" strategic discourse. When the INF decision came along, Giscard treated it as "NATO business," and refused to take sides publicly on the issue.

Ironically, it was during that same period and for the same set of reasons that a much more fundamental evolution began to take place within the left, and particularly within the Socialist Party. By 1977–78, both the PCF and the PS had gone from total opposition to a French deterrent to a position of "recognition" of the nuclear force. They did so, however, for different reasons. The Communists changed their position because they saw the French nuclear force as the key to a policy of genuine armed neutrality and as a bargaining chip for a future French disarmament policy. The Socialists, however, were divided between a strong left-wing minority (CERES), which was staunchly nationalist and neo-Gaullist (in the neutralist sense), and the dominant Mitterrand line, which was both more Atlanticist and more European in its orientation. When the INF issue came along, Mitterrand seized it immediately, sensing not only its importance in the upcoming electoral debate,[8] but also the fundamental political and strategic stakes of the issue for the future of Europe. Even before the May 1981 election, it was clear that Mitterrand had made up his mind: France had to intervene in the European and German debate in order to stop the risk of neutralist drift in non-nuclear Europe, and to help restore a balance of forces "broken by the introduction of the SS-20."[9] Underlying this position was a clearer understanding on the part of Mitterrand than was the case with Giscard of the gravity of the internal evolution in Germany and of the magnitude of Soviet goals—namely a Europe decoupled from America in terms of nuclear deterrence, and gradually sliding from détente to appeasement in the face of the U.S.S.R.'s conventional and nuclear superiority in the region. . . .

This, however, is far from being the end of the story. The various steps taken so far, and particularly Mitterrand's decision to take sides openly in favor of the INF decision, are not politically cost-free for France. As realism forces France to move away from the comfortable ambiguity of the past, France is also beginning to pay the political price of her more "European" security policy, both diplomatically (in the arms control area, in particular) as well as perhaps domestically (in terms of her own internal "consensus" on defense).

On the defense side, however, the situation is still "under control," and despite mounting criticism from the PCF, the government has so far been able to prevent a major disruption of the domestic consensus on security. The government's success on this score has derived from two principal factors. First, in moving toward greater European (and NATO) solidarity, the government has been able to count on the support of the Gaullist party. This is a new situation to the extent that until 1980–81, the Gaullists (and Jacques Chirac himself) vehemently opposed any departure from a strictly national deterrence posture. Second, the government has been extremely careful indeed in redefining its posture vis-à-vis an eventual participation of French forces in the battle in central Europe. In the discussion of the recent five-year program, Defense Minister Charles Hernu and other top officials emphasized that in the event of war the employment of the FAR [Rapid Action Forces] (as well as of the First French Army) would be decoupled from the use of any nuclear weapons, including French battlefield nuclear arms. As mentioned earlier, this does free these forces from their traditional mission of acting as a tripwire for nuclear escalation, and thus makes them more "usable" in Germany. At the same time, however, it may also signal to the Soviets—indirectly at least—that France's participation in the battle is limited to the non-nuclear level only—which Moscow could well interpret as a de facto and unilateral no-first-use pledge on the part of France.

The pressure is much greater, and the risks for France much more considerable, in the arms control area, as a result of the Soviet demand to include French and British nuclear forces in the INF talks. This pressure is all the more serious now that an increasing body of West European opinion—particularly in the left-wing parties—is now openly supporting the Soviet position. . . .[10]

Whatever position France might have taken, the Soviets would in any case have insisted upon the inclusion of French and British systems in the Geneva talks. Including these systems was their best negotiating position given their underlying goal, which was from the very beginning to prevent *any* American deployment. This inclusion also provided them with a handle to prevent (or control) future modernization of French and British deterrent forces (a central Soviet objective dating back to the beginning of the SALT process in 1969). Far from being a tactical negotiating gimmick, moreover, the effort to include French and British forces was aimed at achieving what the Soviets view as a fundamental component of their long-term strategic relationship with the West—namely an overall intercontinental balance with the United States (in which they will of course be entitled to "parity") and a separate "European balance," decoupled from the United States and increasingly free of all U.S. nuclear weapons, in which the Soviets would also be entitled to a separate "parity," this time with France and Britain. This Soviet insistence on both global and regional "parity" actually amounts to giving the U.S.S.R. the right to own as many nuclear weapons as all other non-Soviet nations on the face of the earth, that is to say absolute superiority over each of these nations. Finally, by including French and British systems the Soviets could also hope to drive a wedge, not just between Americans and Europeans, but also between the Europeans themselves—i.e., between the two nuclear powers and the rest of non-nuclear Europe.

The real dilemma for France, therefore, was not whether she could support the decision; in either case, this was (and is) a no-win situation. By keeping silent, France would have projected an image of neutrality and would have only encouraged the neutralist drift at work among her neighbors, without even obtaining the non-inclusion she was seeking from Moscow. But by taking sides on the issue in favor of NATO, the French not only seem to justify the Soviet demand for inclusion, but they are also putting themselves in a position where they are increasingly perceived (thanks to

Soviet propaganda and to the political blindness of many European opinion-makers) as the selfish nuclear power which constitutes the sole obstacle to a rapid and "fair" deal in Geneva. Here again, the end result is deepening discord among the Europeans themselves.

Given this no-win situation, Mitterrand did take the right—and coura-geous—decision. And to some extent, the entry of France into the debate since May 1981 has helped things a great deal in some of the deployment countries—namely Germany and Italy, where the results of the recent elections have turned out to be favorable developments in the context of INF deployment. . . .

NOTES

1. In addition, of course, France, not being a party to the NATO decision of December 1979, is not involved in the deployment of any new American INF on her soil—a situation which simplifies matters greatly.

2. According to a poll published in October 1982 ("Ça m'interesse"), 42% of the French declared that, even if the Red Army were entering France, they would prefer negotiating with the Soviets rather than using nuclear weapons. Thirty-nine percent would favor resisting by conventional means. In all, 44% thought that the *force de frappe* was useless, since its use would entail wiping France off the map.

3. According to the same poll, only 3% of the French would favor resorting to nuclear weapons to defend West Germany (60% would want to remain neutral).

4. This was true, in particular, during the 1976 debate about the concept of "enlarged sanctuarization" put forward under Giscard.

5. The "third" circle being French interests in Africa and the Third World.

6. The Ottawa Declaration recognizes the particular contribution of French and British national deterrents to the Alliance's overall deterrence effort.

7. See Pierre Lellouche, "SALT and European Security: The French Dilemma," *Survival*, January/February 1980; "La France, les SALT et la securité de l'Europe," *Politique Entrangère*, 2/1979.

8. Giscard's May 1980 trip to Warsaw in the aftermath of the Soviet invasion of Afghanistan and his silence on the SS-20 issue became key elements in the Socialist critique of his "soft" foreign policy vis-à-vis Moscow in the spring 1981 electoral campaign.

9. Jacques Chirac is not only in favor of the NATO decision, but he also supports the concept of a "European nuclear deterrent," basically through the extension of France's nuclear deterrent over Germany. (On the latter point, see his recent speech to the Konrad Adenauer Stiftung in Bonn, *Le Monde*, 20 October 1983.)

10. Just recently the SPD formally joined those who demand the inclusion of French and British forces in INF (see Vogel's statement in *Le Monde*, 25–26 September 1983).

30. Summary of Defense of Japan

The Government of Japan

SECTION 2: SOVIET MILITARY BUILDUP AND EXPANSION OF SOVIET INFLUENCE

In its national policy, the Soviet Union has been giving one of the highest priority to military buildup based on the belief that the risk of war is unavoidable as long as imperialism exists. As a result, it has now built up sufficient power to confront the U.S. either in the sphere of nuclear or conventional weapons today. Furthermore, on the one hand, the Soviet Union is attempting to divide the U.S. from its allies by means of its so-called peace offensive, but on the other, it is not showing any indication of relaxing its continued military expansion despite its recent structural economic difficulties including sluggish economic growth, slow growth of oil supplies, and labor shortages. The Chernenko Administration established in February of this year clearly declared both internally and externally that there will be no changes in this policy.

The Soviet Union considers military power indispensable in carrying out its foreign policies and is making considerable efforts to increase its political presence backed by its huge military strength.

SECTION 3: U.S. DEFENSE EFFORTS IN RESPONSE

With a view to defend such values as freedom and democracy, the United States intends to defend the countries of the free world, thus contributing to world peace and stability. The U.S. has consistently followed a strategy of deterrence to prevent aggression of any form by maintaining diverse forces from nuclear to conventional and securing a military posture to effectively respond to any contingency once conflict arises. These national defense efforts by the U.S. are intended to maintain and strengthen the credibility of deterrence in response to the Soviets' military buildup. The Reagan Administration also upholds "deterrence," "defense" and "the restoration of peace" as the basis of U.S. defense policy.

In contrast to the Soviets' move, U.S. national defense efforts were restricted during the period of the so-called detente throughout the 1970s. As the effect of the Soviet military buildup has become clear, however, a sense of crisis concerning changes in the U.S. Soviet military balance and credibility of the deterrent capabilities of the U.S. has arisen in that country. In particular, it is clear from the Soviet military intervention into Afghanistan at the end of 1979 that the Soviet Union will not hesitate to use its military power even against the Third World. With this realization, the U.S. has begun to step up its national defense efforts, and at the same time, to strongly hope that its allies will make appropriate efforts as members of the free nations.

SECTION 4: ARMS CONTROL AND DISARMAMENT
TALKS BETWEEN THE U.S. AND THE SOVIET UNION

Realizing the importance of having credible deterrence of its own as a background and negotiating with the Soviet Union in trying to attain practical arms control and arms reduction with that nation, the U.S. has continued to push the START and INF talks.

. . . However, no agreement was reached between them, and late last year, the U.S began to deploy the Pershing II Missiles and GLCMs in the West European countries to correct the military and political imbalance between the East and West camps as well as to maintain and strengthen the credibility of NATO's deterrent capabilities.

On the other hand, the Soviet Union continued to increase the number of SS-20s even during the above-mentioned talks. And with the deployment of U.S. INF missiles in the West European countries, the Soviet Union unilaterally decided to discontinue the INF talks. As countermeasures, the Soviet Union: 1) lifted the moratorium on its INF deployment on the Soviet European side; 2) promoted the deployment of operational tactical missiles in East Germany and Czechoslovakia; and 3) increased the number of submarines armed with nuclear missiles around the U.S. coastline.

The Soviet Union has deployed 135 SS-20s in the Asian theater. SS-20s have a serious impact on Japan's security. It is Japan's strong hope that SS-20 deployment will be either reduced or abolished on a global basis. If Japan and the NATO nations unite in order to support the U.S., it will lead to the reduction or abolishment of the SS-20s. . . .

———————— **Bibliography: Part Three** ————————

Bertram, Christoph. "The Implications of Theater Nuclear Weapons in Europe." *Foreign Affairs* 60 (Winter 1981/82):305–326.

Buteux, Paul. *The Politics of Nuclear Consultation in NATO 1965–1980.* Cambridge, Mass.: Cambridge University Press, 1983.

Douglass, Joseph D., Jr. *The Soviet Theater Nuclear Offensive.* Vol. 1. Washington, D.C.: Government Printing Office.

Garner, William V. *Soviet Threat Perceptions of NATO's Eurostrategic Missiles.* Paris: Atlantic Institute for International Affairs, 1983.

Hoffman, Stanley. "NATO and Nuclear Weapons: Reasons and Unreason." *Foreign Affairs* 60 (Winter 1981/82):327–346.

Howard, Michael. "Reassurance and Deterrence: Western Defense in the 1980's." *Foreign Affairs* 61 (Winter 1982/83):309–343.

Nitze, Paul, H. "The Relationship of Strategic and Theater Nuclear Forces." *International Security* 2 (Fall 1977):122–132.

Pierre, Andrew J., ed. *Nuclear Weapons in Europe.* New York: Council on Foreign Relations, 1984.

Record, Jeffrey. "Theater Nuclear Weapons: Begging the Soviet Union to Pre-empt." *Survival* 19 (September/October 1977):208–211.

Simes, Dmitri. "Deterrence and Coercion in Soviet Policy." *International Security* 5 (Winter 1980-81):80–103.

Treverton, Gregory F. "Nuclear Weapons and the 'Gray Area.'" *Foreign Affairs* 57 (Summer 1979):1075–1089.

PART FOUR

ARMS CONTROL

CHAPTER 6

The Process

Although arms control deals with some of the world's most esoteric technology, the ideas that move its advocates are simple and obvious ones. They maintain that no nation will gain from nuclear war. They contend that the steady accumulation and refinement of weapons of mass destruction attack the human spirit and undermine the chances of survival not just of the Soviet Union and the United States but of the world and its peoples. Steps must be taken to stop the arms race. These concerns have produced an outpouring of imaginative ideas and proposals for the control, reduction, and elimination of nuclear weapons.

Begun haltingly by the Kennedy and Johnson administrations, arms control became a centerpiece of the détente policies of the 1970s. The presidents of the United States and the USSR, Nixon and Brezhnev, signed the first comprehensive arms control package in 1972, a freeze by executive agreement on the total number of offensive intercontinental nuclear missiles and a treaty requiring the limitation of ABM systems to two small deployments, later reduced to one for each side. The initial Nixon-Brezhnev agreements were nicknamed SALT, for Strategic Arms Limitation Talks, and assigned the number I to indicate there were more agreements to come. Secretary of State Henry Kissinger and others close to the negotiations, such as General Brent Scowcroft, have indicated after leaving office that the overall plan had three phases. In phase one there would be a freeze, necessarily asymmetrical, on both sides' offensive systems. In phase two the arsenals would be brought to a stage of equality, and in the final phase major reductions would be made.

Despite making a poor start with the Soviets, the Carter administration was able to complete the negotiation of an agreement it felt would achieve equality, the goal of the second phase. But before the treaty could be ratified by the U.S. Senate, a series of setbacks affected U.S. foreign policy, particularly the fall of the shah of Iran and the Soviet invasion of Afghanistan. Facing a strong opponent in the presidential elections of 1980 and a deeply unsettled public opinion, President Carter chose to withdraw the SALT II Treaty from the Senate. In short, by 1981 there was no longer any momentum behind arms control. Although President Reagan agreed to abide by the provisions of SALT II and offered to open Strategic Arms Reduction Talks (START) in 1982 with a proposal calling for deep reductions to five thousand warheads on both sides, it seems clear in retrospect that what might have been acceptable to one side would have been unacceptable to the other.

Arms control has two aspects: process and practical agreement. President John Kennedy's American University speech gave the basic rationale of the process. There must be more to international affairs, he observed, than the accumulation of armaments. Whatever the differences between the United States and the Soviet Union, the two nations shared a common interest in avoiding nuclear war. In their books *Cold Dawn* and *Thinking About National Security*, John Newhouse and Harold Brown identified the limited gains one could expect from arms control. As Harold Brown observed, arms control can reduce the damage both sides will suffer in case of war, can improve deterrence by favoring survivable systems and measures that deprive an aggressor of an advantage in striking first, and can reduce the probability of surprise attack or nuclear war by accident or escalation. These are modest claims and even in their pursuit the achievements have been modest.

However modest they may have been, the process and achievements of arms control were hotly rejected by analysts such as Richard Perle and Edward Luttwak in their "Echoes of the 1930s" and "Why Arms Control Has Failed." To Perle, the pursuit of arms control had become self-defeating and smacked of a collapse of will fully comparable to that which befell the European democracies when they faced Hitler's onslaught. Luttwak spoke for many critics of the process when he argued that the logic and process of arms control would actually end by imposing controls only on one side—the United States and its allies—and would favor the Soviet Union. The agreements that have issued from the arms control process have also been searchingly criticized. Senator Henry Jackson's amendment to the Senate resolution approving SALT I politically obligated the Nixon administration and its successors to achieve a goal of strategic equality with the Soviet Union in any subsequent negotiations.

The criticisms of arms control and its modest achievements have caused a loss of enthusiasm and interest in the complex bilateral Soviet-U.S. negotiations that were so fashionable in the 1970s. At the same time, even the Reagan administration, which entered office opposed to the past U.S. record in arms control talks, has chosen to reopen negotiations with the Soviet government and to present itself in the opening year of the president's second term as a government eager for a summit meeting with the new Soviet leader, Mikhail Gorbachev, and a general improvement in Soviet-U.S. relations.

This change owes more to the vision of the future necessary to successful democratic statesmanship than to the possibility of achieving a breakthrough leading to drastic reductions in nuclear weapons. The public backlash in Western Europe and the United States against the Reagan administration when it allowed itself to appear indifferent to arms control drove this point home in the White House. The Reagan administration remains determined to avoid making concessions to the Soviet Union that would, in its view, weaken the United States and its allies. However, by its decision to enter arms control talks with the Soviet Union, the Reagan administration has shown its awareness of the need to hold out to the public on both sides of the Atlantic a reasonable hope that everything that can safely be done will be done to avoid nuclear war and to bring about at least some degree of control over the arms race.

31. Commencement Address at American University in Washington

John F. Kennedy

. . . I speak of peace because of the new face of war. Total war makes no sense in an age when great powers can maintain large and relatively invulnerable nuclear forces and refuse to surrender without resort to those forces. It makes no sense in an age when a single nuclear weapon contains almost ten times the explosive force delivered by all of the allied air forces in the Second World War. It makes no sense in an age when the deadly poisons produced by a nuclear exchange would be carried by wind and water and soil and seed to the far corners of the globe and to generations yet unborn.

Today the expenditure of billions of dollars every year on weapons acquired for the purpose of making sure we never need to use them is essential to keeping the peace. But surely the acquisition of such idle stockpiles—which can only destroy and never create—is not the only, much less the most efficient, means of assuring peace.

I speak of peace, therefore, as the necessary rational end of rational men. I realize that the pursuit of peace is not as dramatic as the pursuit of war—and frequently the words of the pursuer fall on deaf ears. But we have no more urgent task.

Some say that it is useless to speak of world peace or world law or world disarmament—and that it will be useless until the leaders of the Soviet Union adopt a more enlightened attitude. I hope they do. I believe we can help them do it. But I also believe that we must reexamine our own attitude—as individuals and as a Nation—for our attitude is as essential as theirs. And every graduate of this school, every thoughtful citizen who despairs of war and wishes to bring peace, should begin by looking inward— by examining his own attitude toward the possibilities of peace, toward the Soviet Union, toward the course of the cold war and toward freedom and peace here at home.

First: Let us examine our attitude toward peace itself. Too many of us think it is impossible. Too many think it unreal. But that is a dangerous, defeatist belief. It leads to the conclusion that war is inevitable—that mankind is doomed—that we are gripped by forces we cannot control.

We need not accept that view. Our problems are manmade—therefore, they can be solved by man. . . .

Let us focus instead on a more practical, more attainable peace—based not on a sudden revolution in human nature but on a gradual evolution in human institutions—on a series of concrete actions and effective agreements which are in the interest of all concerned. There is no single, simple key to this peace—no grand or magic formula to be adopted by one or two powers. Genuine peace must be the product of many nations, the sum of many acts. It must be dynamic, not static, changing to meet the challenge of each new generation. For peace is a process—a way of solving problems.

With such a peace, there will still be quarrels and conflicting interests, as there are within families and nations. World peace, like community peace, does not require that each man love his neighbor—it requires only that they live together in mutual tolerance, submitting their disputes to a just and peaceful settlement. And history teaches us that enmities between nations, as between individuals, do not last forever. However fixed our likes and dislikes may seem, the tide of time and events will often bring surprising changes in the relations between nations and neighbors.

Second: Let us reexamine our attitude toward the Soviet Union. It is discouraging to think that their leaders may actually believe what their propagandists write. It is discouraging to read a recent authoritative Soviet text on *Military Strategy* and find, on page after page, wholly baseless and incredible claims—such as the allegation that "American imperialist circles are preparing to unleash different types of wars . . . that there is a very real threat of a preventive war being unleashed by American imperialists against the Soviet Union . . . [and that] the political aims of the American imperialists are to enslave economically and politically the European and other capitalist countries . . . [and] to achieve world domination . . . by means of aggressive wars."

Among the many traits the peoples of our two countries have in common, none is stronger than our mutual abhorrence of war. Almost unique, among the major world powers, we have never been at war with each other. And no nation in the history of battle ever suffered more than the Soviet Union suffered in the course of the Second World War. At least 20 million lost their lives. Countless millions of homes and farms were burned or sacked. A third of the nation's territory, including nearly two thirds of its industrial base, was turned into a wasteland—a loss equivalent to the devastation of this country east of Chicago.

Today, should total war ever break out again—no matter how—our two countries would become the primary targets. It is an ironic but accurate fact that the two strongest powers are the two in the most danger of devastation. All we have built, all we have worked for, would be destroyed in the first 24 hours. And even in the cold war, which brings burdens and dangers to so many countries, including this Nation's closest allies—our two countries bear the heaviest burdens. For we are both devoting massive sums of money to weapons that could be better devoted to combating ignorance, poverty, and disease. We are both caught up in a vicious and dangerous cycle in which suspicion on one side breeds suspicion on the other, and new weapons beget counterweapons.

In short, both the United States and its allies, and the Soviet Union and its allies, have a mutually deep interest in a just and genuine peace and in halting the arms race. Agreements to this end are in the interests of the Soviet Union as well as ours—and even the most hostile nations can be relied upon to accept and keep those treaty obligations, and only those treaty obligations, which are in their own interest. . . .

Third: Let us reexamine our attitude toward the cold war, remembering that we are not engaged in a debate, seeking to pile up debating points. We are not here distributing blame or pointing the finger of judgment. We must deal with the world as it is, and not as it might have been had the history of the last 18 years been different.

We must, therefore, persevere in the search for peace in the hope that constructive changes within the Communist bloc might bring within reach solutions which now seem beyond us. We must conduct our affairs in such a way that it becomes in the Communists' interest to agree on a genuine peace. Above all, while defending our own vital interests, nuclear powers must avert those confrontations which bring an adversary to a choice of either a humiliating retreat or a nuclear war. To adopt that kind of course in the nuclear age would be evidence only of the bankruptcy of our policy— or of a collective death-wish for the world.

To secure these ends, America's weapons are nonprovocative, carefully controlled, designed to deter, and capable of selective use. Our military forces are committed to peace and disciplined in self-restraint. Our diplomats are instructed to avoid unnecessary irritants and purely rhetorical hostility.

For we can seek a relaxation of tensions without relaxing our guard. And, for our part, we do not need to use threats to prove that we are resolute. We do not need to jam foreign broadcasts out of fear our faith will be eroded. We are unwilling to impose our system on any unwilling people—but we are willing and able to engage in peaceful competition with any people on earth. . . .

32. Cold Dawn: The Story of SALT

John Newhouse

Bureaucracy speaks of SALT in phases. The first seven rounds were Phase I. We are now in Phase II. It is tempting to say that the Phase I agreements were a major contribution. They may prove to be so, but it is probably too soon to measure the gains of Phase I. The returns may not be in for years. The scope of the achievement will depend on what the two sides do in future about SALT, which in turn will depend in large part on what they do about otherwise unrelated problems affecting themselves and other countries. SALT, like all large projects of history, is linked, however imperceptibly, to many other events and forces.

No one can say which issues will most influence the talks over the years ahead. Most likely, the arguments will become more, not less, complicated, especially as the emphasis shifts away from quantity to quality. New and no less foreign acronyms will emerge. For now, it would be unwise to look upon SALT as other than a semipermanent part of great-power relations: at times a real negotiation, at times a dialogue carried on just to sustain the process. Most people want arms-control negotiations to succeed, especially when nuclear weapons are involved. Within and beyond the government, people have reacted to the twists and turns of such negotiations by allowing themselves to be whipsawed by their fears or misled by their expectations. SALT should help to contain our fears and to hedge our expectations.

Agreement reached for its own sake and affecting little is probably worse than no agreement. Agreements, like those reached in Moscow, that arise from serious negotiation and exact some mutual restraint should become a useful base on which to build. Still, it is well to be cautious. The United States presently has little left to bargain with, save FBS and a threat to withdraw from the ABM Treaty after five years if progress toward a broader agreement on offensive weapons is not achieved.

Both the critics and the SALT bureaucracy, most of which likes the Moscow agreements, take a pessimistic view of the immediate prospects for Phase II. The Soviets will revive the FBS issue. And they will try to avoid being talked into yielding much of their numerical advantage in missiles. They will doubtless reject any proposal to limit MIRV testing, assuming such a proposal is made. The Americans will seek reductions in ICBMs in order to reduce the Soviet advantage in "throw weight," or megatonnage. And Washington, while perhaps ready for a deal of some kind on FBS, will try to minimize its concessions on this very hard issue.

No one expects progress toward a MIRV testing ban, the issue that most interests the arms-control community. At best, it may be possible to limit the number of MIRVs on each ICBM. More than half the Minuteman force will have three MIRVs. Conceivably, the Soviets might accept the same limitation on their big land-based missiles, though only in return for major concessions from the Americans. Whether Washington would pay a heavy price for such a limitation cannot be predicted. Probably neither side will limit the number of MIRVs on their SLBMs, but this is not worrisome, since only the larger and more accurate land-based missiles offer a first-strike threat.

To write off the prospect of setting some limits on innovation and refinement is premature, if not cavalier. After all, a number of so-called qualitative controls were built into the ABM Treaty. Radars were limited in size and power as well as in number. And banning exotic future defensive systems was an enormous contribution. It was also an achievement. Governments sometimes but not often renounce weapons that do not exist.

Skeptics should also understand that SALT is a moderating political force, but one whose effects are neither immediate nor obvious. SALT is likely to affect other things, but slowly. The arms race is not impelled by technology alone, or by technology in tandem with the notorious action-reaction cycle. Both these forces are partially shaped and balanced by politics, by the relations between the powers, by the attitudes and policies of political leaders who may or may not actively seek to improve these relations, for whatever purpose. Normally, it is harder to justify increased military spending during a period of détente than in a period of cold war. This would be especially true if détente appreciably moderated political difficulties between the great powers—reduced the level of their confrontation in Europe, say, and their indirect confrontation elsewhere. Moderating such difficulties in turn would sharpen incentives to apply greater resources to more useful and productive pursuits than improving strategic missiles.

Many Soviet experts, while acknowledging some pressure on Brezhnev to reorder priorities, doubt that it is enough to affect weapons policies. Perhaps. We don't know. We do know that as General Secretary of the Soviet Communist Party, Brezhnev is the boss, but not the chief of state. President Nikolai Podgorny, who is, would normally have signed agreements of this scope and character. Hence, we also know that in choosing to sign the Moscow agreements himself, Brezhnev was proclaiming his identification with SALT, which has become an instrument of his foreign policy—for now, a very much détente-oriented policy.

More and more, the issues that might touch off a political crisis or direct recourse to nuclear weapons are reaching the green baize tables. Fred Iklé, an authority on negotiation, worries that it "has acquired the nimbus of salvation," and warns against the notion that negotiation "will protect the world from destruction and eventually deliver it from the terrible engines of war."

In the West, negotiation is seen as useful because it is a means of working out differences and reaching agreement; negotiation is supposed to force up compromise and settlement. Further east, especially in the Soviet Union, negotiation is often a balloon put into the air, useful only so long as it doesn't come down. In the past, many Soviet offers to talk were designed to buy time, becloud the issue, and acquire propaganda advantage and, hopefully, political leverage.

SALT, so far, appears to be a serious process, as well as a continuing one. But the talks rely on the intent and negotiating style of the parties. The record established over seven rounds of talks, along with the Kissinger-Dobrynin conversations on the margin, is neither reassuring nor disappointing; it is inconclusive. A modest interim agreement has been achieved. From the start, the two sides have engaged in a frank, open, nonpolemical dialogue on the weapons most vital to their security. That itself is a watershed. Certainly, it is not to be taken lightly or for granted.

On the other hand, neither side has really learned much about the other's purposes. The suspicion of each government that the other will use SALT in a self-serving way has not been appreciably narrowed by either the agreements or the process itself.

Negotiating with the Russians requires patience and consistency. In the early rounds of SALT, the Americans often showed neither. After an initial period of studied caution, they did a great deal more talking than listening. In effect, America did the running, too often at the expense of consistency. Washington knew from experience in other negotiations that only by holding to the core of a position would useful agreement emerge. This was the case with the Austrian settlement, with the recent Berlin negotiations, and with other arms-control agreements. But in Vienna and Helsinki, American proposals were sometimes thrown up like pasteboard figures, withdrawn at the first sign of resistance, and replaced by other equally perishable offers.

In fairness, the seemingly chaotic and inconstant character of the talks arises in part from their complexity. Unlike most arms-control negotiations, a basic SALT proposal can have many combinations and permutations; each variant may have merit. The difficulty is that Russian suspicions of entrapment are aroused by this kind of versatility. Americans may understand what they are doing, but Moscow does not. The Soviet leadership normally reacts to unconventional tactics by hardening its own position.

Negotiating against a deadline is always risky. But by arranging to sign a SALT agreement in Moscow, that is what Nixon elected to do. In effect, he placed himself in what French diplomats call the worst of positions: *demandeur*. It is hard for the other side to react to a self-anointed *demandeur* other than to exploit him. Yet, as it turned out, nobody seems to have been exploited. Brezhnev wanted, and apparently felt that he needed, the agreements as much as Nixon.

On May 26, 1972, Nixon and Brezhnev signalled their political constituencies and the world that limiting strategic arms was no longer a rhetorical matter but a fully operational problem. They showed a considerable *sérieux*, as the French would put it, and did so at some risk. Brezhnev, like Khrushchev before him, is seen by Shelest and other Soviet hard-liners to have mounted a high-risk foreign policy. Nixon's triangular politics is popular at home, but worries America's allies and therefore could do some violence to that web of special interests that America has with Western Europe and Japan.

SALT was well served by the Nixon-Kissinger system, which is suited to problems of surpassing sensitivity and long lead time. The back channel is a good place for dealing with rival powers, but allies, like bureaucrats, resent being left out: It is too soon to balance the undoubted gains of a secret trip to China against the long-run effects of not alerting the Japanese.

Triangular politics is useful, if only because the greatest single source of instability in the years ahead probably lies in the relentless hostility that sets China and the Soviet Union against each other. China draws a measure of security from triangular politics, while the United States gains some maneuverability vis-à-vis Moscow and Peking. Moderating its difficulties with the West permits Moscow, however, to focus attention on China. Embedded as it is in triangular politics, SALT cannot be divorced from Moscow's concern with China or Washington's uneasy relations with Japan and Western Europe.

As a political force, SALT has a double edge. It may moderate relations between the great powers, but, depending on how it is handled and what is agreed to, it could also hasten the disrepair of NATO, a prospect to which Moscow must be sensitive. Concessions on the FBS issue could go down badly in Europe. America's forward-based aircraft are the politically credible offset to the Soviet missiles aimed at European cities.

And SALT is the only major East-West transaction that has lacked European participation. Because they are left out, European governments feel more dependent on America and hence more vulnerable. Their security, they know, is as much at stake as America's. Triangular politics, they fear, will restore the world of Palmerston's England, which had "no allies, only interests." Or before him, Metternich, who told Talleyrand: "Don't speak of allies. They no longer exist."

The choices in Phase I involved numbers of missiles and types of ABM defense. The choices in Phase II will be subtler. The lesson of Phase I was the desirability of restraint. Both sides satisfied themselves that they had more than enough strategic launchers, and so they accepted an equivalent of parity. The lesson of Phase II may be the desirability of measure, balance, and proportion. Washington will confront a need to reconcile SALT with other primary interests. It will be a more consciously political negotiation.

Because it is wrapped in the theology and other kindred abstractions of strategy, SALT is an obscure, certainly an elusive, enterprise. Politics, nonetheless, lies at its heart. There is much more to SALT than meets the eye of many a systems analyst. Settling the SLBM replacement issue was not the most important event of the May 1972 summit conference; nor perhaps was the signing of the SALT agreements themselves. The importance of the conference, coming as it did on the heels of Nixon's closure of North Vietnam's ports, lay in the fact that it was held at all.

33. Thinking About National Security

Harold Brown

THE BALANCE AND ARMS CONTROL

Since the mid-1960s, the strategic balance has altered (not continuously, but overall) in a direction less favorable to the United States. There have been and will continue to be interminable arguments about the exact balance of U.S. and Soviet strategic capabilities. There is no single criterion, and it is difficult to make a precise comparison at any one time. It is considerably easier to make a judgment about the trend, which was adverse for the United States from about the mid-1960s through the mid-1970s. This trend has made a difference in perceptions clearly detrimental to the United States in the Third World, among U.S. allies, and with the American people. It is harder to judge, but it is also likely that this trend has affected the perceptions of the leaders of the Soviet Union, although probably they do not consider themselves to have achieved strategic superiority.

Military strength is a central factor in the outcome of a military conflict. The *perception* of military strength can be a critical element in a political confrontation. Perceptions of a military advantage, or even of a trend in relative military capability that reflects a likely future balance, affect the political behavior of potential adversaries and of third parties in contemplating what actions to take in a crisis. The Soviet willingness to indulge in expansionary policies along their borders and in the Third World may in part be the result of a shift in the strategic nuclear balance.

For this reason, as well as to ensure deterrence of nuclear war, the United States must continue to modernize its strategic forces. But the modernization of these weapons will not in itself guarantee that the United States will attain its related national security goals of minimizing the chance of a strategic nuclear conflict and avoiding a situation in which the strategic nuclear posture of the United States is, or is perceived to be, inferior to that of the Soviet Union. The United States must have an associated arms control and diplomatic posture; the unending and unlimited competition likely in their absence clearly carries major risks to U.S. security and indeed to U.S. survival. On the other hand, an arms control and diplomatic program stands no chance of success in the absence of an adequate strategic arms program, because unless the United States has both the appearance and the reality of sufficient strength in strategic matters, neither U.S. allies nor the Soviet Union will be prepared to take those diplomatic and arms control approaches seriously. And any cessation of U.S. efforts in the arms competition that is not accompanied by limitations on Soviet actions would be even more risky to U.S. security and survival than unlimited or resource-limited competition on both sides. Thus, both these approaches together will be needed, and even that may not be enough.

A national security program encompassing both approaches should have as its goal strategic stability, that is, a situation in which neither side can achieve significant gains by a preemptive attack and in which deployment of particular new technologies or weapons systems will not substantially alter the relative strategic positions, at least in the short run. In such a situation, each side would have the strategic forces that it regarded as adequate to deter the other, but it would lack the forces that might make it think it could coerce the other. It is unlikely that meaningful superiority can be achieved against a determined opponent; the Soviet Union is certainly such an opponent, and the United States had better be. There are real dangers that the attempt to achieve superiority will lead to a continued escalation of arms deployment, of rhetoric, and of ill feeling and introduce instabilities into the strategic relationship. But if superiority is an idle goal, inferiority is still a possible outcome—if U.S. will or judgment or technological ability is inadequate. The United States cannot afford inferiority. . . .

Limits on the development and production of nuclear weapons and their delivery systems are possible, as are reductions, even deep ones. Effective steps against further proliferation of nuclear weapons are also feasible. Agreements on arms limitations and reductions can in principle serve any of several different objectives:

- They can improve security and stability, even though there will always be the possibility of covert violations or of a breakout (an abrupt withdrawal from the treaty agreement after making preparations to exceed the limits) by one side.
- They can lead to a more accurate understanding of the size and character of the adversary forces, through such provisions as those that ban interference with national technical means of verification.
- They can increase the ability of each side to calculate and predict the future military capabilities of the other and thus reduce the likelihood of overreaction.
- They can reduce the level of damage in a war—though only extremely deep reductions could mitigate the catastrophic destructiveness of nuclear weapons.
- They can improve deterrence of both nuclear and conventional war—in the first case by increasing the ability of retaliatory forces to survive an attack and in the second by favoring defensive weapons and dispositions of forces, thus tilting the balance against the initiator of an attack.
- They can reduce the likelihood of surprise attack, accidental war, and escalation through misjudgment, by providing for relocation of forces to make surprise less feasible, by including inspection and other confidence-building measures, and by arranging for exchanges of information both in normal times and in times of crisis.

Measured against these glittering possibilities, the achievements of arms negotiations to date have been modest indeed, as are their immediate prospects. The ABM Treaty has helped stabilize deterrence by ensuring the continued mutual vulnerability of the urban and industrial centers of the superpowers. The Nuclear Test Ban Treaty ended nuclear testing in the atmosphere by the superpowers—a move that affected the environment more than it did the arms race. SALT I limited the numbers of some kinds of strategic weapons, but that only channeled the efforts of the superpowers increasingly into qualitative improvements. SALT II imposed important qualitative limits and some reductions, but the failure of the U.S. Senate to approve that treaty has left it in the unstable limbo of voluntary observance. The Mutual and Balanced Force Reduction (MBFR) negotiations have had useful political results within NATO and have favorably influenced decisions for some unilateral reductions on each side. But they have not produced any agreements between the NATO and Warsaw Pact nations. The hot line agreement provides modest emergency communications capability. In all, not much to show for thirty-five years of negotiations and twenty years of treaties. And yet the arms control efforts of the past have not been a trap. They were worth all the effort, and the prospects justify even greater efforts in the future—not as a substitute for defense programs, but as a companion to them. Both are required to improve national security through an adequate and stable military balance.

34. Echoes of the 1930s

Richard Perle

. . . Anyone who has studied the interwar period and reads today's newspapers tends to experience an uneasy sense of *déjà vu*. So it is to the 1930s that this article looks back, and it does so with the present very much in mind. The comparison of the period between the wars with the present is not meant to suggest congruence—only an analogy, and an imperfect one at that. . . .

Indeed, what is striking about the present when placed alongside the past—about the 1970s compared with the 1930s—is not the sense that the Soviet Union's foreign and military policies are a latter-day version of German policies, or that the massive Soviet arms buildup of the last decade is like the German buildup in the decade prior to World War II. Rather, it is the sense that many institutions of government in the United States and among our North Atlantic allies are inadequately inoculated against the disease of appeasement that swept through Europe forty years ago. There is also the troubling sense that among the men who run and staff the institutions the early symptoms of the disease have already begun to appear.

ARMS CONTROL BY EXAMPLE AND RESTRAINT

The pacifist sentiments of the British public in the 1930s are too well known to rehearse here. And they are not, in any case, the central point. After all, the political and bureaucratic elite in prewar Britain believed itself to be clear-eyed and hard-headed in matters of international politics and diplomacy. Its members believed in arms control. They thought that arms control could stabilize the military balance in Europe in the 1930s. But they approached arms control with claims of prudence and caution that ring not so very different from those heard today.

Thus the report of a committee of the Imperial Defence Committee, set up to recommend a British approach to the 1932 Disarmament Conference in (could it be?) Geneva, had this to say: "The military forces of nations . . . should be limited in such a way as to make it unlikely for an aggressor to succeed with a 'knock-out blow.' "[1] The committee went on to urge an end to what is called "disarmament by example," a notion that a Mr. Paul Warnke could be found advocating some forty-five years later. In the British case "disarmament by example" meant a policy that made an example of British disarmament—without affecting in the least the military programs of other nations. So the British committee insisted: "Any further reduction of British armaments could only be undertaken as part of an international agreement containing comparable reductions by other powers."[2]

It is fair to suppose that when SALT got under way, it was a central purpose of the United States to bring about an arms control agreement that would limit strategic forces so that neither the Soviet Union nor the United States could launch a "knock-out blow" against the other's strategic deterrent. And it is surely the case that our arms control negotiators have been steadfast in insisting that the control of American armaments can only be undertaken if there are comparable controls on Soviet forces—no "disarmament by example" for us! Yet, as we approach the conclusion of a SALT II treaty, it seems all but certain that the constraints it will entail will permit the Soviet Union to continue expanding and refining its offensive forces so that, sometime during the life of the treaty, it will have the capacity to strike a "knock-out blow" against the land-based missile leg of the U.S. strategic triad. And it seems equally certain that the American missile force will be unable to threaten the Soviet ICBM force. . . .

TRAGIC PENALTIES OF UNDERESTIMATION

The debate that took place in Britain in 1934 was significantly influenced by official estimates about the relative strength of Britain and Germany—estimates that emanated from a military and intelligence establishment unwilling to acknowledge the emergence of trends that it would later be unable to overcome, but willing enough to obscure them with groundless claims that have a strangely contemporary sound. The following, taken from an Air Staff memorandum circulated in June 1934, is typical: "There is no ground for alarm at the existing situation. Whatever first-line strength Germany might claim, we remain today substantially stronger *if all relevant factors are taken into account*. But the future, as opposed to the present, may cause grave concern."[3]

. . . A careful reading of our own National Intelligence Estimates as they bear on Soviet strategic offensive forces reveals, as Albert Wohlstetter has painstakingly documented, a consistent tendency to underestimate. For no fewer than eleven years the CIA, along with other intelligence agencies, projected substantially smaller strategic forces for the Soviet Union than it in fact deployed.

THE QUICKSAND OF CONCESSIONS

By 1938 Germany's growing military power made British (and French) diplomatic acquiescence all but inevitable. So, at the third of three meetings with Hitler, Chamberlain sealed the fate of Czechoslovakia in accordance with German demands. It was already apparent, after the meetings at Berchtesgarten and Godesberg that preceded Munich, that years of appeasement, accompanied by official near-indifference to the growth of German military power, had narrowed British options to the sacrifice of Czechoslovakia or war with Germany. To those who thought about the matter, it was clear that the sacrifice of Czechoslovakia, however expedient, could only defer the final reckoning with Germany; it could not substitute for it. Yet it is astonishing with what enthusiasm the policy of appeasement was greeted in Britain in the days leading up to and immediately following the Munich agreement. . . .

The British were not alone in their reaction to Munich. Mackenzie King, then Prime Minister of Canada, wrote to Chamberlain on September 30: "A turning point in the world's history will be reached if, as we hope, tonight's agreement means a halt in the mad race of arms, and a new start in building the partnership of all peoples."[4]

Reflecting later on the Munich negotiations, Sir Samuel Hoare had this to say:

Like [Chamberlain] . . . I had been caught up in the toils of a critical negotiation. The longer it went on and the more serious the issue became, the more anxious I grew to see it succeed. This is almost always the course of negotiations. As they proceed, the parties in them become increasingly obsessed with the need to prevent their final failure. If they are to continue, it is necessary to make concessions, and one concession almost invariably leads to another. The time comes when the question has to be faced: Is the substance being sacrificed to the negotiation, and is it not better to admit failure rather than to make further proposals and concessions? Throughout the Munich discussions I often asked myself whether the slide into surrender had not started.[5]

It hardly needs stating that the course of negotiations described by Sir Samuel Hoare applies in great measure to the range of negotiations in which the United States and the Soviet Union have been engaged since the present detente got under way in 1972. It would be hard to imagine a more succinct description of the course of the SALT II negotiations. All the elements are there: the obsession to see them "succeed"; one concession leading to another; sacrificing the substance to the negotiation; the further proposals. Surely it is clear that the original American objectives in the SALT II negotiations have not been realized: preventing the Soviets from achieving a "knock-out blow"; limits on heavy missiles; limits on Backfire; limits on throw-weight; limits on the development of new missiles; an end to missile modernization; and verifiability. . . .

A FAILURE OF INTELLECT

In the largest sense the failure of Britain's interwar leadership reflected an intellectual failing of historic proportions. It was, in the last analysis, a failure to understand one simple proposition—a proposition that Chamberlain himself was to articulate on October 6, 1938, as the policy of appeasement lay in ruins, in a speech to the House of Commons: "Our past experience has shown us only too clearly that *weakness in armed strength means weakness in diplomacy.*"[6]

What troubles one most about the present against the background of the 1930s is the gnawing sense that the current U.S. Administration subscribes to the view that weakness in armed strength *need not* mean weakness in diplomacy—that somehow it will be possible to escape the consequences of military inferiority. What else is one to make of the claim of high officials that the strength of the American economy or the margin of our technological advantage can substitute for strategic power or armor or artillery in Europe?

It will take clarity and judgment to penetrate the "misty atmosphere of peace"; and the judgment we must be clear about is that "weakness in armed strength means weakness in diplomacy."

NOTES

1. Sir Samuel Hoare, Viscount Templewood, *Nine Troubled Years* (London: Collins, 1954), p. 118.
2. Ibid., pp. 118–119.
3. Anthony Eden, Earl of Avon, *Facing the Dictators* (Boston: Houghton Mifflin, 1962), p. 207.
4. Hoare, op. cit., p. 325.
5. Ibid., p. 311.
6. *Hansard Parliamentary Debates*, House of Commons, 5th series, October 6, 1938, p. 551. Emphasis added.

35. Why Arms Control Has Failed

Edward N. Luttwak

Over the last decade, the strategic competition between the United States and the Soviet Union has been transformed. From a clear American superiority, by all criteria of measurement, the balance has tilted to an increasingly precarious parity. A net American inferiority, in *all* dimensions of capability, is projected by 1985. At the same time, the strategic-nuclear arsenals of both the Soviet Union and the United States have increased very greatly in power, and expenditures on strategic forces and their ancillaries have grown considerably. All these changes have taken place under the aegis of an American strategic policy dominated by arms-control objectives. Clearly something must be very wrong with our pursuit of arms control, especially in the Strategic Arms Limitation (SALT) negotiations. But what?

. . . Imagine a world of two identical countries, X and Y, whose array of forces is also identical. Let X make preparations to build a new weapon, say, a bomber. Faced with an emerging bomber force in X, the leaders of Y can choose between two instruments of strategic statecraft, force-building or arms control. . . .

Even if no preventive wars break out, the best result that the force-building response can yield is that the two countries will acquire a matched set of bombers and anti-bomber defenses. While they are reaching this new equilibrium, scarce resources will have been expended, new weapons will have come into existence to add to the violence of an eventual war, and the military and military-industrial interests in each country will have been strengthened, thus making it that much more likely that there will be yet more military competition between X and Y. . . .

The force-building approach guarantees these grim results, while also increasing the risk of war. Arms control, by contrast, offers the possibility of a zero-cost and low-risk solution to the same predicament. . . .

In this elementary and abstract formulation, the merits of arms control as a tool of national strategy are compelling. As compared to force-building, it offers a fully equivalent degree of security, and indeed a higher degree of security in many cases, at a much lower risk of instability and war, at no cost, without any increase in the ultimate destructive potential of war arsenals, and also without the deformations of society caused by the artificial growth of military-industrial interests. Whatever is wrong with arms control, then, it is clearly not its logic.

To be sure, this model of the basic theory is highly artificial, assuming, as it does, only two identical countries, with identical weapons-building capabilities and also good information. When these artificial assumptions are waived, practical obstacles arise. . . .

It is true that even at present the imponderable results of research efforts conducted in the secrecy of military laboratories entail a permanent danger of instability. But the large-scale force-building efforts now conducted openly (because arms control is not effective) provide a double guarantee against destabilizing surprise. First, the sheer scale of the visible efforts implies a corresponding limitation on the resources flowing into secret force-building activities; second, the large and diversified forces now deployed provide a high degree of insurance that the sudden emergence of any revolutionary new weapon will not undermine stability. This indeed is the logic of current American deployment. The land-based missiles, the submarine systems, and the long-range bombers could all be neutralized quite suddenly by the emergence of revolutionary devices. But simultaneous breakthroughs in all the very different scientific areas involved are most unlikely.

Neither guarantee can survive in the wake of prolonged and successful arms-control efforts, through which the forces in place will have been greatly reduced in size and diversification, and through which *overt* force-building efforts will have been greatly diminished or even eliminated, thus releasing abundant resources for less visible activities. True, secrecy is much more easily maintained in laboratory research than in engineering development, when weapons are taken out to be tested, but this does not dispose of the problem. Even if the tests of some revolutionary new weapon that threatens to undermine stability were duly observed and properly understood, it might take years for development to catch up, and in the meantime the prospect of unilateral vulnerability will be an incentive to preventive war. This, then, is the ultimate irony: the inherent limits that information sets on the scope of arms control may mean that the final reward of sustained success in arms control will be the utter defeat of its goals. . . .

As for differences in weapons characteristics, and the built-in asymmetry caused by conservative evaluations on each side, these certainly create many difficulties, but they are after all of a purely technical character, and patient negotiations by technical experts should in principle be able to overcome such problems.

Much the same goes for geographic and demographic differences between the parties. Given good will, or rather a rational appreciation of the mutual advantages of restraint, there is no reason why formulas could not be found to equalize the asymmetries. If Y's light bombers can reach the main cities of X as easily as X's heavy bombers can travel the much greater distance to Y's cities deep inland, then the two physically different forces can be evaluated by their actual effectiveness against their respective targets, and then limited accordingly in a treaty. Similarly, if X is inherently more vulnerable to nuclear bombardment because of its higher population densities, Y's weapons can be more severely restricted so as to yield equality in destructive power under an arms-limitation treaty.

Even the paradox of visibility and vulnerability, as well as the long-term effects of arms control over the more visible weapons which release resources for more unstable invisible weapons, does not make arms control futile. If cruise missiles inherently achieve the substantive aims of arms control, let the negotiations focus on the many other costly and destabilizing weapons now being built. As for the long-term problem, it may well be argued that in the presence of today's luxuriant force-building, the immediate gains of economy and safety that arms control can yield are of much greater consequence than the remote dangers which successful arms control may eventually cause in the very long term. Certainly these hypothetical dangers need not deter the modest efforts now under way. . . .

Whatever is wrong with arms control, therefore, it is not the various practical difficulties, each of which can be overcome, at least in part.

If our arms-control model is brought still nearer to reality by substituting the Soviet Union and the United States for X and Y, a new set of difficulties immediately arises. The Soviet Union undoubtedly lacks the attributes of an ideal partner for arms control; indeed, some might say that it is grotesquely miscast for the part.

The first obstacle is the Soviet negotiating style. If a unilateral concession is made to Soviet negotiators for the sake of "generating a positive atmosphere," the Soviet side will take what is given but will under no circumstances volunteer a reciprocal concession. It is not that Soviet diplomats are necessarily tougher than Western negotiators or even that their conduct is a symptom of inflexibility. It is merely a question of method. Soviet negotiators insist on treating each issue quite separately, making the best bargain they can in each case. They do not try to smooth the path to agreement by yielding on lesser points for the sake of the common interest in the outcome of the negotiations as a whole. . . .

A second and more serious obstacle to arms control with the Soviet Union is the very nature of its politics. The logic of arms control depends on the recognition of a common interest in avoiding force-building which can bring no *national* advantage. In our initial example, country X refrains from building its planned bomber force because it calculates that after Y has duly reacted, it will have gained no net benefit, "it" being the country as a whole, rather than its soldiers and weapons builders. But if country X is the Soviet Union, then the logic may not apply at all. . . .

The politics of the Kremlin are poorly understood, but virtually all our experts agree that the coalition of the KGB, soldiers, and military-production managers is a major force in shaping policy. It follows that within the Soviet Union there is powerful and systematic pressure for more force-building—in fact, there is every reason to believe that the coalition only allowed Brezhnev to embark upon the policy of détente on condition that the military build-up would continue as before, if not actually accelerate. This state of affairs naturally restricts the scope of arms control, since the common interest on which it must be based cannot be the Soviet "society-wide" interest in economy and safety. Instead, there is only the much narrower overlap between the general U.S. interest on the one hand, and the highly specialized interests of the Soviet military-production coalition on the other. . . .

Another serious obstacle to arms control is the traditional Russian passion for secrecy. Even after many years of supposedly intimate negotiations on strategic forces, the Russians refuse to disclose any meaningful information about the characteristics of their weapons and the structure of their forces, let alone about their force-building plans; there is not even any substitute discussion of Soviet strategy and doctrine. . . .

The Soviet refusal to publish future deployment plans—as the United States does—or indeed anything at all specific about its military forces, past, present, or future, therefore restricts the scope of arms control rather seriously.

The Russian passion for secrecy also diminishes the scope of arms control by making verification extremely difficult. Satellite observation is once again crucial, but for all the remarkable detail of the photography, and the valuable performance data that can be obtained from electronic intelligence, the information may nevertheless be too ambiguous to give sufficient confidence for proper verification: data which are technically very good may be worthless because Soviet secrecy denies the wider circumstantial knowledge that can give them meaning. . . .

In practice, this means that in dealing with the Soviet Union, standards of verification must be relaxed, or else many arms-control hopes must be abandoned. In either case, the scope for genuine—that is, properly verified—arms control is considerably restricted. Arms control without high-confidence verification is a contradiction in terms. It does not lessen the risks of conflict but increases them, and it does not diminish incentives to force-building but makes them stronger. Low-confidence verification is not a substitute for proper inspection procedures, and it is not even a good device to mask unilateral disarmament.

Another peculiarity of Soviet statecraft which affects arms control, or rather the observance of its agreed limitations, is the Soviet penchant for the use of probing tactics. Like the hotel thief who will not break a lock but who will walk down the corridors trying each door in the hope of finding one carelessly left open, Soviet policy constantly probes such arms limitations as there are, to exploit any gaps in American vigilance, as well as any loopholes in the agreed texts.

. . . Even the traditionally optimistic analysts of the CIA concede that the Soviet Union is now allocating between 11 percent and 13 percent of its gross national product to military purposes, as compared to roughly 5.5 percent for the United States; other reputable observers reject the CIA estimates as too low and argue for 15 or even 17 percent, almost three times the American proportion. All seem to agree that Soviet defense expenditure has been increasing at a steady rate of roughly 4.5 percent per year, for many years, and this too is much higher than the American rate, which was actually negative between 1968 and 1972, and which even now, when outlays on strategic weapons have increased, stands at *less* than 3 percent, net of inflation.

The contrast in these figures suggests that the Soviet interest in equitable arms control is bound to be rather limited. But even this obstacle is not insurmountable.

. . . While it is true that the earlier plans for a major expansion of consumer-goods production have been dropped, the present leadership is making really huge investments in agriculture and it is steadily increasing the effort going into light industries. Such considerations must moderate in some degree the claims of the "metal eaters" on Soviet resources, and this in turn suggests that there may be room for successful arms control. . . .

Nor should the obstacle of secrecy be overestimated. . . .

The lack of advance information does mean that the huge advantage that comes from negotiating over force-building *plans*, as opposed to actual deployments, is irremediably lost. But the need to wait until weapons are brought out into the open to be tested need not utterly preclude successful arms control. . . .

Nor does the Russian fondness for probing tactics to create gaps in limitation agreements while fully exploiting bona-fide loopholes make a stable arms-control regime impossible. It does mean that the United States must expect probing operations to begin as soon as an agreement is signed, and that it must be ready to take firm action to stop violations—any violations, however small—as soon as they are detected. . . .

It is therefore clear that none of the peculiarities of Soviet conduct is a decisive obstacle to successful arms control. Whatever is wrong with arms control, it is not the fact that dealings with the Soviet Union must loom large in its pursuit.

The process of elimination has left only one possible culprit: the United States.

While arms control can only be effective in limiting specific deployments under specific arrangements, serving thereby as an alternative to force-building, the United States has consistently misused arms control in pursuit of the abstract goal of "strategic parity." When that concept was challenged, it was redefined by the White House as "essential equivalence," which is equally vague. Neither set of words has any meaning in the reality of weapons or forces. Neither set of words can define negotiating objectives.

While arms control can only be effectively pursued if negotiated limitations are defined with extreme precision, the United States has consistently tolerated ambiguities in its urgent pursuit of agreement for its own sake. As a result, a new and entirely artificial source of U.S.-Soviet tensions has been created. Changes in the Soviet strategic forces that would otherwise have passed almost unnoticed have excited suspicion and resentment when seen in high contrast against the poorly drafted texts of the 1972 Moscow accords. One side-benefit of each arms-control agreement should be to build confidence for the next, but the tension-creating ambiguities of SALT-1 have utterly defeated this purpose.

While effective arms control requires that high standards of compliance be enforced, the United States has consistently allowed Soviet probing to develop without effective challenge. . . .

Above all, the United States has misused arms control in the attempt to dampen the strategic competition in itself, as if the growth of strategic arsenals were the cause of Soviet-American rivalry rather than merely one of its symptoms, and incidentally a much less dangerous symptom than the growth of non-nuclear forces, whose warlike use is much more likely. . . .

Arms control can be no more than a tool of national strategy if it is to be effective. It is an alternative to the other tool, the deployment of weapons, and it can be the superior alternative. But this is only true when its goals are the same, that is, the goals of national strategy: to enhance security at the lowest possible cost and risk. In American policy, arms control has usurped the function of strategy and has become an end in itself. The consequences are now manifest: the unilateral arms control pursued by the U.S. since at least 1964, and the bilateral efforts that culminated in the 1972 Moscow agreements, have diminished rather than enhanced American security; they have not contained the growth of the arsenals of both superpowers; they have increased rather than diminished outlays on U.S. strategic forces; and they have now begun to compromise strategic stability, since the increasing obsolescence of an old bomber force and the approaching vulnerability of American land-based missiles will soon leave the submarine-missile force exposed to undivided Soviet counterforce efforts. . . .

CHAPTER 7

The Agreements

In his briefings on SALT, Henry Kissinger answered the critics of arms control by stressing the need to accept a mix of advantage and disadvantage in both the initial and subsequent arms control agreements with the Soviet Union. Over and over, Kissinger made two points. First, arms control had to be seen as part of an effort to establish with the Soviet Union a relationship that would enable the two countries to avoid nuclear war. Second, arms control negotiation and agreement would also help prevent the arms race from gradually overshadowing all other aspects of international affairs. It was not just that the weapons would accumulate without arms control. Worse, the justifications that would be offered to sustain the buildups on both sides would undermine the difference between first- and second-strike capabilities and with it all the effort that had been expended to reduce the incentive to preemptive attack. In the end, Kissinger argued, the political rationale for continuing to expand nuclear arsenals "will, over a period of time, present a major obstacle to the humane or even safe conduct of foreign policy."

The problem for the Nixon agreements, and all arms control efforts ever since, was that to make a beginning meant to accept less U.S. advantage than might be achieved if one or another hypothetical alternative U.S. strategy and defense budget were to be followed. In SALT I, as Kissinger observed, the interim agreement allowed the Soviet Union more missiles than it allowed the United States. This meant that during the brief life of the agreement, neither side could be certain that the other was not merely using the limitations to prepare for a "breakout," or massive buildup when the agreement expired. For that reason, Kissinger said, it was necessary for the two countries over a longer period to achieve an agreement that, by reconciling the differences in the two nuclear arsenals and setting limits on rates of deployment, would avoid a constant escalation of the arms race.

It was an eloquent appeal. In addition, Kissinger was careful to allow the United States a variety of means to modernize its strategic forces, both at land and at sea (a bargaining posture followed by his successors as well). However, the modernizations, such as a new intercontinental bomber and a more powerful and sophisticated ICBM, were either defeated or postponed and thus could have no effect on the Soviet-U.S. balance. This exposed arms control to further attack and helped to poison Soviet-American relations.

The appeal to improve Soviet-American relations and check the arms race was also undercut by public opposition in the United States. Keying on presidential and congressional elections, a variety of lobbying groups formed and launched a massive publicity campaign against the SALT process and agreements. Among these groups, the most influential was the Committee on the Present Danger. Its members were united in a campaign not only to discredit SALT but to force a drastic increase in U.S. military and particularly strategic power. Many of the members of the committee took policymaking positions in the Reagan administration.

Of the efforts by committee members, Paul Nitze's critique of SALT in January 1976 stood out. Nitze argued that SALT was a total failure. Admitting his part in the early phases of arms control, Nitze nonetheless stated flatly that SALT should be abandoned. Under the limitations in SALT the Soviet Union had actually enhanced its nuclear posture to such a degree that nuclear war was more likely than it had been before arms control became a central feature of U.S. foreign policy. Moreover, the Soviet Union could utilize its nuclear advantages to achieve foreign policy gains in critically important regions. "There is every prospect," Nitze argued, "that under the terms of the SALT agreements the Soviet Union will continue to pursue a nuclear superiority that is not merely quantitative but designed to produce a theoretical war-winning capability." According to Nitze, SALT so favored the Soviet Union that it actually increased the likelihood of nuclear war and permitted the Soviet Union to derive significant foreign policy gains from its nuclear advantages.

The efforts of Nitze, his fellow members of the Committee on the Present Danger, and other opposition groups were effective. A marked shift in public perceptions of Soviet and U.S. military capabilities occurred as the 1970s ended. Whether the change would have been sufficient to defeat SALT II had the Soviet Union behaved more prudently in Africa and South Asia is unclear. Strange bedfellows, the anti-SALT Americans and the expansionist Soviets were more than enough to end the decade-long romance of U.S. politicians, policymakers, and analysts with arms control.

36. Press Conference on Soviet-American Relations and Arms Control

Henry A. Kissinger

Secretary Kissinger: I pointed out that there are three fundamental purposes in these summit meetings; one, for the leaders of the Soviet Union and the United States to exchange ideas and to check assessments about international affairs in general. The necessity for this arises because, as the two nations capable of destroying humanity, they have a special obligation to prevent conflicts caused by inadvertence, by miscalculation, by misassessment of each other's motives, examples of which history is replete. The second is to see whether they can, by meeting the needs of their peoples and of mankind, construct a network of positive relationships that will provide an incentive for moderation and for a beneficial and humane conduct of foreign policy.

The second large objective is to prevent the nuclear arms race and the arms race in general from dominating international affairs, and I want to stress again that this objective is no mean goal and one that will occupy American administrations in the absence of comprehensive agreements for as far into the future as we can see. It is not only the complexity of the weapons and their destructiveness; it is also the justifications that will have to be used in each country to sustain large armament programs that will, over a period of time, present a major obstacle to the humane or even safe conduct of foreign policy.

And the third general goal is to identify those areas of common interests, either produced by the nonmilitary aspects of technology or by others or by the nature of modern life, in which the Soviet Union and the United States can cooperate and thereby create a perspective on world affairs that recognizes the interdependence of events and the fact that isolation and confrontation are, over a period of time, inimical to progress and inconsistent with human aspirations.

Now, in terms of these three objectives, a great deal of time was spent by the two leaders in reviewing the international situation. . . . They were the most extensive discussions at that level of the arms race that have ever taken place, and with a frankness that would have been considered inconceivable two years ago, indeed with an amount of detail that would have been considered violating intelligence codes in previous periods.

So, on the issue of SALT, for example, on which I will have more to say in a few minutes, the words of the communiqué, that far-reaching and deep conversations took place, are of very profound significance. And in the next phase of discussions, difficulties cannot be caused by misapprehensions about each other's general intentions and general perceptions of the nature of the strategic environment.

And thirdly, there were a series of agreements, about most of which you have already been briefed, in the field of cooperative relationships.

Now, let me speak for myself about the two areas of arms control and the general review of the international situation.

With respect to arms control, let me cover first the agreements that have been made and then let me talk about the strategic arms limitation talks.

With respect to the agreements that have been made, there are three: the agreement that neither side will build the second ABM site, the agreement on the limited threshold test ban, and thirdly, the agreement to begin negotiations on environmental warfare.

With respect to the first agreement, in which both sides forgo the second ABM site, you remember that the permanent agreement on defensive weapons signed in Moscow in 1972 permitted each of the two countries to maintain two ABM sites, one to defend its capital, the second to defend an ICBM field, provided that field was no closer than 1,300 kilometers to the capital. The United States at that time opted for a defense of an ICBM field. The Soviet Union opted for a defense of its capital. . . .

The United States and the Soviet Union have now decided to forgo that second ABM site and to maintain only the one ABM site that each currently has, which is Moscow for the Soviet Union and an ICBM field for the United States. However, because it was thought desirable to keep some flexibility with respect to which area could be defended, each side is permitted at one time during the course of the agreement, and once in a five-year period, to alter its original decision.

In other words, if the United States should decide that it would prefer to defend Washington rather than the ICBM site, we have the option once in a five-year period to move from the ICBM site to Washington, and equally the Soviet Union has the option of moving once in that five-year period from Moscow to an ICBM site. That option, having once been exercised, cannot be exercised the second time. In other words, countries cannot shuttle their ABM sites back and forth between the capital and an ICBM field. Each side, in short, has the option once to reverse its original decision, and it may do so in any five-year period when the treaty comes up for automatic review.

The significance of this agreement is that it reinforces the original decision implicit in 1972—in fact, explicit in 1972—that neither side would maintain ABM defenses. It makes it even more difficult, if not impossible, to break out of the agreement rapidly, and in turn, the decision to forgo ABM defenses has profound strategic consequences which are sometimes lost sight of.

You must remember that the original impetus for the multiple warheads derived from the desire or the necessity to overcome ABM defenses and to make sure that the required number of missiles would get through. In the absence of ABM defenses, the extraordinary number of foreseeable multiple warheads will create a situation in which such terms as "superiority" should not be lightly thrown around because they may be devoid of any operational meaning.

The notion of nuclear sufficiency, of what is necessary under conditions of no ABM defenses, requires careful correlation with the number of available warheads. For present purposes, I want to say that any idea that any country can easily achieve strategic superiority is almost devoid, under these conditions, of any operational significance and can only have a numerical significance. The ABM agreement reinforces the element of strategic stability that was inherent in the original ABM agreement made in 1972.

The second agreement, on the threshold test ban, prohibits underground nuclear explosions above 150 kilotons and will therefore have the tendency to concentrate competition in the ranges of the lower-yield weapons. The date for its going into effect has been put into the future because a number of additional agreements remain to be worked out.

There remains to have an agreement on the peaceful uses of nuclear explosions, in which adequate assurance will be given that they will not be used to circumvent the intention of the agreement; and there is an agreement in principle that the inspection of peaceful nuclear explosions, among other things, will involve prior notification, precise definition of the time and place, and the presence of observers, which is a major step forward in our discussions. The second subject that will require further discussion is the exchange of geological information which is needed for the adequate verification of this threshold test ban.

The third area in which an agreement was reached was to begin discussions on the dangers of environmental warfare from the point of view of overcoming these dangers. This is a form of warfare that is in its infancy, the nature of which is not properly understood, and which obviously, by definition, can have profound consequences for the future of mankind. The United States and the Soviet Union, in the near future, will open discussions on this problem of environmental warfare.

In addition to these three agreements, two protocols will be signed on the Standing Consultative Commission, and we will certainly make diplomatic history, because it will be the first time that secret agreements are publicly signed. The agreements are being kept secret at the request of the Soviet Union, because they involve dismantling procedures for replacement missiles under the interim agreement and the ABM agreement. However, they will be submitted to the appropriate congressional committees upon our return to the United States.

Let me say a word about the Standing Consultative Commission. The Standing Consultative Commission was created in the 1972 [ABM treaty] in order to implement the provisions for replacement or destruction of weapons under the two agreements on defensive and offensive weapons.

There is a protocol for defensive weapons because the United States will have to dismantle some deployments that have taken place at a site which under the agreement we can no longer maintain, and the Soviet Union will have to dismantle fifteen ABM launchers and associated radars on their test ranges.

Secondly, there is a protocol for offensive weapons, which discusses dismantling and replacement procedure under the provisions of the interim agreement where land-based missiles can be traded in for modern sea-based missiles and where older submarine-launched nuclear missiles can be traded in for newer submarine-launched sea-based missiles. These are the two protocols that have been the subject of an illuminating exchange that took place just before I left the United States. . . .

Now let me say a word about strategic arms limitation talks. As I pointed out prior to our coming here, the administration considers the problem of strategic arms limitation one of the central issues of our time. It is one of the central issues, because if it runs unchecked, the number of warheads will reach proportions astronomical compared to the time—when Armageddon seemed near—when there were something less than 1,000 warheads on both sides.

It is important because a perception may grow that these warheads will provide a capability which will not be sustained by any systematic analysis, but because in any event they bring about a gap between the perceived first- and second-strike capabilities which in itself will fuel a constantly accelerating arms race.

Now, the problem we face in these discussions is that under the interim agreement the Soviet Union possesses more missiles—though if you add together the total number of launchers, that is to say, strategic bombers, there is no significant gap; and after all, it was not the Soviet Union that made us build bombers, that was our own decision—and therefore an attempt has been made to establish a correlation between the number of MIRV missiles and the numbers of launchers, in which perhaps to some extent the larger numbers of missiles on one side can be offset by a larger number of MIRVs on the other.

The difficulty with this approach has been the limited time frame within which it was attempted to be implemented, so that during the maximum deployment period it would not be clear whether any of these limitations would not simply be to provide a base for a breakout when the agreement lapsed.

Therefore the two leaders have decided that the principal focus of the discussions would not be on a brief extension of the interim agreement tied to an equally brief MIRV agreement, but to see whether the three factors—time, quantity of launchers, and quantity of warheads—cannot be related in a more constructive and stabilizing fashion over a longer period of time; that is to say, by 1985. And in that context, some of the difficulty of relating the various asymmetries in number can be taken care of and a stability can be perhaps achieved in deployment rates that would remove, to a considerable extent, the insecurities inherent in an unchecked arms race. . . .

Q: . . . You talked about the technological explosion in Brussels, I think. Does this not suggest that in the period between now and 1985 you will have one hell of an arms race going on?

Secretary Kissinger: No. It depends when the agreement is made. As I said in Brussels, and I maintain, that we have about eighteen months to gain control of the multiple warheads—control not in the sense of eliminating them, but by introducing some stability into the rate and nature of their deployment.

If an agreement is reached within that time frame, more or less—that doesn't mean down to the last month—then it can make a major contribution to turning down the arms race, to including the problem of reduction to which we attach importance, and to bringing stability into the strategic equation.

With every six-month period that it is delayed, the problem becomes more complicated; but the point is precisely to avoid what you called the "hell of an arms race." And the difficulty, as you analyze the problem with cutoff dates of 1978, 1979, is that both sides will be preparing for the break of the agreement while they are negotiating the agreement; and it became clear that one of the obstacles was that both sides, while negotiating limitations, were also putting themselves into the position of the agreement lapsing and therefore having to develop programs that would be pressing against limits of the agreement at the edge of its time period, and for that very reason have another vested interest not to have an agreement.

Q: Dr. Kissinger, General Secretary Brezhnev said last night that these accords could have been still broader than they were. First, I would like your comments on that and also whether it is not correct then from your interpretation that one could not say there are agreed guidelines on the MIRV warhead negotiations. Secondly, on the question of the underground nuclear test ban, could you clarify with some figures what I believe is a fact—that the limit of 150 kilotons would permit all continuing underground testing of MIRVs currently conducted by the United States, which are considerably below that range—and would that not allow the continuance even beyond the target date here of all the projectable multiple warheads likely to be produced by both sides?

Secretary Kissinger: First, the degree of cooperation between the Soviet Union and the United States has not yet reached the point where the General Secretary shows me the text of his speeches before he makes them. [Laughter.] And therefore I am not the best witness of what he may have had in mind.

My impression from what I have observed is that both sides have to convince their military establishments of the benefits of restraint, and that is not a thought that comes naturally to military people on either side.

Now, by definition, the limitations could have been broader. On the one hand, as you know, the Soviet Union has been proposing a complete test ban, but under provisions that are unverifiable and with escape clauses which would make it directed clearly against other countries. And therefore we have deferred a further discussion of the test ban, which we are not rejecting in principle—which, indeed, we are accepting in principle—for a later occasion. So I am assuming this is one thing the General Secretary had in mind.

The second is, from my description of the SALT discussions, obviously a broader agreement is conceivable. With respect to your question—are there agreed guidelines for Geneva?—the idea of extending the time frame arose really only on Monday, and it wasn't possible to work out detailed agreed guidelines in the interval.

On the other hand, certain basic principles do exist, and I believe we have made a major step forward in the approach to the problem.

With respect to the testing, it is not true that all the projected MIRV developments are in the category below 150 [kilotons]. Indeed, the enthusiasm seems to run more in the categories above 150, coupled with improved accuracies, but whenever I link these two I get a rebuttal. So I must be cautious. So if we are concerned that one of the threats to stability is the combination of accuracy and higher yields, then in the next phase of the MIRV warhead race this ban will make a major contribution.

Clearly, for the existing multiple warheads, the testing has been substantially completed on both sides. We are concerned with the next generation of warheads—not this generation of warheads—and with respect to those, it will play a very significant role. . . .

Q: Dr. Kissinger, you sound as though you have, at least for the time being, given up hope for getting a comprehensive SALT agreement with the Soviet Union. Is that correct?

Secretary Kissinger: Not a comprehensive, but a permanent, and this is not a question of giving up hope, it is a question of looking at the realities of how to move matters forward. We have been operating up to now within the constraint of either a very short-term or a sort of permanent agreement.

Now, permanent would have to have review clauses every five to ten years anyway. So when you talk of 1985, that is about as permanent as you can realistically become under present circumstances.

Q: I am kind of puzzled how you can take what happened here on SALT as anything less than a setback. If you have changed from searching for a permanent agreement to searching for one in a finite time period, and you postponed the time you have given yourself, or you have put back the time you have given yourself to find that agreement, it seems to me there are two setbacks there, and I don't see how you can say this hasn't been a failure at the summit.

Secretary Kissinger: If you approach it in a formalistic way, then these are valid arguments. If you approach it from the point of view of what will in fact contribute to slowing down the arms race, then I believe that we have found an approach in which the factors that have inhibited progress can be hopefully overcome.

The difficulty with the previous negotiations has been that it has proved extremely difficult to reconcile the various asymmetries that exist in the design of the forces, in the locations of the forces, and in the relative deployment rates of the forces. And the time limits we have been talking about until this visit created a situation in which both sides would be pressing against the limits of the agreement at the precise moment of its expiration date—the Soviet Union from the point of view of quality, the United States from the point of view of quantity—and therefore there was a great danger that the mere expiration date might fuel, especially in its final phases, a race.

And as a result of the discussion that took place Sunday, where for the first time, I believe, at least where the concerns and the perceptions of both sides were put before each other in what I considered an unusually frank way, and in which it turned out that the perception by each side of the other really was remarkably close—the only difference being that each side of course has to take the worst case of what the other one might do; I think this was the major gap that existed—it became apparent that the time pressure was a greater factor than had been commonly understood by either side.

So I don't want to do this in terms of setback. We are not running a race with ourselves. This is a problem which, I have been stressing, will be with us for a long time, and it shouldn't be seen in terms of hitting a home run on any one occasion.

Q: You, at the Brussels briefing, said there were only eighteen months before their decisions were irrevocable and each six months made it worse in terms of the rate of deployments.

Secretary Kissinger: That is right, and I have reaffirmed that here.

Q: But what I mean is you introduced the time pressure, as you call it.

Secretary Kissinger: There are two time factors, the time factor available for negotiation and the time factor involved in the length of the agreement. I have reaffirmed here that, in my judgment, the time frame in which the problems that I have identified can be constructively settled is in the eighteen-month range—twenty-four months, eighteen months, in that range— and one of the reasons for 1985 is that if this agreement were to be concluded in 1975, it would then take care of the next decade. This was one of the reasons behind it. So that time factor still exists, and that time factor will press on us and must press on us, if we are serious.

Q: Can I follow that up, sir? What would you envision will happen then, if the interim agreement expires or is allowed to expire in 1977 but you have not yet reached a replacement agreement—what will happen between 1977 and 1985 in terms of the arms race psychology?

Secretary Kissinger: If we have not reached an agreement well before 1977, then I believe you will see an explosion of technology and an explosion of numbers at the end of which we will be lucky if we have the present stability, in which it will be impossible to describe what strategic superiority means. And one of the questions which we have to ask ourselves as a country is: What in the name of God is strategic superiority? What is the significance of it, politically, militarily, operationally, at these levels of numbers? What do you do with it?

But my prediction would be that if we do not solve this problem well before, in my judgment, the end of the expiration of the agreement, we will be living in a world which will be extraordinarily complex, in which opportunities for nuclear warfare exist that were unimaginable fifteen years ago at the beginning of the nuclear age, and that is what is driving our concern. . . .

37. Testimony to the U.S. Senate in Favor of the SALT II Treaty: 1

Harold Brown

. . . I have examined this treaty with care and in the light of what I know about the nuclear armaments that exist in the world today and are likely to exist in the future. My judgment is that this treaty will make the people of the United States more secure militarily than we would be without it. For that reason I recommend that the Senate give its approval. . . .

ROLE OF ARMS CONTROL

Turning first to arms control itself, I start with the proposition that we cannot be militarily secure unless our strategic military forces are at least in approximate balance with those of the Soviet Union. The forces of the two countries are in a position of essential equivalence today.

There are two mays to maintain that equivalence. One is for both sides to add to their nuclear arsenals in equivalent or offsetting ways. The other is for both sides to limit their arsenals or to reduce them on a comparable basis. We have the option to follow either course. Either can maintain our security.

The course of limiting arms is preferable, however, for a number of reasons. First, it tends to make the future balance more predictable and stable and less likely to become one sided. Second, it provides more certainty to each side about the current program of the other. Third, it is obviously less costly for both sides. Overall, it is less risky for both sides.

Neither the present balance, nor ongoing Soviet programs, nor the state of arms control agreements, are such that we can avoid substantial defense programs needed for our military security. In fact, we need to increase our present level of such programs overall, regardless of this treaty. But SALT II is a clear and valuable, though limited, step toward curtailing the numbers and types of weapons that can be added by either side, and even toward reducing—by some measures—the number of weapons systems that one side (the Soviet Union) already has on hand.

In short, our military security can be enhanced either by increasing our own defense programs or by limiting the forces of the Soviet Union. Arms control, carried out with balance and care, can add to our military security just as can added defense programs. SALT II takes that approach toward making this country safer. . . .

DETERRENCE

Deterrence of nuclear war is our most fundamental defense objective. For us to achieve this, our potential adversaries must be convinced we possess sufficient military force that, whatever the circumstances, if they were to start a course of action that could lead to war, they would either (1) suffer unacceptable damage or (2) be frustrated in their effort to achieve their objective.

Our strategy requires us to be able to inflict such damage on a potential adversary—and for him to be convinced in advance of our ability—that, regardless of the circumstances, the prospect of that damage will preclude his attack on the United States or our vital interests. To achieve this we need, first of all, a survivable capability to devastate the industry and cities of the Soviet Union, even if the Soviets were to attack first, without warning. That capacity, called assured destruction capability, is the bedrock of nuclear deterrence. It is not, in my judgment, sufficient in itself as a strategic doctrine. Massive retaliation may not be appropriate—nor will its prospect always be sufficiently credible—to deter the full range of actions we seek to prevent. . . .

SALT II TREATY

The treaty is to last through 1985. It does these things:

It limits each side by the end of 1981 to a total of 2,250 intercontinental ballistic missile (ICBM) launchers, submarine-launched ballistic missile (SLBM) launchers, and heavy bombers.

This will require the Soviets to reduce by approximately 250 these strategic nuclear delivery vehicles (called SNDVs). Without SALT II, if the present trend continued, as I believe it would, the Soviet Union would instead have about 3,000 such weapons by 1985, instead of 2,250. No operational U.S. system will have to be reduced.

Within the 2,250 overall limit, the treaty sets a lower limit of 1,200 for launchers of ballistic missiles that carry multiple warheads aimed at more than one target (these systems are called MIRVed). Also, the number of such launchers plus the number of heavy bombers carrying air-launched cruise missiles (ALCM) cannot exceed 1,320.

Again, these limits hold the Soviets down to a level well below what I believe they otherwise could be expected to reach. For example, without SALT II, I would expect them by 1985 to have 1,800 multiple-warhead missiles instead of 1,200.

The limits placed on heavy bombers with cruise missiles will permit us to build the heavy bomber forces we have planned by 1985. To the extent that we may be required to reduce multiple-warhead systems, they will be older ones (chiefly Minuteman III missiles of increasing vulnerability) whose place will be taken by heavy bombers with cruise missiles or Trident submarines, leaving us stronger.

The treaty also sets a sublimit of 820 on launchers for land-based missiles which have multiple warheads that can be aimed at more than one target (MIRVed ICBMs). Those are the most threatening part of the Soviet nuclear arsenal as far as we are concerned.

The 820 figure is at least 100 fewer than we expected a few years ago that the Soviets would reach even under the SALT II limits, and much lower than the number they could reach by 1985 without SALT given their current trend. The 820 figure, however, is well above the number we plan to deploy.

The treaty also contains qualitative limits. It limits each side to only one new type of land-based intercontinental ballistic missile, and it requires that any such new missile not carry more than 10 independently targeted warheads or reentry vehicles (RVs).

This limit permits us to build the only new land-based missile we have planned to develop through 1985, the MX. It permits us to place on it the maximum number of warheads we intended—10.

The Soviets, on the other hand, have been developing several new land-based missiles, their fifth generation of them. The treaty limit means that now all but one of those new missiles will have to be restricted to quite limited modifications of their predecessors. It will also mean that instead of developing a separate specialized new missile for each of several missions, the Soviets will have to make some tough choices. For example, they will have to choose either a replacement for their existing single-warhead land-based missile (the SS-11) or another new missile with up to 10 warheads to replace their SS-17s (4 warheads) and SS-19s (6 warheads). They cannot, under SALT II, develop both of these, nor can they develop a new type missile to replace the SS-18.

The treaty also places a limit on what we call "fractionation." This means that the number of reentry vehicles on existing or modified ICBMs cannot be increased from what it is now and that the permitted new ICBM could not, if MIRVed, have more than 10 reentry vehicles.

We have no plans to increase the number of warheads on our principal land-based missile, the Minuteman III; doing so would not increase its military effectiveness. The Soviets, however, have much larger missiles that could be adapted to carry many more warheads. The treaty takes away the ability of the Soviets to exploit this. For example, except for this limit, the Soviet SS-18 missile could be equipped to carry 20 or even 30 independently targeted warheads in the 1980s. With SALT II, that will not happen.

The treaty also provides measures to permit unimpeded verification by national technical means.

PROTOCOL

The protocol covers a shorter period; it expires on December 31, 1981. It bars operational deployment of ground-launched and sea-launched cruise missiles (GLCMs and SLCMs) with ranges greater than 600 kilometers through 1981, but it permits unimpeded testing and development of such vehicles of any range. This limit will have no impact whatsoever on our present cruise missile testing and development schedule. I might add that the protocol provision was adjusted to our schedules and not vice versa. The protocol also bars deployment of mobile land-based or air-to-surface ballistic missiles through 1981; again, this will not affect our MX development, which will not enter its flight-test phase until 1983.

STATEMENT OF PRINCIPLES

Finally, the agreement includes a statement of principles to guide SALT III, plus Soviet commitments not to produce more than 30 Backfire bombers per year and to limit increases in its capability. The Backfire production restriction means that the Soviets now will not be able to divert Backfires to a strategic role (where they would add only marginally anyway) without greatly reducing Soviet capability for the naval and regional missions to which Backfires are normally assigned. . . .

Without the SALT II agreement, the Soviet Union could have nearly one-third more strategic systems than with the agreement—instead of the 2,250 delivery vehicles of the treaty, they could have 3,000. And there would be corresponding effects on other measures—including overall throw-weight, megatonnage, weapons numbers, and the like. Naturally, we do not know what the Soviets would do in the absence of a treaty, but these higher strategic system levels are well within their capability. . . .

First, the simple addition of numbers would force us to increase our own programs still further to preserve essential equivalence, to maintain areas of U.S. advantage to offset Soviet leads in other areas. . . .

Second, while SALT II won't solve the Minuteman vulnerability problem, it will make the solution of the problem easier than without an agreement. . . .

Third, to the extent that, as seems to me extremely likely, the lack of SALT limits would result in greater Soviet programs, and larger U.S. responses, we would be diverting scarce defense resources. . . .

38. Testimony to the U.S. Senate in Favor of the SALT II Treaty: 2

General David C. Jones

. . . We are unanimous in our view that, although each side retains military advantages, Soviet momentum has brought them from a position of clear inferiority to their present status of at least military equality with the United States. In some areas, they have already surpassed us and we are concerned because their momentum will allow them to gain an advantage over the

United States in most of the major static indicators of strategic force by the early 1980s. There is room for reasoned debate about the practical implications of this prospect, but it is important that we face up to its reality as we consider our own strategic responses.

It is also important to realize that any impending changes in the strategic balance will be the consequence of more than 15 years of unequal rates of investment in force modernization—the product of unilateral choices rather than an outcome of negotiated arms control. Overall, the Soviets have been outinvesting us for 10 years and, for the past few years, their total military investment effort has been about 75% larger than our own. With respect to investments for strategic forces, the disparity has, for many years, been even larger, with the Soviets outspending the United States by a factor of nearly three to one. Moreover, because of lead times in modern weapons programs, this progressive shift in the military balance will continue into the mid-1980s with or without SALT. A major concern my colleagues and I share is how best to minimize the period, extent, and consequences of any Soviet advantages. . . .

SALT'S CONTRIBUTION TO U.S. SECURITY

Having sketched the strategic framework as seen by the Joint Chiefs of Staff, I will now turn to the SALT II agreement itself, and provide an assessment of its contribution to our broader security aims. Such assessments should be based on realistic and reasonable criteria which avoid both unrealistic expectations and overgenerous appraisals. The criteria which my colleagues and I on the Joint Chiefs of Staff have endorsed are threefold.

- The agreement must stand on its own merits regarding equity and mutual interest, to include adequate verification.
- It must accommodate (in fact and in perception) our broader strategic interests, particularly our alliance relationships and the need to preserve our freedom of action in sharing appropriate technology.
- It must be a suitable framework for—and be accompanied by—the national commitment and strategic programs required to arrest the deteriorating state of the military balance. . . .

We should bear in mind that one of the objectives of SALT is to regulate, in a balanced fashion, aspects of two fundamentally dissimilar and asymmetrical force structures. Not only are the force structures different in their composition, but different features on each side's forces are viewed as more threatening by the other side. These different perspectives have produced a negotiating process marked by various compromises and tradeoffs as each side seeks to protect the essential character of its own forces while attempting to minimize the most threatening aspects of the other side's. The result is an agreement with some provisions clearly favoring one side and some clearly favoring the other. The question of equity, then, cannot adequately be addressed by a narrow and selective critique of portions of the SALT II agreement. Only a balanced appraisal of the total will yield an adequate evaluation.

Two issues of particular concern to us with regard to equity have been the Soviets' unilateral right to deploy 308 modern large ballistic missiles (MLBM), which was allowed in SALT I and carried forward to SALT II, and the exclusion of the Backfire from the aggregate totals of strategic nuclear delivery vehicles (SNDVs).

Clearly, the desired result would have been a major reduction in Soviet MLBMs in order to have reduced their very significant throw-weight capability and attendant potential to carry large numbers of warheads. Having failed to achieve that objective, we should accentuate our determination to obtain substantial reductions in future negotiations as a major objective. In the interim, limiting the SS-18 to 10 warheads achieves an important restraint on their MLBM potential.

The second major concern is the failure to count the Backfire bomber in the SNDV aggregate totals. While we are well aware of its employment capabilities in peripheral and maritime roles, the Joint Chiefs of Staff consistently recommended that the Backfire be included in the aggregate because it has an intercontinental range capability.

Nevertheless, the United States did obtain some constraints on the Backfire, the most important of which is a production limit not to exceed 30 per year. Furthermore, the United States retains the right to build and deploy an aircraft with equivalent capabilities. . . .

Among the most important provisions having an impact on Soviet plans for strategic forces are:

- Aggregate limits that will require the Soviets to dismantle (or convert to nonoffensive systems) 250-plus operational systems—these are older and less capable weapons but still a significant fraction of their total systems and megatonnage;
- The various limitations that will enhance the predictability of the range of Soviet force developments, thus assisting us in our force planning;
- The cap on RV fractionation that denies full exploitation of the major Soviet throw-weight advantage for the period of the treaty; and
- Testing, production, and deployment of the SS-16 banned.

On the other hand, the specific limits on the United States are quite nominal and provide the following options in planning our strategic forces.

- We can build an ICBM which fully meets our security requirements.
- We can continue with the modernization of our submarine-launched ballistic missile (SLBM) program at the pace we determine.
- We can continue to modernize our airbreathing systems, including the exploitation of our air-, ground-, and sea-launched cruise missiles. . . .

None of us is totally at ease with all the provisions of the agreement. I expressed our concerns on the Soviet MLBMs and Backfire earlier and we also have significant concerns with regard to our ability to monitor certain aspects of the agreement. We believe, though, that the risks in this area are acceptable provided we pursue vigorously challenges to questionable Soviet practices, improvements in the capability of our monitoring assets, and modernization of our strategic forces. In this context, the Joint Chiefs of Staff believe the agreement is adequately verifiable.

Also, despite differing degrees of concern on specific aspects of SALT II, all of us judge that the agreement which the President signed in Vienna is in the U.S. national interest and merits your support. We believe it is essential that the nation and its leadership view SALT II as a modest but useful step in a long-range process which must include the resolve to provide adequate capabilities to maintain strategic equivalence coupled with vigorous efforts to achieve further substantial reductions. . . .

39. Remarks in the U.S. Senate in Favor of Strategic Equality with the Soviet Union

Henry M. Jackson

Mr. Jackson: Mr. President, the Senate fully appreciates that it has before it an arms limitation measure of very great importance. Equally important, in my view, is how we in the Congress come to judge that agreement and how we convey our judgment to the Soviet Union, to our allies, to the executive branch and to the American people.

International agreements, this one included, have always had a dimension deeper than their words—a dimension that embodies hopes and expectations and reservations, all unwritten, and all of which reflect the blend of risks, doubts and assurances of which treaties are made. The present agreement, which rests to so remarkable a degree on statements with which the other party did not concur, has all of these in full measure.

SOME AMBIGUITIES RESOLVED

In an agreement that goes to the heart of America's security—our capacity to deter nuclear war—nothing is more important than a precise understanding of what the parties have agreed to. Precision can never weaken an arms control agreement; ambiguity can be a source of future tension.

From the moment the agreements were signed in Moscow, there were vague and conflicting reports about the contents of the accords. The press relied on confusing background briefings held in Moscow. Various statements and attachments, integral parts of the agreements, were not made public for weeks. Public figures issued statements of support before the terms of the accords were revealed or analyzed. High administration spokesmen offered contradictory interpretations.

Thus, when the President finally submitted these agreements to the Congress, I was concerned that the Senate and the American people understand what we and the Russians could and could not do within their terms. The Senate Armed Services Committee, under the distinguished chairmanship of my able friend the Senator from Mississippi, held extensive hearings on the military implications of the SALT accords during which we tried very hard to bring before the American people the terms, and the meaning of the terms, of the interim agreement and the ABM treaty. We endeavored to obtain a clear and consistent administration position on key provisions—which was not always easy.

Let me illustrate the kinds of problems the committee hearings had to address. Article I of the interim agreement obligates the parties not to build ICBM launchers after July 1, 1972. But nowhere in the agreement does there appear the number of ICBMs the Soviets are thereby permitted to have. Article II contains a prohibition against substituting "heavy" ICBM launchers for "light" ICBM launchers. But nowhere is there an agreed-upon bilateral definition of what a "heavy" ICBM is. Terms like "modernization" and "significant increase" were used in the agreement, but precise definitions of these terms are not provided.

Mr. President, although the texts of the treaty and the executive agreement contain few numbers, many more numbers have been discussed in connection with the SALT accords. In the last several weeks we have heard these: 1,054; 1,618; 313; 41; 42; 62; 84; 710; 740; 656; 950; and zero. The crucial question for the Nation and for the cause of world peace is whether these numbers add up to stable parity or unstable inferiority. It may clear the air to discuss some of them.

First, 1,054—this is the maximum number of land-based ICBMs that the United States is permitted to retain under the agreement.

Second, 1,618—this is the maximum number of land-based ICBMs that the Soviet Union is permitted to retain given our understanding of the agreement. I emphasize "our understanding" because the figure 1,618 is a U.S. intelligence estimate and not a Soviet-supplied number. This vagueness on the Soviet part is most unfortunate and can only lead to uncertainty and possible tensions. The agreement would be much improved by the use of specific numbers on both sides.

Third, 710—this is the maximum number of submarine-launched ballistic missiles that the United States is permitted to deploy under the agreement. It is arrived at by adding our present 656 Polaris/Poseiden missiles to our potential under the agreement, to replace 54 of our Titan missiles with new submarine-launched ballistic missiles.

Fourth, 950—this is the maximum number of submarine-launched ballistic missiles that the Soviet Union is permitted to deploy under the agreement.

To achieve this total the Soviets would retire their obsolete SS-7 and SS-8 ICBMs and replace them with new submarine-based missiles. This would give the Soviets a total of 62 modern nuclear—Y-class—submarines.

Fifth, 84—the total permissible number of Soviet missile-firing submarines that can be deployed under the agreement. This is derived by adding 22 Soviet G-class submarines to the 62 Y-class submarines that they are permitted to construct.

Sixth, 44—the total permissible number of American missile-firing submarines that can be deployed under the agreement. However, only 41 in actual fact are part of the U.S. deterrent in this period.

Seventh, 313—this is the maximum number of "heavy" ICBMs that the Soviet Union is permitted to deploy under the agreement. This again is a U.S. intelligence estimate, not a confirmed Soviet figure. Each of these missiles can carry at least a 25-megaton warhead and perhaps, eventually, as much as 50 megatons. Of course the Soviets are free to replace single large warheads with many MIRV warheads per missile when they are able to do so.

Eighth, 0—this is the maximum number of "heavy" ICBMs that the United States is permitted to deploy under the agreement.

These numbers, Mr. President, are merely representative of the thrust of the agreements—which is to confer on the Soviet Union the authority to retain or deploy a number of weapons based on land and at sea that exceeds our own in every category, and by a 50 percent margin.

Now there will be some who argue that "numbers do not matter"—that both sides have "sufficiency" and that therefore the strategic balance is stable. How curious it is that the people who hold to the "numbers do not matter" doctrine are the same ones who believe that without an immediate arms control agreement the world is in danger of a great nuclear war. Either numbers matter or agreements do not—you cannot have it both ways.

I have never seen an international agreement that depends so greatly on the attachment of unilateral statements. Clearly each of these unilateral statements reflects, not U.S.-Soviet agreement but, on the contrary, a failure to reach agreement. In my view the interim agreement is substantially weakened as a result of the failures indicated by the resort to unilateral assertions with no legal standing. No long-term treaty covering these vital matters would be acceptable to me and, I suspect, to a majority of my colleagues, if it depended to the extent of the present agreement on unilateral statements. The fact that the present agreement runs for only 5 years mitigates a situation that would be intolerable in a future treaty. . . .

I remember the argument over the H-bomb. Some argued that it would take the Russians a decade to get the H-bomb after we did. We finally got it, as I recall, only 7 months ahead of the Russians.

I just do not want the numbers in the interim agreement with the throw-weight advantage a permanent arrangement in a treaty at SALT II. That would destabilize the strategic balance rather than stabilize it. . . .

Under this interim agreement, they get 4 times our throw weight. That is the key. They get four times the throw weight of our forces. As the Senator knows, they are permitted 50 percent more launchers. So they will have the ability to have more warheads with greater megatonnage if we do not reverse this trend in SALT II. It is that simple. I think that is an undisputed fact. If the Senator has facts to contradict it, I will be glad to hear them. . . .

As I said earlier this afternoon, we had a great Navy, the largest in the world, on December 6, 1941. On December 7, we had no Navy at all. This is what we are talking about. I listened over the years to the debate on the ABM. We were told, "Look, the Russians really are not going for more than 1,200 ICBMs, and they are not about to go for Polaris-type submarines." We went through that in 1969 and 1970 and 1971. Now, look at the figures. Senators argued: "All they want is parity with the United States. They want an equal number of submarines and an equal number of ICBMs."

What did the Soviets do. They have gone ahead and have 50 percent more intercontinental ICBM launchers, and they have a 4-to-1 throw-weight advantage. Under the interim agreement they will have 62 submarines to our theoretical 44. In actuality, it will be 41.

I am just a country boy, but I think I know what is equality and what is not, and the interim agreement is not equality. The President recognizes it, and the President, who has the job of directing the negotiations, has indicated that the interim agreement is not the basis for the SALT II talks. . . .

THE CASE FOR THE AMENDMENT

Mr. Jackson: Mr. President, my amendment, which is broadly cosponsored by a bipartisan group of Senators, deals with three issues: First, the threat to the survivability of the U.S. strategic deterrent under the interim agreement; second, the need for equality in any follow-on agreement on offensive intercontinental strategic weapons; and third, the need for research, development and force modernization. These are issues that I have thoroughly discussed with the administration, with the witnesses before the Armed Services Committee and, in many cases, with my colleagues in the Senate. It is my firm conviction that we ought to state our views with respect to these issues and I believe my amendment is a medium for the expression of views that I am confident are shared by a majority of my colleagues. . . .

EQUALITY IN SALT II

Mr. President, I have elsewhere described the present agreement as providing the United States with "interim subparity." The agreement confers on the Soviets a 50-percent advantage in numbers of land- and sea-based launchers and a 400-percent advantage in throw weight. Now, the argument is made that this enormous disparity in numbers of launchers and throw weight is offset by superior technology and numbers of warheads on our side. There is a certain limited truth to this claim. It is not an enduring truth: for while numbers are limited under the agreement, technology is not. It stands to reason, therefore, that in the long run "superior" technology cannot be relied upon to offset inferior numbers.

The inability of technology to compensate for numbers is not only true in general but is, in the present case, true for specific reasons as well. The greatest part of our presumed technological advantage lies in our lead over the Soviets in the development and deployment of MIRV warheads on our missile forces.

This lead is not one that can be maintained at anything approaching our current margin. On the contrary, when the Soviets develop a MIRV capability—and they are expected to do so at "any moment"—the combination of that capability and their vastly superior throw weight will give them, given time and effort on their part, superiority in numbers of warheads. . . .

There is an enormous volume of misinformation on the subject of alleged U.S. advantages arising from technology and geography. There is no doubt that in the long run technology will tend toward equalization. How well I remember those who argued that the Soviets would require a decade or more to catch up with the United States in developing hydrogen weapons. The same sort of scientists who today argue that we can rest comfortably with inferior numbers of launchers because of an unbridgeable advantage in technology miscalculated by about 9½ out of 10 years back in 1947. The Russians, of course, were only months behind us, and our scientists were behind the eight ball.

As to geography, I have heard it argued—the chairman of the Foreign Relations Committee made the case himself last week—that owing to our possession of forward bases for our submarine fleet we need fewer submarines than the Soviets in order to maintain on-station times equal to theirs. Now, sea-based strategic forces are assuming increasing importance; so it is essential that we be correct on this point. Despite some statements to the contrary, the geographical asymmetries favor the Soviet submarine fleet and not our own. With the increased range such as that of the Soviet SSNX-8 submarine-launched missile, the importance of forward bases is greatly diminished. Russian submarines will be on station with respect to a large number of U.S. targets within 1 day's travel time from Murmansk or Petropavlovsk. This is not substantially different from the situation of our submarines operating out of their forward bases. What is more important, however, is that the Russians have a very large land mass between our submarines and their vulnerable points while we do not. Most of the U.S. points that are targets for Soviet submarine-launched missiles are coastal or near-coastal.

So there is little substance to the claim that we are in a favorable geographical situation. . . .

My amendment provides the Senate with an opportunity to declare itself in favor of equality in a follow-on agreement; and I am certain that in view of the basic good sense of that position and the overwhelming testimony before us, we will act to affirm it.

Mr. President, the question of what is to be included in the computation of equal forces in a follow-on agreement is related to the difficult issue of our forward deployments in Europe which are dedicated to the defense of our European allies and which are at sea.

The intent of my amendment as it bears on this matter is, I believe, perfectly clear and straightforward. In stating that "the Congress recognizes the principle of United States–Soviet Union equality reflected in the antiballistic missile treaty" and that accordingly "the Congress requests the President to seek a future treaty that, inter alia, would not limit the United States to levels of intercontinental strategic forces inferior to the limits provided for the Soviet Union" it is unmistakably clear that so-called forward based systems, which are not intercontinental, should not be included in that calculation of equality. It is my view, and the intent of the pending amendment, that any eventual treaty must recognize the necessity that the intercontinental strategic forces of the U.S. and the U.S.S.R., by which I mean to include ICBMs, submarine-launched nuclear missiles, and intercontinental range bombers of the two powers, should bear an equal relationship to one another. This says nothing about the eventual role of or disposition of the issue of forward-based systems. . . .

I fully expect that our negotiators at SALT II will insist upon equality just as the Soviets insisted upon equality in the ABM treaty. The issue of whether the present agreement adds up to equality is beside the point; and there will be differences of opinion on that. But what I am certain we can agree on is the necessity that we not accept in SALT II levels of intercontinental strategic weapons that are inferior to the levels of intercontinental forces permitted for the Soviet Union. My amendment does that.

Finally, I am confident that the Senate would wish to reaffirm its confidence in the importance of our research and development efforts. . . .

40. Public Law 92-448, Requiring the U.S. Government to Seek Strategic Equality Between the United States and the Soviet Union

JOINT RESOLUTION

Approval and authorization for the President of the United States to accept an Interim Agreement between the United States of America and the Union of Soviet Socialist Republics on certain measures with respect to the limitation of strategic offensive arms.

Resolved by the Senate and House of Representatives of the United States of America in Congress assembled,

. . . Sec. 2. The President is hereby authorized to approve on behalf of the United States the interim agreement between the United States of America and the Union of Soviet Socialist Republics on certain measures with respect to the limitation of strategic offensive arms, and the protocol related thereto, signed at Moscow on May 26, 1972, by Richard Nixon, President of the United States of America, and Leonid I. Brezhnev, General Secretary of the Central Committee of the Communist Party of the Soviet Union.

Sec. 3. The Government and the people of the United States ardently desire a stable international strategic balance that maintains peace and deters aggression. The Congress supports the stated policy of the United States that, were a more complete strategic offensive arms agreement not achieved within the five years of the interim agreement, and were the survivability of the strategic deterrent forces of the United States to be threatened as a result of such failure, this could jeopardize the supreme national interests of the United States; the Congress recognizes the difficulty of maintaining a stable strategic balance in a period of rapidly developing technology; the Congress recognizes the principle of United States–Soviet Union equality reflected in the antiballistic missile treaty, and urges and requests the President to seek a future treaty that, inter alia, would not limit the United States to levels of intercontinental strategic forces inferior to the limits provided for the Soviet Union; and the Congress considers that the success of these agreements and the attainment of more permanent and comprehensive agreements are dependent upon the maintenance under present world conditions of a vigorous research and development and modernization program as required by a prudent strategic posture.

41. Assuring Strategic Stability in an Era of Détente

Paul H. Nitze

Even though the translation of the Vladivostok Accord on strategic arms into a SALT II treaty has not yet been resolved, I believe it is now timely to take stock of the strategic arms balance toward which the United States and the Soviet Union would be headed under the terms of such a treaty. To that end it is necessary to raise certain basic questions about the maintenance of strategic stability—in terms of minimizing both the possibility of nuclear war and the possibility that nuclear arms may be used by either side as a means of decisive pressure in key areas of the world.

It appears to be the general belief that while such strategic stability may not be assured by the SALT agreements, it is not and will not be substantially endangered—that on the contrary it has been furthered by the SALT negotiations and agreements since 1969—and that in any event the best hope of stability lies in further pursuit of negotiations with the aim of reducing the level of strategic weapons and delivery systems on both sides. Unfortunately—and to the profound regret of one who has participated both in the SALT negotiations and in a series of earlier U.S. decisions designed to stabilize the nuclear balance—I believe that each of these conclusions is today without adequate foundation.

On the contrary, there is every prospect that under the terms of the SALT agreements the Soviet Union will continue to pursue a nuclear superiority that is not merely quantitative but designed to produce a theoretical war-winning capability. Further, there is a major risk that, if such a condition were achieved, the Soviet Union would adjust its policies and actions in ways that would undermine the present détente situation, with results that could only resurrect the danger of nuclear confrontation or, alternatively, increase the prospect of Soviet expansion through other means of pressure.

While this highly disturbing prospect does not mean that strategic arms limitation should for a moment be abandoned as a U.S. (and world) goal, the practical fact we now face is that a SALT II treaty based on the Vladivostok Accord would *not* provide a sound foundation for follow-on negotiations under present trends. If, and only if, the United States now takes action to redress the impending strategic imbalance, can the Soviet Union be persuaded to abandon its quest for superiority and to resume the path of meaningful limitations and reductions through negotiation.

Finally, I believe that such corrective action *can* be taken: (a) within the framework of the Vladivostok Accord; (b) with costs that would increase the strategic arms budget marginally above present levels (themselves less than half the strategic arms budget we supported from 1956 through 1962, if the dollar values are made comparable); (c) with results that would encourage the diversion of the Soviet effort from its present thrust and in directions compatible with long-range strategic stability. At the close of this article I shall outline the key elements in such a corrective program. . . .

Let us begin by discussing the similarities and contrasts between Soviet and American views on certain strategic questions.

"Is the avoidance of war—particularly a nuclear war—between the two countries desirable?" On this question I think both sides are in agreement. However, there is a certain difference of approach. Clausewitz once said that the aggressor never wants war; he would prefer to achieve his objectives without having to fight for them. The Soviets take seriously their doctrine that the eventual worldwide triumph of socialism is inevitable; that they are duty bound to assist this process; and that, as the process progresses, the potential losers may stand at some point and feel impelled to fight back. On the U.S. side some say that there is no alternative to peace and therefore to détente. This attitude misses two points. The first is that capitulation is too high a price for free men. The second is that high-quality deterrence, not unilateral restraint to the point of eroding deterrence, is the surest way of avoiding a nuclear war.

This thus leads to a second pair of questions: "Is nuclear war unthinkable? Would it mean the end of civilization as we know it?" We in the United States tend to think that it is, and this view prevailed (except for a small group of believers in preventive war, who never had strong policy influence) even in the periods when the United States enjoyed a nuclear monopoly and, at a later time, a clear theoretical war-winning capability.[1] When the effort was made in the late 1950s and early 1960s to create a significant civil defense capability, public resistance soon aborted the effort, so that today the United States has only the most minute preparations in this area. Rather, Americans have thought throughout the last 30 years in terms of deterring nuclear war, with the debate centering on how much effort is necessary to maintain deterrence, to keep nuclear war unthinkable.

In the Soviet Union, the view has been quite different. Perhaps initially because of the U.S. monopoly, Soviet leaders from the outset discounted the impact of nuclear weapons to their people. But as the Soviet nuclear capability grew, the Soviet leaders still declined to depict nuclear war as unthinkable or the end of civilization. On the contrary, they directed, and still direct, a massive and meticulously planned civil defense effort, with expenditures that run at approximately a billion dollars a year (compared to U.S. civil defense expenditures of approximately $80 million a year).[2] The average Soviet citizen is necessarily drawn into this effort, and the thinking it represents appears to permeate the Soviet leadership. In the Soviet Civil Defense Manual issued in large numbers beginning in 1969 and 1970, the estimate is made that implementation of the prescribed evacuation and civil defense procedures would limit the civilian casualties to five to eight percent of urban population or three to four percent of the total population—even after a direct U.S. attack on Soviet cities. The Soviets may well overestimate the effectiveness of their civil defense program, but what is plain is that they have made, for 20 years or more, an approach to the problem of nuclear war that does assume, to a degree incomprehensible to Americans (or other Westerners), that nuclear war could happen, and that the Soviet Union could survive.

These differences in approach and attitude appear to be basic and deeply rooted. In essence, Americans think in terms of deterring nuclear war almost exclusively. The Soviet leaders think much more of what might happen in such a war. To the extent that humanitarian and moral objections to the use of nuclear weapons exist in the Soviet Union—as of course they do—such objections are subordinated for practical planning purposes to what Soviet leaders believe to be a realistic view.

I have explained elsewhere at greater length the decisions of the early 1960s, in which I was one of those who participated with Robert McNamara, then Secretary of Defense.[3] In essence, the United States opted at that point to stress technological improvement rather than expanded force levels. While numerical comparisons were not ignored, the basic aim was an underlying condition of what may be called "crisis stability," a situation where neither side could gain from a first strike, and of "mutual assured destruction," where each side would have a fully adequate second-strike capability to deter the other. In such a condition it was believed that neither could realistically threaten the other in the area of strategic weapons, and that the result would be much greater stability and higher chances of the peaceful resolution of crises if they did occur. While nuclear weapons would always be a major deterrent, the conventional arms balance at any point of confrontation would remain important (as it had been in the Berlin crisis of 1958-62 and also in the Cuban missile crisis itself). In short, the aim was to downgrade nuclear weapons as an element in U.S.-Soviet competition and to prepare the way for systematic reductions in nuclear arms. If both sides were to adopt such a concept, it should be possible, over time, to move from what might be called a "high deterrent" posture to a "low deterrent" posture, with the deterrent remaining essentially equivalent on both sides but at successively lower levels.

As the United States thus adjusted its posture, the invitation for the Soviet Union likewise to seek a similar posture—and stop there—was patent both from statements of American policy and from the always-visible American actions. Unfortunately, however, the Soviet Union chose to pursue a course that was ambiguous: it could be interpreted as being aimed at overtaking the United States but then stopping at parity; it could, however, be interpreted as being aimed at establishing superiority in numbers of launchers and in throw-weight[4] and, perhaps ultimately, a nuclear-war-winning capability on the Soviet side. . . .

Harking back to the Soviet penchant for actually visualizing what would happen in the event of nuclear war, it seems highly likely that the Soviet leaders, in those hectic October days of 1962, did something that U.S. leaders, as I know from my participation, did only in more general terms—that is, ask their military just how a nuclear exchange would come out. They must have been told that the United States would be able to achieve what they construed as victory, that the U.S. nuclear posture was such as to be able to destroy a major portion of Soviet striking power and still itself survive in a greatly superior condition for further strikes if needed. And they must have concluded that such a superior capability provided a unique and vital tool for pressure in a confrontation situation. It was a reading markedly different from the American internal one, which laid much less stress on American nuclear superiority and much more on the fact that the United States controlled the sea lanes to Cuba and could also have expected to prevail in any conflict over Cuba waged with conventional arms.

One cannot prove that this was the Soviet reasoning. But the programs they set under way about 1962—above all the new family of weapons systems, embodying not only numbers and size but also greatly advanced technology, the development and deployment of which began to be evident beginning in 1971 but which must have been decided upon some years earlier—seem to reflect a fundamental state of mind on the Soviet side that contains no doubt as to the desirability of a war-winning capability, *if feasible*. Believing that evacuation, civil defense and recuperation measures can minimize the amount of damage sustained in a war, they conclude that they should be prepared if necessary to accept the unavoidable casualties. On the other hand, the loss of a war would be irretrievable. Therefore, the best deterrent is a war-winning capability, if that is attainable.

There have been, and I believe still are, divisions of opinion on the Soviet side as to whether such a capability *is* feasible. There are those who have argued that the United States is a tough opponent with great technical expertise and that the United States can be expected to do whatever is necessary to deny such a war-winning capability to the Soviet side. Others have taken the view that the developing correlation of forces—social, economic and political as well as military and what they call the deepening crisis of capitalism—may prevent the United States and its allies from taking the necessary countermeasures and that the target of a war-winning capability, therefore, is both desirable and feasible. Again, this is not to say that Soviet leaders would desire to initiate a nuclear war even if they had a war-winning capability. They would, however, consider themselves duty bound by Soviet doctrine to exploit fully that strategic advantage through political or limited military means.

The SALT negotiations got under way in late 1969. As a participant in those talks from then until mid-1974, I have described elsewhere some of the difficulties that attended the U.S. side. What was most fundamental was that the U.S. delegation sought at every level and through every form of contact to bring home to the Soviet delegation, and the leaders behind it, the desirability of limitations which would assure "crisis stability" and "essential equivalence"—and that the Soviet side stoutly resisted these efforts.

Indeed, the negotiations very early revealed other major stumbling blocks. One, in particular, revolved around the Soviet conception of "strategic parity." In the SALT negotiations the U.S. delegation consistently argued for the acceptance by both sides of the concept of "essential equivalence." By that we meant that both sides did not have to be exactly equal in each component of their nuclear capabilities but that overall the nuclear strategic capability of each side should be essentially equal to that of the other and at a level, one could hope, lower than that programmed by the United States. The Soviets have never accepted this concept, but have argued instead for the concept of "equal security taking into account geographic and other considerations." In explaining what they meant by "geographic and other considerations," they said that, "The U.S. is surrounded by friendly countries. You have friends all around the oceans. We, the U.S.S.R., are surrounded by enemies. China is an enemy and Europe is a potential enemy. What we are asking for is that our security be equal to yours taking into account these considerations." They never went so far as to say that this really amounts to a requirement for Soviet superiority in capabilities over the United States, the U.K., France and China simultaneously, but watching the way they added things up and how they justified their position, this is what it boiled down to.

Yet the two sides were able to reach agreement in May of 1972 on stringent limitations on the deployment of ABM interceptor missiles, ABM launchers and ABM radars and on an Interim Agreement temporarily freezing new offensive missile-launcher starts.

After the May 1972 signing of the ABM Treaty and the Interim Agreement, it turned out that the two sides had quite different views as to how the negotiating situation had been left. On the U.S. side, we told the Congress that the Interim Agreement was intended to be merely a short-term freeze on new missile-launcher starts, and that this, together with the ABM Treaty, should create favorable conditions for the prompt negotiation of a more complete and balanced long-term agreement on offensive strategic arms to replace the Interim Agreement and be a complement to the ABM Treaty. . . .

However, the Soviet Union had a quite different view. Its negotiators held that in accepting the Interim Agreement we had conceded that the Soviet Union was entitled to an advantage for an indefinite time of some 40 percent in the number of missile launchers and something better than double the average effective size, or throw-weight, of their missiles over ours. In working out a more complete and longer term agreement, in their view, all that was necessary was to add strict and equal limits on bombers and their armaments, provide for the withdrawal of our nuclear forces deployed in support of our allies capable of striking Soviet territory, and halt our B-1 and Trident programs but not the "modernization" of their systems. The difference of position between the two sides was such that it was difficult to see how agreement could be reached.

In the Vladivostok Accord of December 1974 the Soviets did make concessions from their past extremely one-sided negotiating demands. Those concessions were greater than many in the U.S. executive branch expected. However, does the Accord promise to result in achieving the objectives which the United States has for many years thought should be achieved by a long-term agreement on offensive forces? Those objectives were parity, or essential equivalence, between the offensive capabilities on the two sides, the maintenance of high-quality mutual deterrence and a basis for reducing strategic arms expenditures. I believe it does not.

The Vladivostok Accord, in essence, limits the total number of strategic launchers—ICBMs, submarine-launched ballistic missiles (SLBMs) and heavy strategic bombers, to 2,400 on both sides, and the number of MIRVed missile launchers to 1,320 on both sides. It limits the Soviet Union to the number of modern large ballistic launchers (MLBMs) that they now have, while prohibiting the United States from deploying any modern launchers in this category.[5] The Accord calls for air-to-surface missiles with a range greater than 600 kilometers, carried by heavy bombers, to be counted against the 2,400 ceiling. The treaty would allow freedom to mix between the various systems subject to these limitations. . . .

A notable feature of the Vladivostok Accord is that it does not deal with throw-weight. The agreement would not effectively check the deployment of the new Soviet family of large, technically improved and MIRVed offensive missiles. While both sides are permitted equal numbers of MIRVed missiles, the new Soviet SS-19s have three times the throw-weight of the U.S. Minuteman III, and the new SS-18s, seven times. What this comes down to is that under the Accord the Soviets can be expected to have a total of about 15 million pounds of missile throw-weight and bomber throw-weight equivalent. If the Congress goes forward with the B-1 and the Trident system but the United States does not add further strategic programs, the Soviets can be expected to end up with an advantage of at least three-to-one in missile throw-weight and of at least two-to-one in overall throw-weight, including a generous allowance for the throw-weight equivalent of heavy bombers, and two-to-one or three-to-one in MIRVed missile throw-weight. This disparity leaves out of consideration the Backfire, the FB-III, and the highly asymmetrical advantage in air defenses that the Soviet Union enjoys.[6]

Thus, the Vladivostok Accord, while a considerable improvement upon the prior negotiating positions presented by the Soviet Union, continues to codify a potentially unstable situation caused by the large disparity in throw-weight, now being exploited by Soviet technological improvements. . . .

The country as a whole has looked at strategic nuclear problems during the last six years in the context of SALT, hoping to make the maintenance of our national security easier through negotiations. It now appears, however, for the reasons outlined above, that we are not likely to get relief from our nuclear strategic problems through this route. Therefore, we have to look at our strategic nuclear posture in much the way we used to look at it before the SALT negotiations began and determine what is needed in the way of a nuclear strategy for the United States and what kind of posture is needed to support it. A fundamental aim of nuclear strategy and the military posture to back it up must be deterrence: the failure to deter would be of enormous cost to the United States and to the world.

Once again, two important distinctions should be borne in mind: the distinction between the concept of "deterrence" and the concept of "military strategy," and the accompanying distinction between "declaratory policy" and "action policy." Deterrence is a political concept; it deals with attempts by indications of capability and will to dissuade the potential enemy from taking certain actions. Military strategy deals with the military actions one would, in fact, take if deterrence fails. A responsible objective of military strategy in this event would be to bring the war to an end in circumstances least damaging to the future of our society.

From the U.S. standpoint, just to level a number of Soviet cities with the anticipation that most of our cities would then be destroyed would not necessarily be the implementation of a rational military strategy. Deterrence through the threat of such destruction thus rests on the belief that in that kind of crisis the United States would act irrationally and in revenge. Yet serious dangers can arise if there is such a disparity between declaratory deterrence policy and the actual military strategy a nation's leaders would adopt if deterrence fails—*or* if there is a belief by the other side that such a disparity would be likely. I think former Secretary James Schlesinger's flexible response program was, in effect, an attempt to get our declaratory policy closer to a credible action policy and thus improve deterrence.

Ultimately, the quality of that deterrence depends importantly on the character and strength of the U.S. nuclear posture versus that of the Soviet Union. In assessing its adequacy, one may start by considering our ability to hold Soviet population and industry as hostages, in the face of Soviet measures to deter or hedge against U.S. retaliation directed at such targets.

In 1970 and 1971—when the focus was almost exclusively on "mutual assured destruction"—the congressional debates on whether or not to deploy a U.S. anti-ballistic missile system recognized clearly the importance to deterrence of hostage populations. Critics of the ABM argued—and with decisive impact on the outcome of the debate—that an effective ABM defense of urban/industrial centers could be destabilizing to the nuclear balance: if side A (whether the United States or the U.S.S.R.) deployed an ABM defense of its cities, side B could no longer hold side A's population as a hostage to deter an attack by A on B. And in 1972 the same argument carried weight in the negotiation and ratification of the ABM limits in the SALT I agreements.

Yet today the Soviet Union has adopted programs that have much the same effect on the situation as an ABM program would have. And as the Soviet civil defense program becomes more effective it tends to destabilize the deterrent relationship for the same reason: the United States can then no longer hold as significant a proportion of the Soviet population as a hostage to deter a Soviet attack. Concurrently, Soviet industrial vulnerability has been reduced by deliberate policies, apparently adopted largely for military reasons, of locating three-quarters of new Soviet industry in small and medium-sized towns. The civil defense program also provides for evacuation of some industry and materials in time of crisis.

In sum, the ability of U.S. nuclear power to destroy without question the bulk of Soviet industry and a large proportion of the Soviet population is by no means as clear as it once was, even if one assumes most of U.S. striking power to be available and directed to this end. . . .

In sum, the trends in relative military strength are such that, unless we move promptly to reverse them, the United States is moving toward a posture of minimum deterrence in which we would be conceding to the Soviet Union the potential for a military and political victory if deterrence failed. While it is probably not possible and may not be politically desirable for the United States to strive for a nuclear-war-winning capability, there are courses of action available to the United States whereby we could deny to the Soviets such a capability and remove the one-sided instability caused by their throw-weight advantage and by their civil defense program.

To restore stability and the effectiveness of the U.S. deterrent: (1) the survivability and capability of the U.S. strategic forces must be such that the Soviet Union could not foresee a military advantage in attacking our forces, and (2) we must eliminate or compensate for the one-sided instability caused by the Soviet civil defense program. Specifically, we must remove the possibility that the Soviet Union could profitably attack U.S. forces with a fraction of their forces and still maintain reserves adequate for other contingencies.

As to the civil-defense aspect, the absence of a U.S. capability to protect its own population gives the Soviet Union an asymmetrical possibility of holding the U.S. population as a hostage to deter retaliation following a Soviet attack on U.S. forces. Although the most economical and rapidly implementable approach to removing this one-sided instability would be for the United States to pursue a more active civil defense program of its own, such a program does not appear to be politically possible at this time. Its future political acceptability will be a function of the emerging threat and its appreciation by U.S. leadership and by the public.

Two more practicable avenues of action suggest themselves. First, all of the options which would be effective in diminishing the one-sided Soviet advantage involve some improvement in the *accuracy* of U.S. missiles. Differential accuracy improvements can, at least temporarily, compensate for throw-weight inequality. . . .

Second, the prospective Soviet advantage could be offset by measures to decrease the *vulnerability* of U.S. strategic nuclear forces. Here there are several ongoing programs already under way, notably the development of the Trident submarine and the B-1 bomber; both these delivery systems will be inherently less vulnerable to a counterforce attack than fixed ICBM installations, the submarine by reason of its mobility at sea and the B-1 by virtue of its mobility and escape speed as well as the potential capacity to maintain a portion of the B-1 force airborne in time of crisis. In addition, programs to increase the pre-launch survivability of U.S. bomber forces generally, as well as programs to increase air defense capability through the so-called AWACS system, operate to reduce vulnerability of the total U.S. force.

. . . Unfortunately, I believe the record shows that neither negotiations nor unilateral restraint have operated to dissuade Soviet leaders from seeking a nuclear-war-winning capability—or from the view that with such a capability they could effectively use pressure tactics to get their way in crisis situations.

Hence it is urgent that the United States take positive steps to maintain strategic stability and high-quality deterrence. If the trends in Soviet thinking continue to evolve in the manner indicated by the internal statements of Soviet leaders, and if the trends in relative military capability continue to evolve in the fashion suggested by the prior analysis, the foundations for hope in the evolution of a true relaxation of tensions between the U.S.S.R. and much of the rest of the world will be seriously in doubt.

NOTES

1. To see how top officials viewed American nuclear power even in the period of American monopoly, one can now consult the recently declassified text of the NSC 68 policy paper dated in the spring of 1950. Even though Soviet nuclear

capacity (after the first Soviet test of August 1949) was assessed as small for some years to come, that paper rejected any idea of reliance on American nuclear power for the defense of key areas. To be sure, in the 1950s under John Foster Dulles, the United States had a declaratory policy of "massive retaliation." But in the actual confrontations of that period, this declaratory policy was not in fact followed; instead, conventional force was used, for example in the Lebanon crisis of 1958 and, less directly, in the Offshore Islands crisis of the same year. After 1961 massive retaliation was abandoned.

2. Eugene Wigner, "The Atom and the Bomb," *Christian Science Monitor*, November 13, 1974, p. 4.

3. See Paul H. Nitze, "The Vladivostok Accord and SALT II," *The Review of Politics* (University of Notre Dame), April 1975, pp. 147–60, especially pp. 149–50.

4. "Throw-weight" is a measure of the weight of effective payload that can be delivered to an intended distance. In the case of intercontinental ballistic missiles (ICBMs) and submarine-launched ballistic missiles (SLBMs), the throw-weight is a direct measure of such a payload in terms of the potential power of the missiles' boosters. In view of the more variable loads carried by heavy bombers, a formula for equivalence is needed to take account of all factors including explosive power.

5. There has been no agreed definition of a heavy ballistic missile. However, both sides acknowledge that the SS-9 and the SS-18 are MLBMs and that the U.S. Titan missile, while it is considered heavy, does not fall within the definition of "modern." The U.S. has no launchers for MLBMs and is prohibited from converting any of its silos to such launchers. The Soviets are estimated to have had 308 launchers for MLBMs and are permitted to convert the SS-9 launchers into launchers for the even larger and much more capable SS-18s.

6. In mid-1973 the United States had 602 fighter interceptors and 481 surface-to-air missiles, compared to the Soviet Union's 3,000 fighter interceptors and 10,000 surface-to-air missiles. Edward Luttwak, *The U.S.-U.S.S.R. Nuclear Weapons Balance*, The Washington Papers (Beverly Hills, Calif.: Sage Publications, 1974).

—————————— **Bibliography: Part Four** ——————————

Bertram, Christoph. "Rethinking Arms Control." *Foreign Affairs* 59 (Winter 1980/ 81):352–365.

Blacker, Coit D., and Duffy, Gloria, eds. *International Arms Control*. 2d ed. Stanford, Calif.: Stanford University Press, 1984.

Burt, Richard. "Reassessing the Strategic Balance." *International Security* 5 (Summer 1980):37–52.

deRose, Francois. "SALT and Western Security in Europe." *Foreign Affairs* 57 (Summer 1979):1065–1074.

Frye, Alton. "How to Fix SALT." *Foreign Policy* 39 (Summer 1980):58–73.

Garn, Jake. "The Salt II Verification Myth." *Strategic Review* 7 (Summer 1979):16–24.

"The Great SALT Debate." *Washington Quarterly* 2 (Winter 1979):3–103.

Labrie, Roger P., ed. *SALT Handbook: Key Documents and Issues 1972–1979*. Washington, D.C.: American Enterprise Institute, 1979.

Luttwak, Edward N. "Ten Questions About SALT II." *Commentary* 8 (Summer 1979):21–32.

Nitze, Paul. "The Merits and Demerits of a SALT II Agreement." Chap. 2 in Paul Nitze, James E. Dougherty, and Francis X. Kane, *The Fateful Ends and Shades of SALT.* New York: Crane, Russak & Co., 1979.

Panofsky, Wolfgang. *Arms Control and SALT II.* Seattle and London: University of Washington Press, 1979.

Rosen, Stephen P. "Safeguarding Deterrence." *Foreign Policy* 35 (Summer 1979):109–123.

Rostow, Eugene. "SALT II—A Soft Bargain, A Hard Sell." *Policy Review* 6 (Fall 1978):41–56.

Smith, Gerard. *Doubletalk: The Story of SALT I.* New York: Doubleday, 1980.

Talbott, Strobe. *Endgame.* New York: Harper & Row, 1979.

Wolfe, Thomas. *The SALT Experience.* Cambridge, Mass.: Ballinger, 1979.

PART FIVE

MORALITY AND NUCLEAR WEAPONS

CHAPTER 8

Secular and Religious Views on the Moral Dimensions of Nuclear Strategy

In their anxiety to prevail, the opponents of SALT oversold their case. Far from being a preferred stance, arms control became a pariah in the Reagan administration. So great was its aversion to arms control that the new U.S. administration permitted itself to appear indifferent to or even pleased by the consequences of a new round in the arms race, with its attendant risks of nuclear war. As a result, the administration lost control of public opinion in North America and in a number of countries in Western Europe, particularly the Protestant and northern ones. An antinuclear movement flourished once again, reminiscent of the 1950s Campaign for Nuclear Disarmament. Massive public demonstrations, some becoming violent, took place. Militant feminists keened over cruise missiles on Greenham Commons in Britain; Germans, Danes, and Dutch took to the streets forming miles-long human chains. In the United States, cities declared themselves nuclear-free zones, and town meetings in New England voted overwhelmingly for a freeze on the production and deployment of nuclear weapons. Resolutions endorsing a mutual verifiable freeze were passed by a number of state legislatures and the U.S. House of Representatives.

Through an excess of zeal, the administration had left the stage to the protesters. They obliged with countless manifestos, such as E. P. Thompson's "For Disarming Europe," and called for a welter of solutions, from Thompson's thoroughgoing pacifism to cautious unilateral measures taken to build mutual confidence between the Soviet Union and the United States.

As a result, the Reagan administration found itself scrambling not only to continue its efforts to increase U.S. military and strategic power but simultaneously to manage a public relations and policy campaign that would split the protesters and regain a hold on public and political opinion in the NATO countries. It is probably not unfair to suggest that the principal arms control proposals of the first Reagan administration were composed and presented with this goal as much in mind as the aim of achieving a new arms control agreement with the Soviet Union.

One of the more interesting manifestations of the greater public concern with nuclear weapons at this time was the appearance of the U.S. bishops' pastoral letter on nuclear issues. The bishops conducted a prolonged and very public debate among themselves about nuclear strategy and nuclear war. With the help of skilled antinuclear academic and political advisers, the bishops invited academic and government experts to appear before them and gained a remarkable amount of publicity for their deliberations. Basing their position on the teachings of the Catholic church about Just War and obligations of conscience, the bishops strongly condemned all use of nuclear weapons and planning for any kind of nuclear war. Nuclear war would inevitably escalate, they concluded. Therefore, all use of nuclear weapons must be avoided, even at the price of surrender.

At most, the majority of the bishops were prepared to admit the utility of nuclear deterrence as a means of preventing war. But their acceptance of deterrence was conditional. Nations might depend on deterrence, the pastoral letter said, but only if they sought with all the means at their command to bring about nuclear disarmament. Any other course, the bishops' position implied, was not only ruinous but immoral and contrary to the tenets of church teaching. Because the pastoral letter passed through a series of drafts, in the final version opponents of this approach were able to insert several passages intended to preserve the possibility and legitimacy of the use of force to resist evil. In several of the passages, one may see clearly the hand of Pope John Paul, a man who knew, along with millions of his Polish countrymen, of the true price of the passive submission to evil.

The antinuclear movement in general and the pastoral letter in particular evoked a number of telling responses. The Council of French Bishops addressed the same subject as the U.S. bishops and from within the same religious tradition. Their opinions, however, differed markedly from those of their U.S. colleagues. The French bishops corrected an omission in the U.S. pastoral letter by describing the heavy weight of Soviet pressure on the Western democracies. To the French churchmen, the Soviets sought the benefits of war without its costs. In addition, the French bishops stressed (as their U.S. peers had not) that states were charged with the protection of justice and liberty as well as peace. What an individual could risk in obedience to nonviolence might be forbidden a state, if it were to be true to all its responsibilities. Third, the French bishops observed that the threat of use in no way constituted the employment of nuclear weapons. To deny deterrence to free societies was to prevent them from dissuading an aggressor from attacking. The central question was "Given the context of the current geopolitical situation, does a country that is threatened in its existence, its liberty, or its identity have a moral right to meet the threat with an effective counterthreat, even if that counterthreat is nuclear?" Provided the threatened country acted in self-defense and followed a peaceful policy, the French bishops emphatically answered "Yes!"

In the United States an effective response came from Albert Wohlstetter, in "Bishops, Statesmen, and Other Strategists on the Bombing of Innocents." Drawing on a lifetime of study and government service concerned with nuclear strategy, Wohlstetter raised two fundamental objections to the U.S. bishops' position: To adopt a position of "use never" actually increases the risk of war; following a strategy of "deterrence only" would mean that if nuclear war should occur there would be attacks on cities. In Wohlstetter's eyes, there was a direct connection between the bishops' position and the views of secular advocates of arms control and the "deterrence only" strategy of Mutual Assured Destruction. The marriage of "use never" and "deterrence only," Wohlstetter concluded, utterly failed to recognize that there was no distinction between deterring and fighting: "Our *threat* to fight back will dissuade an opponent only if he thinks we are able and if necessary willing to fight back. . . ."

42. Pastoral Letter on War and Peace—The Challenge of Peace: God's Promise and Our Response

U.S. Catholic Bishops

THE JUST-WAR CRITERIA

(80) The moral theory of the "just-war" or "limited-war" doctrine begins with the presumption which binds all Christians: We should do no harm to our neighbors; how we treat our enemy is the key test of whether we love our neighbor; and the possibility of taking even one human life is a prospect we should consider in fear and trembling. How is it possible to move from these presumptions to the idea of a justifiable use of lethal force?

(81) Historically and theologically the clearest answer to the question is found in St. Augustine. Augustine was impressed by the fact and the consequences of sin in history—the "not yet" dimension of the kingdom. In his view war was both the result of sin and a tragic remedy for sin in the life of political societies. War arose from disordered ambitions, but it could also be used in some cases at least to restrain evil and protect the innocent. The classic case which illustrated his view was the use of lethal force to prevent aggression against innocent victims. Faced with the fact of attack on the innocent, the presumption that we do no harm even to

our enemy yielded to the command of love understood as the need to restrain an enemy who would injure the innocent.

(82) The just-war argument has taken several forms in the history of Catholic theology, but this Augustinian insight is its central premise.[1] In the 20th century, papal teaching has used the logic of Augustine and Aquinas[2] to articulate a right of self-defense for states in a decentralized international order and to state the criteria for exercising that right. The essential position was stated by Vatican II: "As long as the danger of war persists and there is no international authority with the necessary competence and power, governments cannot be denied the right of lawful self-defense, once all peace efforts have failed."

(84) The determination of *when* conditions exist which allow the resort to force in spite of the strong presumption against it is made in light of *jus ad bellum* criteria. The determination of *how* even a justified resort to force must be conducted is made in light of the *jus in bello* criteria. We shall briefly explore the meaning of both.[3]

(85) *Jus ad Bellum*: Why and when recourse to war is permissible.

(86) a. Just Cause: War is permissible only to confront "a real and certain danger," i.e., to protect innocent life, to preserve conditions necessary for decent human existence and to secure basic human rights. As both Pope Pius XII and Pope John XXIII made clear, if war of retribution was ever justifiable, the risks of modern war negate such a claim today.

(87) b. Competent Authority: In the Catholic tradition the right to use force has always been joined to the common good; war must be declared by those with responsibility for public order, not by private groups or individuals. . . .

(92) c. Comparative Justice: Questions concerning the *means* of waging war today, particularly in view of the destructive potential of weapons, have tended to override questions concerning the comparative justice of the positions of respective adversaries or enemies. In essence: Which side is sufficiently "right" in a dispute, and are the values at stake critical enough to override the presumption against war? The question in its most basic form is this: Do the rights and values involved justify killing? For whatever the means used, war by definition involves violence, destruction, suffering and death. . . .

(95) d. Right Intention: Right intention is related to just cause—war can be legitimately intended only for the reasons set forth above as a just cause. During the conflict, right intention means pursuit of peace and reconciliation, including avoiding unnecessarily destructive acts or imposing unreasonable conditions (e.g., unconditional surrender).

(95) e. Last Resort: For resort to war to be justified, all peaceful alternatives must have been exhausted. There are formidable problems in this requirement. No international organization currently in existence has exercised sufficient internationally recognized authority to be able either to mediate effectively in most cases or to prevent conflict by the intervention of U.N. or other peacekeeping forces. Furthermore, there is a tendency for nations or peoples

which perceive conflict between or among other nations as advantageous to themselves to attempt to prevent a peaceful settlement rather than advance it. . . .

(98) f. Probability of Success: This is a difficult criterion to apply, but its purpose is to prevent irrational resort to force or hopeless resistance when the outcome of either will clearly be disproportionate or futile. The determination includes a recognition that at times defense of key values, even against great odds, may be a "proportionate" witness.

(99) g. Proportionality: In terms of the *jus ad bellum* criteria, proportionality means that the damage to be inflicted and the costs incurred by war must be proportionate to the good expected by taking up arms. Nor should judgments concerning proportionality be limited to the temporal order without regard to a spiritual dimension in terms of "damage," "cost" and "the good expected." In today's interdependent world even a local conflict can affect people everywhere; this is particularly the case when the nuclear powers are involved. Hence a nation cannot justly go to war today without considering the effect of its action on others and on the international community.

JUS IN BELLO

Even when the stringent conditions which justify resort to war are met, the conduct of war (i.e., strategy, tactics and individual actions) remains subject to continuous scrutiny in light of two principles which have special significance today precisely because of the destructive capability of modern technological warfare. These principles are proportionality and discrimination. In discussing them here we shall apply them to the question of *jus ad bellum* as well as *jus in bello*; for today it becomes increasingly difficult to make a decision to use any kind of armed force, however limited initially in intention and in the destructive power of the weapons employed, without facing at least the possibility of escalation to broader, or even total, war and to the use of weapons of horrendous destructive potential. . . .

(103) Response to aggression must not exceed the nature of the aggression. To destroy civilization as we know it by waging a "total war" as today it *could* be waged would be a monstrously disproportionate response to aggression on the part of any nation.

(104) Moreover, the lives of innocent persons may never be taken directly, regardless of the purpose alleged for doing so. To wage truly "total" war is by definition to take huge numbers of innocent lives. Just response to aggression must be discriminate; it must be directed against unjust aggressors, not against innocent people caught up in a war not of their making. . . .

When confronting choices among specific military options, the question asked by proportionality is: Once we take into account not only the military advantages that will be achieved by using this means, but also all the harms reasonably expected to follow from using it, can its use still be justified? We know, of course, that no end can justify means evil in themselves, such as the executing of hostages or the targeting of non-combatants. Nonetheless, even if the means adopted is not evil in itself, it is necessary to take into account the probable harms that will result from using it and the justice of accepting those harms. . . .

In terms of the arms race, if the *real* end in view is legitimate defense against unjust aggression and the means to this end are not evil in themselves, we must still examine the question of proportionality concerning attendant evils. . . . Do the exorbitant costs, the general climate of insecurity generated, the possibility of accidental detonation of highly destructive weapons, the danger of error and miscalculation that could provoke retaliation and war—do such evils or others attendant upon and indirectly deriving from the arms race make the arms race itself a disproportionate response to aggression? Pope John Paul II is very clear in his insistence that the exercise of the right and duty of a people to protect their existence and freedom is contingent on the use of proportionate means.[4]

(107) Finally, another set of questions concerns the interpretation of the principle of discrimination. The principle prohibits directly intended attacks on non-combatants and non-military targets. It raises a series of questions about the term "intentional," the category of "non-combatant" and the meaning of "military. . . ."

These two principles in all their complexity must be applied to the range of weapons—conventional, nuclear, biological and chemical—with which nations are armed today. . . .

(139) . . . As bishops we see a specific task defined for us in Pope John Paul II's 1982 World Day of Peace Message:

> Peace cannot be built by the power of rulers alone. Peace can be firmly constructed only if it corresponds to the resolute determination of all people of good will. Rulers must be supported and enlightened by a public opinion that encourages them or, where necessary, expresses disapproval.[5]

The pope's appeal to form public opinion is not an abstract task. Especially in a democracy, public opinion can passively acquiesce in policies and strategies or it can through a series of measures indicate the limits beyond which a government should not proceed. The "new moment" which exists in the public debate about nuclear weapons provides a creative opportunity and a moral imperative to examine the relationship between public opinion and public policy. We believe it is necessary for the sake of prevention to build a barrier against the concept of nuclear war as a viable strategy for defense. There should be a clear public resistance to the rhetoric of "winnable" nuclear wars, or unrealistic expectations of "surviving" nuclear exchanges and strategies of "protracted nuclear war." We oppose such rhetoric. . . .

(141) Charting a moral course in a complex public policy debate involves several steps. We will address four questions, offering our reflections on them as an invitation to a public moral dialogue:

1. The use of nuclear weapons:

2. The policy of deterrence in principle and in practice;
3. Specific steps to reduce the danger of war;
4. Long-term measures of policy and diplomacy.

C. *The Use of Nuclear Weapons:* Establishing moral guidelines in the nuclear debate means addressing first the question of the use of nuclear weapons. . . .

(147) 1. *Counterpopulation Warfare:* Under no circumstances may nuclear weapons or other instruments of mass slaughter be used for the purpose of destroying population centers or other predominantly civilian targets. Popes have repeatedly condemned "total war," which implies such use. For example, as early as 1954 Pope Pius XII condemned nuclear warfare "when it entirely escapes the control of man" and results in "the pure and simple annihilation of all human life within the radius of action."[6]

(148) Retaliatory action, whether nuclear or conventional, which would indiscriminately take many wholly innocent lives, lives of people who are in no way responsible for reckless actions of their government, must also be condemned. This condemnation, in our judgment, applies even to retaliatory use of weapons striking enemy cities after our own have already been struck. No Christian can rightfully carry out orders or policies deliberately aimed at killing non-combatants. . . .

(150) 2. *The Initiation of Nuclear War:* We do not perceive any situation in which the deliberate initiation of nuclear warfare on however restricted a scale can be morally justified. Non-nuclear attacks by another state must be resisted by other than nuclear means. Therefore, a serious moral obligation exists to develop non-nuclear defensive strategies as rapidly as possible. . . .

(152) Whether under conditions of war in Europe, parts of Asia or the Middle East, or the exchange of strategic weapons directly between the United States and the Soviet Union, the difficulties of limiting the use of nuclear weapons are immense. A number of expert witnesses advise us that the commanders operating under conditions of battle probably would not be able to exercise strict control; the number of weapons used would rapidly increase, the targets would be expanded beyond the military and the level of civilian casualties would rise enormously.[7] No one can be certain that this escalation would not occur even in the face of political efforts to keep such an exchange "limited." The chances of keeping use limited seem remote, and the consequences of escalation to mass destruction would be appalling. Former public officials have testified that it is improbable that any nuclear war could actually be kept limited. Their testimony and the consequences involved in this problem lead us to conclude that the danger of escalation is so great that it would be morally unjustifiable to initiate nuclear war in any form. The danger is rooted not only in the technology of our weapons systems, but in the weakness and sinfulness of human communities. We find the moral responsibility of beginning nuclear war not justified by rational political objectives. . . .

(154) At the same time we recognize the responsibility the United States has had and continues to have in assisting allied nations in their defense

against either a conventional or a nuclear attack. Especially in the European theater, the deterrence of a *nuclear* attack may require nuclear weapons for a time, even though their possession and deployment must be subject to rigid restrictions.

(155) The need to defend against a conventional attack in Europe imposes the political and moral burden of developing adequate, alternative modes of defense to present reliance on nuclear weapons. Even with the best coordinated effort—hardly likely in view of contemporary political division on this question—development of an alternative defense position will still take time.

(156) In the interim, deterrence against a conventional attack relies upon two factors: the not inconsiderable conventional forces at the disposal of NATO and the recognition by a potential attacker that the outbreak of large-scale conventional war could escalate to the nuclear level through accident or miscalculation by either side. We are aware that NATO's refusal to adopt a "no first use" pledge is to some extent linked to the deterrent effect of this inherent ambiguity. Nonetheless, in light of the probable effects of initiating nuclear war, we urge NATO to move rapidly toward the adoption of a "no first use" policy, but doing so in tandem with development of an adequate alternative defense posture.

(157) 3. *Limited Nuclear War:* It would be possible to agree with our first two conclusions and still not be sure about retaliatory use of nuclear weapons in what is called a "limited exchange." The issue at stake is the *real* as opposed to the *theoretical* possibility of a "limited nuclear exchange."

(158) We recognize that the policy debate on this question is inconclusive and that all participants are left with hypothetical projections about probable reactions in a nuclear exchange. While not trying to adjudicate the technical debate, we are aware of it and wish to raise a series of questions which challenge the actual meaning of "limited" in this discussion.

- Would leaders have sufficient information to know what is happening in a nuclear exchange?
- Would they be able under the conditions of stress, time pressures and fragmentary information to make the extraordinarily precise decision needed to keep the exchange limited if this were technically possible?
- Would military commanders be able in the midst of the destruction and confusion of a nuclear exchange to maintain a policy of "discriminate targeting"? Can this be done in modern warfare waged across great distances by aircraft and missiles?
- Given the accidents we know about in peacetime conditions, what assurances are there that computer errors could be avoided in the midst of a nuclear exchange?
- Would not the casualties, even in a war defined as limited by strategists, still run in the millions?
- How "limited" would be the long-term effects of radiation, famine, social fragmentation and economic dislocation?

(159) Unless these questions can be answered satisfactorily, we will continue to be highly skeptical about the real meaning of "limited." One of the criteria of the just-war tradition is a reasonable hope of success in bringing about justice and peace. We must ask whether such a reasonable hope can exist once nuclear weapons have been exchanged. The burden of proof remains on those who assert that meaningful limitation is possible.

(160) A nuclear response to either conventional or nuclear attack can cause destruction which goes far beyond "legitimate defense." Such use of nuclear weapons would not be justified. . . .

Without making a specific moral judgment on deterrence, the council clearly designated the elements of the arms race: the tension between "peace of a sort" preserved by deterrence and "genuine peace" required for a stable international life; the contradiction between what is spent for destructive capacity and what is needed for constructive development.

In the post-conciliar assessment of war and peace and specifically of deterrence, different parties to the political-morale debate within the church and in civil society have focused on one or another aspect of the problem. For some, the fact that nuclear weapons have not been used since 1945 means that deterrence has worked, and this fact satisfies the demands of both the political and the moral order. Others contest this assessment by highlighting the risk of failure involved in continued reliance on deterrence and pointing out how politically and morally catastrophic even a single failure would be. Still others note that the absence of nuclear war is not necessarily proof that the policy of deterrence has prevented it. Indeed, some would find in the policy of deterrence the driving force in the superpower arms race. Still other observers, many of them Catholic moralists, have stressed that deterrence may not morally include the intention of deliberately attacking civilian populations or non-combatants.

The statements of the NCCB-USCC [National Conference of Catholic Bishops–U.S. Catholic Conference] over the past several years have both reflected and contributed to the wider moral debate on deterrence. In the NCCB pastoral letter "To Live in Christ Jesus" (1976), we focused on the moral limits of declaratory policy while calling for stronger measures of arms control.[8] In 1979 Cardinal John Krol, speaking for the USCC in support of SALT II ratification, brought into focus the other element of the deterrence problem: The actual use of nuclear weapons may have been prevented (a moral good), but the risk of failure and the physical harm and moral evil resulting from possible nuclear war remained.

"This explains," Cardinal Krol stated, "the Catholic dissatisfaction with nuclear deterrence and the urgency of the Catholic demand that the nuclear arms race be reversed. It is of the utmost importance that negotiations proceed to meaningful and continuing reductions in nuclear stockpiles and eventually to the phasing out altogether of nuclear deterrence and the threat of mutual-assured destruction."[9]

These two texts, along with the conciliar statement, have influenced much of Catholic opinion expressed recently on the nuclear question.

In June 1982, Pope John Paul II provided new impetus and insight to the moral analysis with his statement to the U.N. Second Special Session on Disarmament. The pope first situated the problem of deterrence within the context of world politics. No power, he observes, will admit to wishing to start a war, but each distrusts others and considers it necessary to mount a strong defense against attack. He then discusses the notion of deterrence:

> Many even think that such preparations constitute the way—even the only way—to safeguard peace in some fashion or at least to impede to the utmost in an efficacious way the outbreak of wars, especially major conflicts which might lead to the ultimate holocaust of humanity and the destruction of the civilization that man has constructed so laboriously over the centuries.
>
> In this approach one can see the "philosophy of peace" which was proclaimed in the ancient Roman principle: *Si vis pacem, para bellum.* Put in modern terms, this "philosophy" has the label of "deterrence" and one can find it in various guises of the search for a "balance of forces" which sometimes has been called, and not without reason, the "balance of terror."[10]

(173) Having offered this analysis of the general concept of deterrence, the Holy Father introduces his considerations on disarmament, especially, but not only, nuclear disarmament. Pope John Paul II makes this statement about the morality of deterrence:

> In current conditions "deterrence" based on balance, certainly not as an end in itself but as a step on the way toward a progressive disarmament, may still be judged morally acceptable. Nonetheless in order to ensure peace, it is indispensable not to be satisfied with this minimum, which is always susceptible to the real danger of explosion.[11]

(174) In Pope John Paul II's assessment we perceive two dimensions of the contemporary dilemma of deterrence. One dimension is the danger of nuclear war with its human and moral costs. The possession of nuclear weapons, the continuing quantitative growth of the arms race and the danger of nuclear proliferation all point to the grave danger of basing "peace of a sort" on deterrence. The other dimension is the independence and freedom of nations and entire peoples, including the need to protect smaller nations from threats to their independence and integrity. Deterrence reflects the radical distrust which marks international politics, a condition identified as a major problem by Pope John XXIII in "Peace on Earth" and reaffirmed by Pope Paul VI and Pope John Paul II. Thus a balance of forces, preventing either side from achieving superiority, can be seen as a means of safeguarding both dimensions.

(175) The moral duty today is to prevent nuclear war from ever occurring *and* to protect and preserve those key values of justice, freedom and independence which are necessary for personal dignity and national integrity. . . .

(178) Targeting doctrine raises significant moral questions because it is a significant determinant of what would occur if nuclear weapons were ever

to be used. Although we acknowledge the need for deterrent, not all forms of deterrence are morally acceptable. There are moral limits to deterrence policy as well as to policy regarding use. Specifically, it is not morally acceptable to intend to kill the innocent as part of a strategy of deterring nuclear war. The question of whether U.S. policy involves an intention to strike civilian centers (directly targeting civilian populations) has been one of our factual concerns.

(179) This complex question has always produced a variety of responses, official and unofficial in character. The NCCB committee has received a series of statements of clarification of policy from U.S. government officials.[12] Essentially these statements declare that it is not U.S. strategic policy to target the Soviet civilian population as such or to use nuclear weapons deliberately for the purpose of destroying population centers.

These statements respond, in principle at least, to one moral criterion for assessing deterrence policy: the immunity of non-combatants from direct attack either by conventional or nuclear weapons.

(180) These statements do not address or resolve another very troublesome moral problem, namely, that an attack on military targets or militarily significant industrial targets could involve "indirect" (i.e., unintended) but massive civilian casualties. We are advised, for example, that the U.S. strategic nuclear targeting plan (SIOP—Single Integrated Operational Plan) has identified 60 "military" targets within the city of Moscow alone, and that 40,000 "military" targets for nuclear weapons have been identified in the whole of the Soviet Union.[13] It is important to recognize that Soviet policy is subject to the same moral judgment; attacks on several "industrial targets" or politically significant targets in the United States could produce massive civilian casualties. The number of civilians who would necessarily be killed by such strikes is horrendous. This problem is unavoidable because of the way modern military facilities and production centers are so thoroughly interspersed with civilian living and working areas. It is aggravated if one side deliberately positions military targets in the midst of a civilian population.

In our consultations, administration officials readily admitted that while they hoped any nuclear exchange could be kept limited, they were prepared to retaliate in a massive way if necessary. They also agreed that once any substantial numbers of weapons were used, the civilian casualty levels would quickly become truly catastrophic and that even with attacks limited to "military" targets the number of deaths in a substantial exchange would be almost indistinguishable from what might occur if civilian centers had been deliberately and directly struck. These possibilities pose a different moral question and are to be judged by a different moral criterion: the principle of proportionality.

(181) While any judgment of proportionality is always open to differing evaluations, there are actions which can be decisively judged to be disproportionate. A narrow adherence exclusively to the principle of non-combatant immunity as a criterion for policy is an inadequate moral posture for it ignores some evil and unacceptable consequences. Hence, we cannot be

satisfied that the assertion of an intention not to strike civilians directly or even the most honest effort to implement that intention by itself constitutes a "moral policy" for the use of nuclear weapons.

(182) The location of industrial or militarily significant economic targets within heavily populated areas or in those areas affected by radioactive fallout could well involve such massive civilian casualties that in our judgment such a strike would be deemed morally disproportionate, even though not intentionally indiscriminate.

(183) The problem is not simply one of producing highly accurate weapons that might minimize civilian casualties in any single explosion, but one of increasing the likelihood of escalation at a level where many, even "discriminating," weapons would cumulatively kill very large numbers of civilians. Those civilian deaths would occur both immediately and from the long-term effects of social and economic devastation.

(184) A second issue of concern to us is the relationship of deterrence doctrine to war-fighting strategies. We are aware of the argument that war-fighting capabilities enhance the credibility of the deterrent, particularly the strategy of extended deterrence. But the development of such capabilities raises other strategic and moral questions. The relationship of war-fighting capabilities and targeting doctrine exemplifies the difficult choices in this area of policy. Targeting civilian populations would violate the principle of discrimination—one of the central moral principles of a Christian ethic of war. But "counterforce targeting," while preferable from the perspective of protecting civilians, is often joined with a declaratory policy which conveys the notion that nuclear war is subject to precise rational and moral limits. We have already expressed our severe doubts about such a concept. Furthermore, a purely counter-force strategy may seem to threaten the viability of other nations' retaliatory forces, making deterrence unstable in a crisis and war more likely.

(185) While we welcome any effort to protect civilian populations, we do not want to legitimize or encourage moves which extend deterrence beyond the specific objective of preventing the use of nuclear weapons or other actions which could lead directly to a nuclear exchange.

(186) These considerations of concrete elements of nuclear deterrence policy, made in light of John Paul II's evaluation, but applying it through our own prudential judgments, lead us to a strictly conditioned moral acceptance of nuclear deterrence. We cannot consider it adequate as a long-term basis for peace.

(187) This strictly conditioned judgment yields *criteria* for morally assessing the elements of deterrence strategy. Clearly, these criteria demonstrate that we cannot approve of every weapons system, strategic doctrine or policy initiative advanced in the name of strengthening deterrence. On the contrary, these criteria require continual public scrutiny of what our government proposes to do with the deterrent.

(188) *On the basis of these criteria we wish now to make some specific evaluations:*

1. If nuclear deterrence exists only to prevent the *use* of nuclear weapons by others, then proposals to go beyond this to planning for prolonged periods of repeated nuclear strikes and counterstrikes, or "prevailing" in nuclear war, are not acceptable. They encourage notions that nuclear war can be engaged in with tolerable human and moral consequences. Rather, we must continually say no to the idea of nuclear war.

2. If nuclear deterrence is our goal, "sufficiency" to deter is an adequate strategy; the quest for nuclear superiority must be rejected.

3. Nuclear deterrence should be used as a step on the way toward progressive disarmament. Each proposed addition to our strategic system or change in strategic doctrine must be assessed precisely in light of whether it will render steps toward "progressive disarmament" more or less likely.

(189) Moreover, these criteria provide us with the means to make some judgments and recommendations about the present direction of U.S. strategic policy. Progress toward a world freed of dependence on nuclear deterrence must be carefully carried out. But it must not be delayed. There is an urgent moral and political responsibility to use the "peace of a sort" we have as a framework to move toward authentic peace through nuclear arms control, reductions and disarmament. Of primary importance in this process is the need to prevent the development and deployment of destabilizing weapons systems on either side; a second requirement is to ensure that the more sophisticated command and control systems do not become mere hair triggers for automatic launch on warning; a third is the need to prevent the proliferation of nuclear weapons in the international system.

(190) In light of these general judgments *we oppose* some specific proposals in respect to our present deterrence posture:

1. The addition of weapons which are likely to be vulnerable to attack, yet also possess a "prompt hard-target kill" capability that threatens to make the other side's retaliatory forces vulnerable. Such weapons may seem to be useful primarily in a first strike;[14] we resist such weapons for this reason and we oppose Soviet deployment of such weapons which generate fear of a first strike against U.S. forces.

2. The willingness to foster strategic planning which seeks a nuclear war-fighting capability that goes beyond the limited function of deterrence outlined in this letter.

3. Proposals which have the effect of lowering the nuclear threshold and blurring the difference between nuclear and conventional weapons.

(191) In support of the concept of "sufficiency" as an adequate deterrent and in light of the present size and composition of both the U.S. and Soviet strategic arsenals, *we recommend:*

1. Support for immediate, bilateral, verifiable agreements to halt the testing, production and deployment of new nuclear weapons systems.[15]

2. Support for negotiated bilateral deep cuts in the arsenals of both superpowers, particularly those weapons systems which have destabilizing characteristics; U.S. proposals like those for START (Strategic Arms Reduction Talks) and INF (Intermediate-Range Nuclear Forces) negotiations

in Geneva are said to be designed to achieve deep cuts; our hope is that they will be pursued in a manner which will realize these goals.

3. Support for early and successful conclusion of negotiations of a comprehensive test ban treaty.

4. Removal by all parties of short-range nuclear weapons which multiply dangers disproportionate to their deterrent value.

5. Removal by all parties of nuclear weapons from areas where they are likely to be overrun in the early stages of war, thus forcing rapid and uncontrollable decisions on their use.

6. Strengthening of command and control over nuclear weapons to prevent inadvertent and unauthorized use.

(192) These judgments are meant to exemplify how a lack of unequivocal condemnation of deterrence is meant only to be an attempt to acknowledge the role attributed to deterrence, but not to support its extension beyond the limited purpose discussed above. Some have urged us to condemn all aspects of nuclear deterrence. This urging has been based on a variety of reasons, but has emphasized particularly the high and terrible risks that either deliberate use or accidental detonation of nuclear weapons could quickly escalate to something utterly disproportionate to any acceptable moral purpose. That determination requires highly technical judgments about hypothetical events. Although reasons exist which move some to condemn reliance on nuclear weapons for deterrence, we have not reached this conclusion for the reasons outlined in this letter. . . .

NOTES

1. Augustine called it a Manichean heresy to assert that war is intrinsically evil and contrary to Christian charity, and stated: "War and conquest are a sad necessity in the eyes of men of principle, yet it would be still more unfortunate if wrongdoers should dominate just men" (*The City of God*, Book IV, C. 15). Representative surveys of the history and theology of the just-war tradition include: F. H. Russell, *The Just War in the Middle Ages* (N.Y.: 1975); P. Ramsey, *War and the Christian Conscience* (Durham, N.C.: 1961), *The Just War: Force and Political Responsibility* (N.Y.: 1968); J. T. Johnson, *Ideology, Reason and the Limitation of War* (Princeton: 1975), *Just War Tradition and the Restraint of War: A Moral and Historical Inquiry* (Princeton: 1981); L. B. Walters, "Five Classic Just-War Theories" (Ph.D. Dissertation, Yale University, 1971); W. O'Brien, *War and-or Survival* (N.Y.: 1969), *The Conduct of Just and Limited War* (N.Y.: 1981); J. C. Murray, "Remarks on the Moral Problem of War," *Theological Studies* 20 (1959) pp. 40-61.

2. Aquinas treats the question of war in the *Summa Theologica*, II-IIae, q. 40; also cf. II-IIae, q. 64.

3. For an analysis of the content and relationship of these principles cf: R. Potter, "The Moral Logic of War," *McCormick Quarterly* 23 (1970) p. 203-233; J. Childress in Shannon, cited, p. 40-58.

4. John Paul II, World Day of Peace Message, 1982, 12.

5. John Paul II, World Day of Peace Message, 1982, 6, cited, p. 476.

6. Pius XII, Address to the VIII Congress of the World Medical Association, in Documents, p. 131.

7. Testimony given to the NCCB Committee during preparation of this pastoral letter.

8. USCC, "To Live in Christ Jesus" (Washington: 1976) p. 34.

9. Cardinal John Krol, Testimony on SALT II, *Origins* (1979) p. 197.

10. John Paul II, Message U.N. Special Session, 1982; 3.

11. Same, 8.

12. Particularly helpful was the letter of Jan. 15, 1983, of William Clark, national security adviser, to Cardinal Bernardin. Clark stated: "For moral, political and military reasons, the United States does not target the Soviet civilian population as such. There is no deliberately opaque meaning conveyed in the last two words. We do not threaten the existence of Soviet civilization by threatening Soviet cities. Rather, we hold at risk the war-making capability of the Soviet Union—its armed forces, and the industrial capacity to sustain war. It would be irresponsible for us to issue policy statements which might suggest to the Soviets that it would be to their advantage to establish privileged sanctuaries within heavily populated areas, thus inducing them to locate much of their war-fighting capability within those urban sanctuaries." A reaffirmation of the administration's policy is also found in Secretary Weinberger's Annual Report to the Congress (Caspar Weinberger, Annual Report to the Congress, Feb. 1, 1983, p. 55): "The Reagan administration's policy is that under no circumstances may such weapons be used deliberately for the purpose of destroying populations." Also the letter of Weinberger to Bishop O'Connor of Feb. 9, 1983, has a similar statement.

13. S. Zuckerman, *Nuclear Illusion and Reality,* (N.Y.: 1982); D. Ball, cited, p. 36; T. Powers, "Choosing a Strategy for World War III," *The Atlantic Monthly* (November 1982) p. 82–110.

14. Several experts in strategic theory would place both the MX missile and Pershing II missiles in this category.

15. In each of the successive drafts of this letter we have tried to state a central moral imperative: that the arms race should be stopped and disarmament begun. The implementation of this imperative is open to a wide variety of approaches. Hence we have chosen our own language in this paragraph, not wanting either to be identified with one specific political initiative or to have our words used against specific political measures.

43. Bishops, Statesmen, and Other Strategists on the Bombing of Innocents

Albert Wohlstetter

. . . Reckless nuclear threats and the intimidating growth of both Soviet conventional and nuclear strength have had much to do with the rise of the anti-nuclear movement here and in Protestant Northern Europe. By revising many times in public their pastoral letter on war and peace, American Catholic bishops have dramatized the moral issues which statesmen, using empty threats to end the world, neglect or evade. For the bishops stand in a long moral tradition which condemns the threat to destroy innocents as well as their actual destruction. They try but do not escape reliance on threatening bystanders. Ironically, the view dominating all their revisions reflects an evasive secular extreme which, instead of speeding improvements in the ability to avoid bystanders, has tried to halt or curb them. But because the bishops must take threats seriously, they make more visible the essential evasions of Western statesmen. That, however, is a kind of virtue. The letter offers a unique opportunity to examine the moral, political, and military issues together, and to show that . . . threatening to bomb innocents is not part of the nature of things. Nor has it been, as is now widely claimed, an essential of deterrence from the beginning. Nor is it the inevitable result of "modern technology. . . ."

The bishops have been sending a message to strategists in Western foreign-policy establishments—and to strategists in Western anti-nuclear counter-establishments. It seems unequivocal: "Under no circumstances may nuclear weapons or other instruments of mass slaughter be used for the purpose of destroying population centers or other predominantly civilian targets." Though that only restates an exemplary part of Vatican II two decades earlier, it is far from commonplace. Nonetheless it should be obvious to Catholics and non-Catholics alike. Informed realists in foreign-policy establishments as well as pacifists should oppose aiming to kill bystanders with nuclear or conventional weapons: indiscriminate Western threats paralyze the West, not the East. We have urgent political and military as well as moral grounds for improving our ability to answer an attack on Western military forces with less unintended killing, not to mention deliberate mass slaughter.

The bishops *seem* to be countering the perverse dogma which, after the Cuban missile crisis, came increasingly to be used by Western statesmen eager to spend less on defense: that the West should rely for deterring the Soviets on the ability to answer a nuclear military attack by assuring the deliberate destruction of tens or even hundreds of millions of Soviet civilians; and that the United States should also, for the supposed sake of "stability," give up any defense of its own civilians and any attack on military targets in order to assure the Soviets that they could, in response, destroy a comparable number of American civilians. The long humanist as well as the religious tradition on "just war" stresses especially the need to avoid attacks on "open," that is undefended, cities. The new doctrine exactly reversed this; it called both for leaving cities undefended and threatening to annihilate them. John Newhouse succinctly stated this dogma, to which he was sympathetic, in the "frosty apothegm": "Offense is defense, defense is offense. Killing people is good, killing weapons is bad." The late Donald Brennan, a long-term advocate of arms control to defend people and restrain offense from killing innocents, was not sympathetic. He noted that the acronym for Mutual Assured Destruction—MAD—described that Orwellian dogma.

Having observed long ago that not even Genghis Khan avoided combatants in order to focus solely on destroying noncombatants, I was grateful, on a first look at this issue in the evolving pastoral letter, to find the bishops on the side of the angels. Unfortunately, a closer reading suggested that they were also on the other side. For, while they sometimes say that we should not threaten to destroy civilians, they say too that we may continue to maintain nuclear weapons—and so implicitly threaten their use as a deterrent—while moving toward permanent verifiable nuclear and general disarmament; *yet we may not meanwhile plan to be able to fight a nuclear war even in response to a nuclear attack.*

Before that distant millennial day when all the world disarms totally, verifiably, and irrevocably—at least in nuclear weapons—if we should not intend to attack noncombatants, as the letter says, what alternative is there to deter nuclear attack or coercion? Plainly only to be able to aim at the combatants attacking us, or at their equipment, facilities, or direct sources of combat supply. That, however, is what is meant by planning to be able to fight a nuclear war—which the letter rejects. . . .

On many matters of technical military and political fact the bishops derive their views not from sacred authority but from a more doubtful range of secular strategists than they realize. Much of the letter, for example, stems from the strategists who hold that defense is offense and that killing people is good and killing weapons bad—the very strategists who would rely exclusively on threatening to destroy cities.

In invoking divine authority to sustain such lay strategies, the bishops' power seems dangerous to many Catholics who disagree. But their moral prestige alone gives weight to the bishops' strategic views with non-Catholics and Catholics. They reinforce the impassioned pacifist and neutralist movements that have been growing in Europe and in the United States, as well as the establishment strategies which helped to generate these protest movements.

For the bishops pass lightly over or further confound many already muddled and controversial questions of fact and policy. In a world where so many intense, deep, and sometimes mutually reinforcing antagonisms divide regional as well as superpowers, are there serious early prospects for negotiating the complete, verifiable, and permanent elimination of nuclear or conventional arms? If antagonists don't agree, should we disarm unilaterally? If we keep nuclear arms, how should we use them to deter their use against us or an ally? Might an adversary in some plausible circumstance make a nuclear attack on an element (perhaps a key non-nuclear element) of our military power or that of an ally to whom we have issued a nuclear guarantee? Might such an enemy nuclear attack (for example, one generated in the course of allied conventional resistance to a conventional invasion of NATO's center or of a critical country on NATO's northern or southern flank) have decisive military effects yet restrict side effects enough to leave us, and possibly our ally, a very large stake in avoiding "mutual mass slaughter"? Could some selective but militarily useful Western response to such a restricted nuclear attack destroy substantially fewer innocent bystanders than a direct attack on population centers? Would any discriminate Western response to a restricted nuclear attack—even one in an isolated area on a flank—inevitably (or more likely than not, or just possibly, or with some intermediate probability) lead to the destruction of humanity, or "something little better"? Or at least to an unprecedented catastrophe? Would it be less or more likely than an attack on population to lead to unrestricted attacks on populations? Can we deter a restricted nuclear attack better by threatening an "unlimited," frankly suicidal, and therefore improbable attack on the aggressor's cities, or by a limited but much more probable response suited to the circumstance?

The bishops' authorities slip by or confuse almost all these questions. The bishops sometimes seem only to be saying that the extent of direct collateral harm done by a particular restricted attack is uncertain, quite apart from the possibilities of "escalation." At other times they are certain that restricted attacks will lead to an entirely unrestricted war. And they then suggest that the chance is "so infinitesimal" that any Western nuclear response to a restricted attack would end short of ending humanity itself, that we might better threaten directly to bring on the apocalypse. The bishops cite experts as authority for their judgment that any use whatever of nuclear weapons would with an overwhelming probability lead to unlimited destruction. And some of their experts do seem to say just that. But some they cite appear only to say that we cannot be quite sure (that is, the probability is not equal to one) that any use of nuclear weapons would stay limited. If any response other than our surrender is to be believed, it makes a difference whether we talk of a probability that is not quite zero or a probability that is not quite equal to one that any nuclear response would bring on a suicidally total disaster. Yet two successive paragraphs in the 1982 *Foreign Affairs* article by McGeorge Bundy, George F. Kennan, Robert S. McNamara, and Gerard Smith proposing "no first use" of nuclear weapons, which the bishops cite, assert each of a wide range of such differing possibilities without distinction. Most authorities relied on by the bishops are themselves not very discriminating about which point they are trying to make.

Some important components of conventional military power vulnerable to nuclear attack are close to population centers. Others, however, may be very far from them—for example, naval forces at sea; or satellites in orbit hundreds or even a hundred thousand miles above the earth, that may be expected to perform the essential tasks during a conventional war of reconnaissance, surveillance, navigation, guidance, and communications. These are more vulnerable to nuclear than conventional attack. If we have no way of discouraging a limited nuclear attack except by extracting a promise from an adversary that he will not attack, or by threatening that we will respond to such isolated attacks with a suicidal retaliation on his cities, an adversary might, in the course of a conventional war, chance a small but effective nuclear attack against such isolated military targets. Such an attack would do incomparably less damage to civilians in the West than any of the "limited" attacks discussed by the bishops' authorities. Is it really so evident that a similarly restricted Western nuclear response to such a nuclear attack would be nearly certain to escalate to the end of humanity? Wouldn't a restricted response doing minimal damage to civilians on either side be much less likely to escalate than an attack on cities? And wouldn't the ability to respond in a proportionate way be a better deterrent to an adversary's crossing the gap between nuclear and conventional weapons? The bishops' lay experts tend to see the Soviets as mirror iimages of themselves, but sometimes diabolize them. They argue as if the Soviets would not continue during a war to have the strongest possible incentives to keep escalation within bounds; and as if the Soviets would love every killing of a Western bystander exactly as much as the West values his survival; as if the Soviet interest were in annihilating rather than dominating Western society. . . .

Moralists who have chosen to emphasize the shallow paradoxes associated with deterrence by immoral threats against population have been at their worst when they have opposed any attempts to improve the capability to attack targets precisely and discriminately. While they have thought of themselves as aiming their opposition at the dangers of bringing on nuclear mass destruction, they have often stopped research and engineering on ways to destroy military targets without mass destruction; and they have done collateral damage to the development of precise, long-range conventional weapons. . . .

They have tried to stop, and have slowed, the development of technologies which can free us from the loose and wishful paradoxes involved in efforts to save the peace with unstable threats to terrorize our own as well as adversary civilians. . . .

Declaratory doctrine for the American defense of Europe started in the 1950s with the belief that strategic and tactical nuclear weapons could replace the conventional firepower which our NATO allies hesitated to supply against conventional invasion. It went through a phase in which many of the present advocates of MAD entertained exaggerated hopes for limiting the harm done by the large-scale use of tactical nuclear weapons on European battlefields; and for using massive active and civil defense, limiting to quite small amounts the damage done by a large raid on U.S. cities. When their hopes began to seem excessive, they switched to the view that the *threat* of unlimited mutual destruction was actually good, since it was nearly sure to deter even a conventional invasion. The last year or two have seen signs of renewed serious interest in improving NATO's ability to meet a conventional invasion in Europe on its own terms. Manfred Woerner, the current Minister of Defense in the German Federal Republic, has set forth a program which is designed not only to discourage a Soviet conventional invasion, but to do it responsibly in a way that will also put to rest the growing West German anti-nuclear movement. He would exploit the advanced technologies that are coming to be available for that purpose.

Woerner's view stands in contrast to that of his predecessor, who held that even a conventional war in Europe would be "the end of Europe," and that it was essential that tactical nuclear weapons be used quickly but only as a link to the "intercontinental exchange"—which would be "the end of the world." But anyone who relies on such threats to deter a conventional attack is likely to threaten up to the last minute and then, when it would have become clear that the Soviets did not believe that NATO leaders would consciously bring on the end of Europe and then the end of the world, rush to reassure the Soviets that they did not really mean to execute the "threat." Such a policy, Herman Kahn accurately labeled "preemptive surrender." It differs from the policy advocated by West Germany's party of the Greens in the anti-nuclear movement who would make their accommodation with the Soviets now, in time of peace, safely in advance of a threatened Soviet attack. Pierre Hassner has characterized the difference between the leaders of the anti-nuclear movement and some leading figures in the West European establishment who rely on suicidal threats: it is the difference between "preventive surrender" and "preemptive surrender."

DETERRING NUCLEAR ATTACK ON AN ALLY

Bundy, McNamara, Kennan, and Smith have lost their faith in suicidal threats as a way of deterring a conventional invasion. They believe in the necessity and adequacy of such threats to deter nuclear attacks. However, a hope that an adversary can be safely deterred by our threat to blow him up along with ourselves is unfounded not only for a conventional attack but also for a *nuclear* attack on the ally.

Consider a strategically placed ally like Norway with an American nuclear guarantee and no nuclear weapons of its own. How would a capability to destroy Soviet civilians, along with American civilians and possibly the civilization of Europe itself, discourage Soviet use of nuclear weapons against military targets in the course of an attack aimed at seizing the sparsely populated but strategic northernmost counties of Norway? No one—no Norwegian, no American leader, and no Soviet leader—would seriously expect us to respond to such an attack by consciously initiating the killing of 100 million or so innocent Soviet civilians and a corresponding number of Americans and/or West Europeans. That is one reason why some believers in MAD are explicitly *for* threats and *against* their execution. But a capability which plainly will never be used to initiate a chain of events we believe would lead to the end of civilization will terrify an adversary no more than a capability that would destroy half, or a tenth, or a millionth the number of civilians, or no civilians at all. The only way weapons can inspire concern is by the likelihood that they will be used. The residual fear that the West might deliberately blow up the world tends to terrify some in our own elites much more than the Soviets who chatter less on this subject.

The recognition at the end of 1953 that fusion warheads might be made small enough to be carried in ballistic missiles by the 1960s might have seemed to hold out the prospect for reducing collateral damage somewhat.

For these first ballistic-missile warheads were expected to be substantially smaller than the gravity bombs carried in aircraft. (Later Navy SLBM warheads were about the same size as some early A-bombs, 40 kilotons. Even the first SLBM and ICBM warheads were about a half-megaton, much smaller than the H-bombs contemplated in the initial debate.) In fact, however, the prospect of the ballistic missile worsened expectations about collateral damage because the first generation of missiles was expected to be much more inaccurate than aircraft. The median miss distance then expected for the first ballistic missiles was anywhere from two to five miles. A five-mile median radius of inaccuracy meant that half the bombs would strike outside of an 80-square-mile area!

But inaccuracy determines the unintended harm done in destroying a small target more basically than does the explosive yield of individual bombs. It is the lack of technology smart enough, rather than the availability of large brute-force single weapons, that lies at the root of the problem of collateral damage. One makes up for incompetence in aiming by filling an enormous area of uncertainty either with a few large-yield nuclear weapons or, as the British did in World War II, with many thousands of small conventional bombs. When the British discovered in June 1941 that only a third of the bomber crews who thought they had bombed the target were within 80 square miles of it, they resorted to huge raids involving thousands of bombers with results that became visible in Hamburg and in Dresden. David Irving's estimate of the dead in Dresden came to 135,000— much more than the official estimates of the Hiroshima dead. A single American conventional raid on Tokyo in March 1945 destroyed an area over three times that destroyed by the Hiroshima bomb (15.8 compared to 4.7 square miles) and nearly nine times that destroyed by the Nagasaki bomb (1.8 square miles). The average area destroyed in 93 conventional attacks against Japanese cities amounted to the same as that in Nagasaki. . . . The upholders of the bishops' doctrine of "Use, Never" (i.e., No Use— First-Second-Or Ever) seem unaware that an adversary might be concerned not only about the magnitude of the harm we threaten but about the likelihood that we will inflict it.

However, it is a familiar fact of everyday life that we consider implicitly in our behavior not only the size of the assorted catastrophes we might conceivably face when we get up each morning but also their likelihood. Blizzards in August might find us peculiarly unequipped to survive them. So also sunstroke in December. Neither bothers us much, nor leads us to wear furs in summer and carry parasols in winter. Even when we face adversaries and not merely environmental dangers, we have a way of arraying threats according to the probability that they will be carried out and not only in terms of the damage they would do if they were. When a threatener can execute a terrible threat to us with little harm to himself, we worry more than when he would suffer at least as much as we would. Moreover, when a threatener, who expects to destroy himself and his allies along with the aggressor, says that he has no intention whatsoever and, in fact, would regard it as immoral to execute his threat, this can only be reassuring to a potential aggressor. It is an invitation rather than a deterrent. Somehow it does not occur to those who hope to deter by a suicidal threat (which they loudly proclaim they will never execute) that they may be doing the opposite of deterring. Their policy is—to use that dread catchword—"destabilizing."

Soviet leaders who were not deterred by a threat they knew would never be executed would not, as Cardinal Krol suggests, have to be insane. It seems more nearly insane, as O'Brien says, to hold that in all circumstances, even during a stalled conventional invasion when all alternatives looked risky to them, the Soviets would be deterred "beyond question" from using nuclear weapons by our self-confessed suicidal bluff. Nonetheless, the doctrine of "Use, Never" advanced by the bishops merely makes more explicit the operational meaning of secular strategies of Deterrence Only. The Stanford physicist, Sidney Drell, recently has repeated the standard jumble about deterrence and fighting: instead of observing that our *threat* to fight back will dissuade an opponent only if he thinks we are able and if necessary willing to fight back, Drell says deterring and fighting are incompatible goals.

Deterrence only focuses on deterring Western responses rather than Soviet attacks. It assumes that it is really the West, and especially the United States, in its misunderstanding of the Soviets, that menaces the nuclear peace and not the Soviets. This is an assumption widely held, even by those who oppose the disarmers. Michael Howard of Oxford tells us that the Soviets are entirely satisfied with the present division of Europe and that only Western extremists are not. He grants that the Soviets would revise the rest of the world, but doesn't notice that in that process they might effectively alter the division of power within West Europe too. It would be hard for the Soviet Union to avoid altering the division of power in Europe, even if unintentionally, if it seized some future opportunity to satisfy its long expressed interest in expanding toward the Persian Gulf and the Eastern Mediterranean. (England is said to have acquired its empire in a fit of absentmindedness.) Moreover, from the Soviet point of view, the destruction of the Western alliance that would result would surely be a bonus in defense of Soviet Western borders. George Kennan draws rather more satisfaction than is warranted from Soviet paranoid defensiveness. Paranoids can be dangerous.

But Michael Howard isn't terribly worried about the Soviets beginning a war. He worries about Americans. Though he has been subject to attack by E. P. Thompson and the nuclear disarmers, he sometimes sounds a little like them. He says: "Whether I could encounter the same phenomenon in the Soviet Union, I do not know. But wars begin in the minds of men, and in many American minds the flames of war seem already to have taken a very firm hold." And: "When I hear some of my American friends speak of that country [the Soviet Union], when I note how their eyes glaze over, their voices drop an octave, and they grind out the words, '*the Soviets*' in tones of gravelly hatred, I become really frightened; far more frightened than I am by the nuclear arsenals themselves or the various proposals for their use." I know some of Howard's American friends (indeed have counted myself as one), but none resembling that description. If such glazed-eyed monsters controlled the U.S. arsenal, instead of planning proportionate Western responses that might credibly discourage Soviet attack, the West might focus its attention entirely on stopping us and let the credibility of U.S. guarantees erode. . . .

In a war, when all alternatives may be extremely risky to an adversary, we may not convince him that the alternative of nuclear attack is riskier than the others if we have persuaded him also that it can be done safely because we won't retaliate for fear of the unlimited harm we would bring on ourselves. We only complete the absurdity and undermining of deterrence when we *say* that we have no intention to fight, that is, to use nuclear weapons if deterrence fails. Unfortunately, the principle of deterrence and the principle of "Use, Never" mutually annihilate each other. . . .

Declaring—or telling oneself—that one does not really mean to use nuclear weapons if deterrence fails is one way of stilling uneasiness about threatening to kill innocents in order to deter. Another standard way of softening guilt is to say that the West should continue to raise such a threat even implicitly only if it is making serious progress toward the total elimination of nuclear weapons. That, however, does not lie solely within the West's power. It depends on others who have or may acquire nuclear weapons, and in particular it depends on the disposition of the deeply suspicious, hostile leadership of the Soviet Union. . . .

The Soviets see the lasting independence of Western democracies side by side with their own system as a permanent danger to its maintenance, not to say its expansion toward an international utopia. Meanwhile, there is little evidence that some plausible arrangement would lead them to surrender so powerful an instrument of coercion or defense. That, after all, was indicated in their rejection of the Baruch-Acheson-Lilienthal plan for international control of atomic energy. . . .

If Western political as well as religious leaders take Western possession of nuclear weapons as justified only if there is progress toward agreement with the Russians to eliminate them altogether, they place in Soviet hands the decision as to whether the West will continue to maintain a nuclear deterrent.

Not all differences are negotiable. Pretending that they are suggests a willingness to disarm unilaterally—either because the Soviets prevent agreement or because they agree only to a disarmament which would be purely nominal for them but real for the West. . . .

We should recognize that utopian hopes for total nuclear disarmament cannot excuse a Western failure to defend its independence soberly without using reckless threats. Unfortunately, our elites now link the phrase "arms control" not only to millennial dreams of early complete nuclear disarmament, but to the strategy of using threats to annihilate cities as a way of deterring attack; and to a perverse myth of the "arms race" that suggests that nuclear war is imminent because our nuclear arms have been spiraling exponentially and will continue to do so unless we limit our objectives to the destruction of a fixed small number of vulnerable population centers. (No one has ever suggested that the only way to avoid an exponential race in conventional arms is to train our fire on villages rather than enemy tanks. But when it comes to nuclear arms our elites will believe almost anything.) That is not the "arms control" Donald Brennan had in mind. "Arms control," as he and the Princeton physicist, Freeman Dyson, have understood it, should aim at the more traditional and more sensible goal of restraining the bombardment of civilians. But the phrase is now loaded with wishful and mistaken prejudices. It suggests that without arms agreements our spending on defense inevitably will rise exponentially and uncontrollably; and that with arms agreements Soviet arms efforts will diminish. Experience for nearly two decades after the Cuban missile crisis illustrates the opposite. . . .

We assumed that the Soviets, like ourselves, had as a principal objective the desire to reduce the percentage of their resources devoted to arms spending and that they would choose "arms control" rather than arms competition. The record plainly shows that Western assumptions were wishful. The Soviets pursued arms agreements as a method of limiting Western spending—which did decline as a proportion of GNP by nearly half in the period after the missile crisis—while they themselves steadily increased their spending and did succeed in changing the balance. Now the West has the problem of catching up and that is especially hard to negotiate.

Serious negotiations today must recognize the limits to what they can accomplish. We and the Soviets share an interest in avoiding mutual suicide, an interest which each of us will pursue whether or not we reach genuine agreement in various understandings and formal treaties. But the Soviets also have interests in expanding their influence and control and, in the process, destabilizing the West, if necessary by the use of external force rather than simply by manipulating internal dissension. Arms agreements might temper, but are unlikely to eliminate, this reality. In particular, there seems scant basis to hope for major economies in our security effort through negotiated limits or reductions.

Experience suggests that when the Soviets agree to close off one path of effort, they redirect their resources to other projects posing differing but no lesser dangers. On the other hand, many of the ostensible goals of arms agreements are best achieved through measures which we can and should implement on our own. Our current efforts—which a freeze would stop—to design and deploy nuclear weapons which are more accident-proof and more secure against theft or unauthorized use, are a good example. Measures to improve the safety, security, and invulnerability of nuclear weapons can be implemented by both sides individually because they make sense for each side independently of formal treaties or elaborate verification measures. These need not mean a net increase in the numbers or destructiveness of nuclear weapons in our stockpile. The United States has already greatly reduced both the megatonnage and the numbers of its nuclear weapons. It recently removed 1,000 weapons from Europe and has said that if, in accordance with NATO's decision in 1979, it installs 572 intermediate-range nuclear missiles, it will withdraw an equal number of warheads. If we increase precision further, we can drastically further reduce the number and destructiveness of our nuclear weapons. Increased precision can also improve the effectiveness of conventional weapons so that they may increasingly replace nuclear brute force. And it would improve our ability to avoid the unintended bombing of innocents with nuclear or conventional warheads. It would enlarge rather than foreclose our freedom to choose. . . .

44. Win the Peace: Declaration of the Plenary Assembly of French Bishops

French Catholic Bishops

Because the survival of humanity is at stake, there is no cause that can justify the outbreak of a nuclear war. The same applies to other forms of suicidal warfare which are less often discussed, even though they too are being prepared for: chemical and biological warfare. Moreover, by centering too much attention on nuclear war we run the risk of minimizing "conventional" war. . . . And everyone knows that in a direct military confrontation between the two nuclear powers conventional armaments could serve as detonators for nuclear weapons. . . .

Nobody wants war. . . . Yet some countries are bent on reaping the benefits of warfare without paying the price of war: By brandishing its threat they make permanent use of blackmail. . . . While former democracies are kept by force within the Soviet bloc, the Western democracies are subject to constant pressure to neutralize them and to bring them, if possible, into the sphere of influence of Marxist-Leninist ideology. The latter, convinced that it holds the secret to the total liberation of mankind and of nations, thinks itself mandated to impose upon all what it believes to be the greatest good.

There is no question here of cultivating a Manichaean view of the world—all the evil on one side, all the good on the other. The West is also ailing. Materialism—whether theoretical, as in communist societies, or practical, as in the West—is a deadly disease of humanity. And Marxist-Leninist states do not hold a monopoly on imperialism. Sometimes they gain a following even within the systems that oppose them most strenuously. But it would be unfair to simply state and accept the conflict of ideologies while closing one's eyes to the domineering and aggressive character of

Marxism-Leninism, which holds that everything, even a nation's hopes for peace, must be used as a tool for world conquest.

Given those conditions, does an absolute condemnation of all warfare not put peace-loving nations at the mercy of those animated by an ideology of domination? In their efforts to avoid war, peaceful nations could fall prey to other forms of violence and injustice: colonization, alienation, deprivation of freedom and identity. Pushed to its ultimate consquences, peace at any price leads a nation to all sorts of capitulations. Unilateral disarmament could even encourage aggressive behavior on the part of neighbors by presenting them with the temptation of an easy prey. . . .

The church has always recognized the right of political power to counter force by the use of force. . . . Nonviolence is a risk a person can take. Can states, whose function it is to preserve the peace, take that risk? In our world of violence and injustice, it is the duty of politicians to safeguard the peace of the community for which they are responsible. That community is made of peace, but it is also made of justice, of solidarity, of liberty. To protect it, politicians must have the means of deterring, as far as possible, a potential aggressor. . . .

We know all too well the injustice and disorder that arise when a state of law gives way to the law of the stronger. . . . In international relations, unfortunately, there is no authority powerful and effective enough to impose that state of law. Therefore individual countries cannot be denied the right of legitimate defense against external threats as well as internal perils. . . .

We shall not involve ourselves here in the technical debates among experts on the credibility of our defense. . . . In those highly technical questions, which give rise to considerations of ethics, one must beware of two types of excess: (1) the suspension of ethical judgment, as if something as heavy with human significance could be left to simple technical logic; (2) preemptive judgments of a deductive kind that make light of technical considerations. . . .

The central question thus becomes the following: Given the context of the current geopolitical situation, does a country that is threatened in its existence, its liberty or its identity have a moral right to meet the threat with an effective counterthreat, even if that counterthreat is nuclear?

Until now, while stressing the possible consequences of such a parry and the terrible risks it entails, the Catholic Church has not felt the necessity to condemn it. . . .

Such logic, of course, is the logic of distress, and its weakness is obvious. Of course, it is to avoid having to wage war that one wants to show oneself capable of waging it. One does serve the cause of peace in deterring an aggressor by inspiring in him, through fear, a minimum of wisdom. The threat of violence does not constitute violence. That is the basis of dissuasion, and it is something we often forget when we attribute the same moral status to the threat as to the use of violence.

Nevertheless the dangers of the logic of dissuasion are obvious. To leave the potential aggressor no doubt as to the credibility of one's defense, one must show oneself to be firm in one's resolve to resort to action if dissuasion fails. Now the moral legitimacy of resort to action is more than unclear—especially in France, where our deterrence is that of the weak facing the powerful, a "poor man's deterrence" which relies on a wholesale threat: for lack of means, it is compelled to threaten cities: and that is a strategy the council condemns, clearly and finally. . . .

Yet the threat does not constitute use. Does the immorality of use entail the immorality of the threat? Not necessarily. For according to the council, "We cannot set aside the complexities of the situation as it stands." Given the state of violence and sin in which the world exists, it is the duty of politicians and military officials to defuse the blackmail to which the nation could be subjected. . . .

It is clear that to be morally acceptable, recourse to nuclear deterrence must presuppose:

- that it applies only to self-defense;
- the avoidance of overarmament: Deterrence is effective as soon as it represents a threat sufficient to discourage aggression;
- all the necessary precautions against "mistakes," against the eventuality of interference by terrorists, by a madman, etc.;
- that the nation which assumes the risks inherent in nuclear deterrence will pursue a constructive policy that serves the cause of peace. . . .

The church does not encourage unconditional pacifism. The church has never advocated unilateral disarmament, for it is aware that unilateral disarmament could serve to kindle the violence of an aggressive military, political and ideological complex. But the church recognizes the evangelical message present in calls to nonviolence: They are prophetic reminders of the destructiveness of violence. . . . While recognizing the current need for armed defense, the church calls for that need to be overcome. . . .

A nation cannot live with its eyes glued to the radar screens that survey its territory. Nor can it stare forever at the charts of its economists. All those things are important: but they are only means. Beyond the means of life stands the question of the reason for living. For people, and also for nations and for mankind itself. That is a cultural problem and therefore a spiritual one.

45. For Disarming Europe

E. P. Thompson

For three years, on both sides of the Atlantic, there have been conferences and church meetings and marches and agonizing over the fate of the Earth. Well, are we serious or not? Somewhere, at some time, some people have got to decide that they will say no. We in the European peace movement have decided that the place is Europe and the time is now and that the people will have to be us. Europe is the no-man's-land between the superpowers where at last the nuclear arms race might be stopped.

We are not interested in President Reagan's "zero option"—to forgo deployment of American intermediate-range nuclear missiles in exchange for a Soviet agreement to dismantle its intermediate-range missiles—or in his compromise "zero plus" proposal for limiting these missiles. The European peace movement decided long ago that the matter is not open to negotiation. Even one cruise missile on our territory, owned and operated by a foreign power, takes us across an unacceptable threshold.

Mr. Reagan is supposed to have made the "zero plus" proposal reluctantly in response to European pressure. Maybe we should examine the term "European pressure." In reality, European peace movements brought pressure on their governments, which in turn brought pressure on the United States. European leaders ran to Washington and pleaded with Mr. Reagan to get them off the hook of their own restive public opinion. The purpose of the "zero plus" proposal was less to achieve a settlement with Moscow than to aid leaders of the North Atlantic Treaty Organization countries to reassert control over their domestic constituencies.

After the demonstrations two weeks ago in Britain, the Netherlands and West Germany, it must be clear to the President that something has gone wrong. He, and the American public, should be warned that, just as they got into the present mess in part by listening too closely to Helmut Schmidt, the former Chancellor, and a narrow circle of Atlanticist "defense experts," so they are signing on for another period of purgatory by following the same kind of ill advice from the same political circles.

The President is badly briefed. In his calculation, the Soviet Union has a "monopoly" of missiles in Europe, at a ratio of some 1,000 warheads to nil. That depends on what categories you count and what you hold your hand over while counting. What was being hidden was all the tactical stuff and all the aircraft-launched stuff (on both sides), the American Poseidons under NATO command, the British Polaris submarine-launched missiles, the whole French armory and the multitude of American air-launched and sea-launched cruise missiles now on their way. All this can be added up and balanced in different ways, but the sum is never 1,000 to nil.

In any case, it is disputable whether the Pershing 2s and ground-launched cruise missiles can be properly described as "intermediate" missiles at all. They are forward-based American strategic missiles, to be owned and operated by American personnel on European territory, that can strike deeply into the Soviet Union—whereas the SS-20, nasty as it is, cannot cross the Atlantic.

Europeans who have been counting and recounting these beads for three years find the President's calculations only boring. The "zero plus" proposal was boring also, for it, too, offers to match missile numbers only in this one notional category, while ignoring all the rest. Besides, it would amount to only a small reduction in the missiles planned for Europe.

I doubt that the Russians are interested in the "zero plus" proposal, but it matters little, since there are three parties who must consent in this negotiation. And certainly the European peace movement will not be co-signatories. The missiles will have to be brought by helicopter into installations at Greenham Common and Comiso in the face of our refusal at the gates.

The only numbers game that would interest us would be a Dutch auction, downward. What we want is a freeze, of course. As for Euromissiles, we want no phony counting—and Poseidon submarines and Polaris missiles should be thrown into the calculus as sweeteners to bring Moscow to eliminate all SS-20s.

We propose also a European disarmament conference with this agenda: removing all nuclear weapons (including British and French) from Europe (measured from the eastern Atlantic waters to the Urals), mutual deep reductions in conventional forces, and guarantees of the human rights clauses of the Helsinki Final Act. Finally, we call for a summit meeting between the President and the leader of the Soviet Union to bring World War II to a final settlement, with the mutual withdrawal, by phases, of all American and Soviet forces in Europe.

It may seem to the superpower leaders that nuclear weapons are for security, stability, certainty. Their terminus is certain also: The stability of the tomb.

Bibliography: Part Five

Bennett, John C. "The Deterioration of Deterrence." *Christianity and Crisis* 44 (August 13, 1984):296–301.

Bundy, McGeorge. "The Bishops and the Bomb." *The New York Review of Books* (June 16, 1983):2–3.

Dougherty, James E. *The Bishops and Nuclear Weapons: The Catholic Pastoral Letter on War and Peace.* Hamden, Conn.: Archon Books, 1984.

Johnson, James Turner. *Can Modern War Be Just?* New Haven, Conn.: Yale University Press, 1984.

Novak, Michael. "Moral Clarity in the Nuclear Age." *National Review* 35 (April 1, 1983):354–388.

_____. *Moral Clarity in the Nuclear Age.* Nashville, Tenn.: Thomas Nelson Publishers, 1983.

Ramsey, Paul. *The Just War: Force and Political Responsibility.* New York: Charles Scribner's Sons, 1968.

Tucker, Robert W. "The Nuclear Debate." *Foreign Affairs* 63 (Fall 1984):1–32.

Van Voorst, Bruce. "The Churches and Nuclear Deterrence." *Foreign Affairs* 61 (Spring 1983):827–852.

Woolsey, James R. *Nuclear Arms: Ethics, Strategy, Politics.* San Francisco: Institute for Contemporary Studies, 1984.

PART SIX

STRATEGY AND ARMS CONTROL IN THE FUTURE

CHAPTER 9

Strategy

Both irony and innovation may be found in the attempts by strategists and policymakers to look to the future. The irony emerged clearly in the Scowcroft Commission report. While demonstrating that the Soviet Union had not gained a decisive nuclear advantage over the United States, the commission recommended a return to single-warheaded missiles. The idea was to make these Midgetmen safe against attack through mobility. They would not invite attack, because they could not be located as precisely as silo-based ICBMs and because their single warheads would not provide a rewarding target against which to use two or more attacking warheads. It was jarring, at best, after a decade of immense effort and expenditure devoted to developing, deploying, and refining multiple warheads, to learn that they ought to be replaced by a better single-warheaded missile.

The Scowcroft Commission made two other recommendations: (1) the United States should produce and deploy one hundred MX missiles in Minuteman silos, and (2) the United States should engage in arms control negotiations to move both the United States and Soviet forces toward more crisis-stable deployments. The commission's recommendations had been carefully negotiated in advance with congressional leaders in an at least partially successful attempt to reconcile strongly conflicting views about the the need to deploy the MX. The eloquent warning letter from Soviet physicist and Nobel Peace Prize winner Andrei Sakharov to Sidney Drell on the danger of thermonuclear war provided a strong case against failing to deploy the MX and to make the other efforts needed for continued strategic modernization.

President Reagan provided the greatest spur to innovation by urging the United States and particularly its scientific community to work toward the day when the nation could replace its offense-dominated nuclear strategy with a defense-dominated approach based on a capability to destroy attacking missiles. Dubbed Star Wars by its opponents, the president's Strategic Defense Initiative (SDI) held more promise than reality. Its purpose appeared to have been twofold: to enable the administration to augment research into defensive technology and to hold out to the U.S. people and to the peoples of the world the prospect of at least a partial escape from reliance on the threat of nuclear destruction to avoid major war.

Opponents of SDI argued that there were strong technical and political reasons why it should not be pursued. On technical grounds there was no reason to expect the development of an antimissile defense capable of replacing the offense-dominated deterrent strategies of past decades. Cities could not be protected against nuclear attack, because even a few missiles that "leaked" through the defense would cause catastrophic damage. Politically, the wholesale pursuit of strategic defense would destroy the ABM treaty, an arms control agreement that had prevented a defensive arms race and had contributed to at least a minimum of trust between the Soviet Union and the United States.

As Soviet-American relations soured in the last years of the Carter administration and throughout the first Reagan term, two changes occurred among those concerned with arms control. The first was a growing conviction that future steps in arms control probably would not take the form of comprehensive agreements laboriously negotiated by the Soviet Union and the United States.[1] The most that could be expected would be tacit agreements and unilateral measures that would enhance stability and restrain the arms competition. The second change was the appearance of a variety of imaginative proposals by strategists, former policymakers, and other analysts designed to allow progress on arms control even without the negotiation of a comprehensive Soviet-American agreement.

NOTES

1. See Kenneth L. Adelman, "Arms Control with and Without Agreements," *Foreign Affairs* 63 (Winter 1984/85):240–263.

46. The Danger of Thermonuclear War

Andrei Sakharov

An Open Letter to Dr. Sidney Drell

Dear Friend: . . . I fully agree with your assessment of the danger of nuclear war. In view of the critical importance of this thesis, I will dwell on it in some detail, perhaps repeating what is already well known.

Here, and later on, I use the terms "nuclear war" and "thermonuclear war" nearly interchangeably. Nuclear weapons mean atomic and thermonuclear weapons; conventional weapons mean any weapons with the exception of three types with the capability of mass destruction—nuclear, chemical, and bacteriological weapons.

A large nuclear war would be a calamity of indescribable proportions and absolutely unpredictable consequences, with the uncertainties tending toward the worse. . . .

In sum, it should be said that all-out nuclear war would mean the destruction of contemporary civilization, hurl man back centuries, cause the deaths of hundreds of millions or billions of people, and, with a certain degree of probability, would cause man to be destroyed as a biological species and could even cause the annihilation of life on earth.

Clearly it is meaningless to speak of victory in a large nuclear war which is collective suicide.

I think that basically my point of view coincides with yours as well as with the opinion of a great many people on earth.

I am also in complete agreement with your other conclusions. I agree that if the "nuclear threshold" is crossed, i.e., if any country uses a nuclear weapon even on a limited scale, the further course of events would be difficult to control and the most probable result would be swift escalation leading from a nuclear war initially limited in scale or by region to an all-out nuclear war, i.e., to general suicide.

It is relatively unimportant how the "nuclear threshold" is crossed—as a result of a preventive nuclear strike or in the course of a war fought with conventional weapons, when a country is threatened with defeat, or simply as a result of an accident (technical or organizational).

In view of the above, I am convinced that the following basic tenet of yours is true: *Nuclear weapons only make sense as a means of deterring nuclear aggression by a potential enemy,* i.e., a nuclear war cannot be planned with the aim of winning it. Nuclear weapons cannot be viewed as a means of restraining aggression carried out by means of conventional weapons.

Of course you realize that this last statement is in contradiction to the West's actual strategy in the last few decades. For a long time, beginning as far back as the end of the 1940s, the West has not been relying on its "conventional" armed forces as a means sufficient for repelling a potential aggressor and for restraining expansion. There are many reasons for this— the West's lack of political, military, and economic unity; the striving to avoid a peacetime militarization of the economy, society, technology, and science; the low numerical levels of the Western nations' armies. All that at a time when the U.S.S.R. and the other countries of the socialist camp have armies with great numerical strength and are rearming them intensively, sparing no resources. It is possible that for a limited period of time the mutual nuclear terror had a certain restraining effect on the course of world events. But, at the present time, the balance of nuclear terror is a dangerous remnant of the past! In order to avoid aggression with conventional weapons one cannot threaten to use nuclear weapons if their use is inadmissible. One of the conclusions that follows here—and a conclusion you draw—is that it is necessary to restore strategic parity in the field of conventional weapons. . . .

The restoration of strategic parity [in this field] is only possible by investing large resources and by an essential change in the psychological atmosphere in the West. There must be a readiness to make certain limited economic sacrifices and, most important, an understanding of the seriousness of the situation and of the necessity for some restructuring. In the final analysis, this is necessary to prevent nuclear war, and war in general. Will the West's politicians be able to carry out such a restructuring? Will the press, the public, and our fellow scientists help them (and not hinder them as is frequently now the case)? Can they succeed in convincing those who doubt the necessity of such restructuring? A great deal depends on it—the opportunity for the West to conduct a nuclear arms policy that will be conducive to the lessening of the danger of nuclear disaster. . . .

I should stress especially that a restructuring of strategy could of course only be carried out gradually and very carefully in order to prevent a loss of parity in some of the intermediate phases.

For talks on nuclear disarmament you propose that one quite simple— and, within the limits of the possible, fair—criterion for assessing nuclear strength be worked out. As that criterion you propose taking the sum total of the number of delivery vehicles and the total number of nuclear charges which can be delivered (probably one should assume the maximal number of certain standard or conventional charges which can be delivered by a given type of missile with a corresponding division of the usable weight).

I will begin by discussing that latter proposal of yours (made jointly with your student, Kent Wisner). This proposal seems practical to me. Your criterion takes into account delivery vehicles of various throw-weights by assigning them various weight factors. This is very important—the assigning of an equal weight factor to both the small American missiles and the large Soviet missiles was one of the points for which I, at one time, criticized the SALT I Treaty (while in general viewing the very fact of the talks and the concluding of the Treaty in a positive light). Here, in distinction to criteria using the power of the charge, as a rule not published officially, the number of deliverable charges is easy to determine. Your criterion also takes into account the fact that, for example, five missiles each carrying one warhead have a significant tactical advantage over one large missile carrying five warheads. Of course, the criterion you propose does not encompass all the parameters like distance, accuracy, or degree of vulnerability—they will have to be allowed for supplementarily or, in some cases, not taken into account so as to facilitate agreements.

I hope that your (or some analogous) criterion will be accepted as the basis for negotiations both on intercontinental missiles and (independently) on medium-range missiles. . . .

From this relatively specific question I will move to one more general, more complex and controversial. Is it actually possible when making decisions in the area of nuclear weapons to ignore all the considerations and requirements relevant to the possible scenarios for a nuclear war and simply limit oneself to the criterion of achieving a reliable deterrent—when that criterion is understood to mean an arsenal sufficient to deal a devastating blow in response? Your answer to this question—while perhaps formulating it somewhat differently—is positive and you draw far-reaching conclusions.

There is no doubt that at present the United States already possesses a large number of submarine-based missiles and charges carried by strategic bombers which are not vulnerable to the U.S.S.R. and, in addition, has silo-based missiles though they are smaller than the U.S.S.R.'s—all these in such amounts that, were those charges used against the U.S.S.R., nothing, roughly speaking, would be left of it. You maintain that this has *already* created a reliable deterrent—independently of what the U.S.S.R. and the United States have and what they lack! Therefore, you specifically consider the building of the MX missile unnecessary and similarly consider irrelevant the arguments which are advanced in support of developing it—the U.S.S.R.'s substantial arsenal of intercontinental missiles with large throw-weight which the United States does not have; and the fact that Soviet missiles and MX missiles have multiple warheads so that one missile can destroy several enemy silos during a missile duel. Therefore you consider it acceptable (with certain reservations) for the United States to freeze the nuclear arsenals of the United States and the U.S.S.R. at their current numerical levels.[1]

Your line of reasoning seems to me very strong and convincing. But I think that the concept presented fails to take into account all the complex realities of the opposition that involves two world systems and that there is the necessity (despite your stance) for a more specific and comprehensive unbiased consideration than a simple orientation toward a "reliable deterrent" (in the meaning of the word as formulated above, i.e., the possibility of dealing a devastating retaliatory strike). I will endeavor to explain this statement.

Precisely because an all-out nuclear war means collective suicide, we can imagine that a potential aggressor might count on a lack of resolve on the part of the country under attack to take the step leading to that suicide; i.e., it could count on its victim capitulating for the sake of saving what could be saved. Given that, if the aggressor has a military advantage in some of the variants of conventional warfare or—which is also possible *in principle*—in some of the variants of partial (limited) nuclear war, he would attempt to use the fear of further escalation to force the enemy to fight the war on his (the aggressor's) own terms. There would be little cause for joy if, ultimately, the aggressor's hopes proved false and the aggressor country perished along with the rest of mankind.

You consider it necessary to achieve a restoration of strategic parity in the field of conventional arms. Now take the next logical step—while nuclear weapons exist it is also necessary to have strategic parity in relation to those variants of limited or regional nuclear warfare which a potential enemy could impose; i.e., it is really *necessary* to examine in detail the various scenarios for both conventional and nuclear war and to analyze the various contingencies. It is of course not possible to analyze fully all these possibilities or to ensure security entirely. But I am attempting to warn of the opposite extreme—"closing one's eyes" and relying on one's potential enemy to be perfectly sensible. As always in life's complex problems, some sort of compromise is needed.

Of course I realize that in attempting not to lag behind a potential enemy in any way, we condemn ourselves to an arms race that is tragic in a world with so many critical problems admitting of no delay. But the main danger is slipping into an all-out nuclear war. If the probability of such an outcome could be reduced at the cost of another ten or fifteen years of the arms race, then perhaps that price must be paid while, at the same time, diplomatic, economic, ideological, political, cultural, and social efforts are made to prevent a war.

Of course it would be wiser to agree now to reduce nuclear and conventional weapons and to eliminate nuclear weapons entirely. But is that now possible in a world poisoned with fear and mistrust, a world where the West fears aggression from the U.S.S.R., the U.S.S.R. fears aggression from the West and from China, and where China fears it from the U.S.S.R., and no verbal assurances and treaties can eliminate those dangers entirely?

I know that pacifist sentiments are very strong in the West. I deeply sympathize with people's yearning for peace, for a solution to world problems by peaceful means; I share those aspirations fully. But, at the same time, I am certain that it is absolutely necessary to be mindful of the specific political, military, and strategic realities of the present day and to do so objectively without making any sort of allowances for either side; this also means that one should not proceed from an a priori assumption of any special peace-loving nature in the socialist countries due to their supposed progressiveness or the horrors and losses they have experienced in war. Objective reality is much more complicated and far from anything so simple. People both in the socialist and the Western countries have a passionate inward aspiration for peace. This is an extremely important factor, but, I repeat, itself alone does not exclude the possibility of a tragic outcome.

What is necessary now, I believe, is the enormous practical task of education so that specific, exact, and historically and politically meaningful objective information can be made available to all people, information that will enjoy their trust and not be veiled with dogma and propaganda. Here one must take into account that, in the countries of the West, pro-Soviet propaganda has been conducted for quite a long time and is very goal-oriented and clever, and that pro-Soviet elements have penetrated many key positions, particularly in the mass media.

The history of the pacifist campaigns against the deployment of missiles in Europe is telling in many respects. After all, many of those participating in those campaigns entirely ignore the initial cause of NATO's "dual decision"—the change in strategic parity in the 1970s in favor of the U.S.S.R.—and, when protesting NATO's plans, they have not advanced any demands on the U.S.S.R. Another example: President Carter's attempt to take a minimal step toward achieving balance in the area of conventional arms, i.e., to introduce draft registration, met with stiff resistance. Meanwhile, balance in the area of conventional arms is a necessary prerequisite for reducing nuclear arsenals. For public opinion in the West to assess global problems correctly, in particular the problems of strategic parity both in conventional and in nuclear weapons, a more objective approach, one which takes the real world strategic situation into account, is vitally needed.

A second group of problems in the field of nuclear weapons about which I should make a few supplementary remarks here concerns the talks on nuclear disarmament. For these talks to be successful the West should have something that it can give up! The case of the "Euromissiles" once again demonstrates how difficult it is to negotiate from a position of weakness. Only very recently has the U.S.S.R. apparently ceased to insist on its unsubstantiated thesis that a rough nuclear parity now exists and therefore everything should be left as it is.

Now, the next welcome step would be the reduction of the number of missiles—which must include a fair assessment of the *quality* of missiles and other means of delivery (i.e., the number of charges deliverable by each carrier, its range and accuracy, and its degree of vulnerability—the last being greater for aircraft and less for missiles;[2] most likely, it would be expedient to use your criterion, or analogous ones). And what is absolutely at issue here is not moving the missiles beyond the Urals but *destroying* them. After all, rebasing is too "reversible." Of course, one also must not consider powerful Soviet missiles, with mobile launchers and several warheads, as being equal to the now-existing Pershing I, the British and French missiles, or the bombs on short-range bombers—as the Soviet side sometimes attempts to do for purposes of propaganda.

No less important a problem is that of the powerful silo-based missiles. At present the U.S.S.R. has a great advantage in this area. Perhaps talks about the limitation and reduction of these most destructive missiles could become easier if the United States were to have MX missiles, albeit only potentially (indeed, that would be best of all). . . .

A specific danger associated with silo-based missiles is that they can be destroyed relatively easily as a result of enemy attack, as I have just demonstrated. At the same time, they can be used to destroy enemy launch sites in an amount four to five times larger than the number of missiles used for the attack. A country with large numbers of silo-based missiles (at the present time this is primarily the U.S.S.R., but if the United States carries out a major MX program, then it too) could be "tempted" to use such missiles first before the enemy destroys them. In such circumstances the presence of silo-based missiles constitutes a destabilizing factor.

In view of the above, it seems very important to me to strive for the abolition of powerful silo-based missiles at the talks on nuclear disarmament. While the U.S.S.R. is the leader in this field there is very little chance of its easily relinquishing that lead. If it is necessary to spend a few billion dollars on MX missiles to alter this situation, then perhaps this is what the West must do. But, at the same time, if the Soviets, in deed and not just in word, take significant verifiable measures for reducing the number of land-based missiles (more precisely, for destroying them), then the West should not only abolish MX missiles (or not build them!) but carry out other significant disarmament programs as well.

On the whole I am convinced that nuclear disarmament talks are of enormous importance and of the highest priority. They must be conducted continuously—in the brighter periods of international relations but also in the periods when relations are strained—and conducted with persistence, foresight, firmness and, at the same time, with flexibility and initiative. In so doing, political figures should not think of exploiting those talks, and the nuclear problem in general, for their own immediate political gains but only for the long-term interests of their country and the world. And the planning of the talks should be included in one's general nuclear strategy as its most important part—on this point as well I am in agreement with you!

VIII

The third group of problems which should be discussed here is political and social in nature. A nuclear war could result from a conventional war, while a conventional war is, as is well known, a result of politics. We all know that the world is not at peace. There are a variety of reasons for this—national, economic, and social reasons, as well as the tyranny of dictators.

Many of the tragic events now occurring have their roots in the distant past. It would absolutely be wrong to see only Moscow's hand everywhere. Still, when examining the general trend of events since 1945 there has been a relentless expansion of the Soviet sphere of influence—objectively, this is nothing but Soviet expansion on a world scale. This process has spread as the U.S.S.R. has grown stronger economically (though that strength is one-sided), and in scientific, technological and military terms, and has today assumed proportions dangerously harmful to international equilibrium. The West has grounds to worry that the world's sea routes, Arab oil, and the uranium, diamonds, and other resources of South Africa are now threatened.

One of the basic problems of this age is the fate of the developing countries, the greater part of mankind. But, in fact, for the U.S.S.R., and to some degree for the West as well, this problem has become exploitable and expendable in the struggle for dominance and strategic interests. Millions of people are dying of hunger every year, hundreds of millions suffer from malnutrition and hopeless poverty. The West provides the developing countries with economic and technological aid, but this remains entirely insufficient due largely to the rising price of crude oil. Aid from the U.S.S.R. and the socialist countries is smaller in scale and, to a greater degree than the West's aid, military in nature and bloc-oriented. And, very importantly, that aid is in no way coordinated with world efforts.

The hot spots of local conflicts are not dying but are rather threatening to grow into global wars. All this is greatly alarming.

The most acutely negative manifestation of Soviet policies was the invasion of Afghanistan which began in December 1979 with the murder of the head of state. Three years of appallingly cruel anti-guerrilla war have brought incalculable suffering to the Afghan people, as attested by the more than four million refugees in Pakistan and Iran.

It was precisely the general upsetting of world equilibrium caused by the invasion of Afghanistan and by other concurrent events which was the fundamental reason that the SALT II agreement was not ratified. I am with you in regretting this but I cannot disregard the reasons I have just described.

Yet another subject closely connected to the problem of peace is the openness of society and human rights. I use the term the "openness of society" to mean precisely what the great Niels Bohr meant by it when introducing it more than 30 years ago.

In 1948, the U.N.'s member states adopted the Universal Declaration of Human Rights and stressed its significance for maintaining peace. In 1975, the relationship of human rights and international security was proclaimed by the Helsinki Final Act, which was signed by 35 countries including the U.S.S.R. and the United States. Among those rights are: the right to freedom of conscience; the right to receive and impart information within a country and across frontiers; the right to a free choice of one's country of residence and domicile within a country; freedom of religion; and freedom from psychiatric persecution.

Finally, citizens have the right to control their national leaders' decision-making in matters on which the fate of the world depends. But we don't even know how, or by whom, the decision to invade Afghanistan was made! People in our country do not have even a fraction of the information about events in the world and in their own country which the citizens of the West have at their disposal. The opportunity to criticize the policy of one's national leaders in matters of war and peace as you do freely is, in our country, entirely absent. Not only critical statements but those merely factual in nature, made on even much less important questions, often entail arrest and a long sentence of confinement or psychiatric prison.

In conclusion I again stress how important it is that the world realize the absolute inadmissibility of nuclear war, the collective suicide of mankind. It is impossible to win a nuclear war. What is necessary is to strive, systematically though carefully, for complete nuclear disarmament based on strategic parity in conventional weapons. As long as there are nuclear weapons in the world, there must be a strategic parity of nuclear forces so that neither side will venture to embark on a limited or regional nuclear war. Genuine security is possible only when based on a stabilization of international relations, a repudiation of expansionist policies, the strengthening of international trust, openness and pluralization in the socialist societies, the observance of human rights throughout the world, the rapprochement—convergence—of the socialist and capitalist systems, and worldwide coordinated efforts to solve global problems. . . .

NOTES

1. *Editor's Note* [*Foreign Affairs*]: Professor Drell notes that maintaining the U.S. and Soviet nuclear arsenals at their present numerical levels is not the same as the kind of "freeze" usually discussed today—in that it would not preclude changes in

the types of weapons within the numerical level. As to a strict "freeze" as usually discussed, Professor Drell's position, stated in his Grace Cathedral speech, is that "the freeze movement has been very helpful in creating . . . a constituency for arms control. Though I recognize some deficiencies of the freeze as literal policy, I support it and will vote for it as a mandate for arms control. . . ."

2. *Editor's Note [Foreign Affairs]:* This phrase is a literal translation from the Russian. It apparently refers to the shape and size of the area in which a given missile is likely to land in accordance with its accuracy characteristics. The comparable American term is "circular error probable," or "CEP," defined as the area within which a given missile has a 50-percent chance of landing. Such an area is in fact usually elliptical in shape rather than circular.

47. Report of the President's Commission on Strategic Forces

Scowcroft Commission

DETERRENCE AND ARMS CONTROL

The responsibility given to this Commission is to review the purpose, character, size, and composition of the strategic forces of the United States. The members of the Commission fully understand not only the purposes for which this nation maintains its deterrent, but also the devastating nature of nuclear warfare, should deterrence fail. The Commission believes that effective arms control is an essential element in diminishing the risk of nuclear war—while preserving our liberties and those of like-minded nations. At the same time the Commission is persuaded that as we consider the threat of mass destruction we must consider simultaneously the threat of aggressive totalitarianism. Both are central to the political dilemmas of our age. For the United States and its allies the essential dual task of statecraft is, and must be, to avoid the first and contain the second.

It is only by addressing these two issues together that we can begin to understand how to preserve both liberty and peace. Although the United States and the Soviet Union hold fundamentally incompatible views of history, of the nature of society, and of the individual's place in it, the existence of nuclear weapons imbues that rivalry with peril unprecedented in human history. The temptation is sometimes great to simplify—or oversimplify—the difficult problems that result, either by blinking at the devastating nature of modern full-scale war or by refusing to acknowledge the emptiness of life under modern totalitarianism. But it is naive, false, and dangerous to assume that either of these, today, can be ignored and the other dealt with in isolation. We cannot cope with the efforts of the Soviet Union to extend its power without giving thought to the way nuclear weapons have sharply raised the stakes and changed the nature of warfare. Nor can we struggle against nuclear war or the arms race in some abstract sense without keeping before us the Soviet Union's drive to expand its power, which is what makes those struggles so difficult.

Deterrence is central to the calm persistence we must demonstrate in order to reduce these risks. American strategic forces exist to deter attack on the United States or its allies—and the coercion that would be possible if the public or decisionmakers believed that the Soviets might be able to launch a successful attack. Such a policy of deterrence, like the security policy of the West itself, is essentially defensive in nature. The strategic forces that are necessary in order to support such a policy by their very existence help to convince the Soviet Union's leaders: that the West has the military strength and political will to resist aggression; and that, if they should ever choose to attack, they should have no doubt that we can and would respond until we have so damaged the power of the Soviet state that they will unmistakably be far worse off than if they had never begun.

There can be no doubt that the very scope of the possible tragedy of modern nuclear war, and the increased destruction made possible even by modern non-nuclear technology, have changed the nature of war itself. This is not only because massive conventional war with modern weapons could be horrendously destructive—some fifty million people died in "conventional" World War II before the advent of nuclear weapons—but also because *conventional* war between the world's major power blocs is the most likely way for *nuclear* war to develop. The problem of deterring the threat of nuclear war, in short, cannot be isolated from the overall power balance between East and West. Simply put, it is war that must concern us, not nuclear war alone. Thus we must maintain a balance between our nuclear and conventional forces and we must demonstrate to the Soviets our cohesion and our will. And we must understand that weakness in any one of these areas puts a dangerous burden on the others as well as on overall deterrence.

Deterrence is not, and cannot be, bluff. In order for deterrence to be effective we must not merely have weapons, we must be perceived to be able, and prepared, if necessary, to use them effectively against the key elements of Soviet power. Deterrence is not an abstract notion amenable to simple quantification. Still less is it a mirror image of what would deter ourselves. Deterrence is the set of beliefs in the minds of the Soviet leaders, given their own values and attitudes, about our capabilities and our will. It requires us to determine, as best we can, what would deter them from considering aggression, even in a crisis—not to determine what would deter us.

Our military forces must be able to deter war even if the Soviets are unwilling to participate with us in equitable and reasonable arms control agreements. But various types of agreements can, when the Soviets prove willing, accomplish critical objectives. Arms control can: reduce the risk of war; help limit the spread of nuclear weapons; remove or reduce the risk of misunderstanding of particular events or accidents; seal off wasteful, dangerous, or unhelpful lines of technical development before either side gets too committed to them; help channel modernization into stabilizing rather than destabilizing paths; reduce misunderstanding about the purpose of weapons developments and thus reduce the need to over-insure against worst-case projections; and help make arsenals less destructive and costly. To achieve part or all of these positive and useful goals, we must keep in mind the importance of compliance and adequate verification—difficult problems in light of the nature of the Soviet state—and the consequent importance of patience in order to reach fair and reasonable agreements. . . .

SOVIET OBJECTIVES AND PROGRAMS

Effective deterrence and effective arms control have both been made significantly more difficult by Soviet conduct and Soviet weapons programs in recent years. The overall military balance, including the nuclear balance, provides the backdrop for Soviet decisions about the manner in which they will try to advance their interests. This is central to our understanding of how to deter war, how to frustrate Soviet efforts at blackmail, and how to deal with the Soviets' day-to-day conduct of international affairs. The Soviets have shown by word and deed that they regard military power, including nuclear weapons, as a useful tool in the projection of their national influence. In the Soviet strategic view, nuclear weapons are closely related to, and are integrated with, their other military and political instruments as a means of advancing their interests. The Soviets have concentrated enormous effort on the development and modernization of nuclear weapons, obviously seeking to achieve what they regard as important advantages in certain areas of nuclear weaponry. . . .

PREVENTING SOVIET EXPLOITATION OF
THEIR MILITARY PROGRAMS

In our effort to make a strategy of deterrence and arms control effective in preventing the Soviets from political or military use of their strategic forces, we must keep several points in mind.

The Soviets must continue to believe what has been NATO's doctrine for three decades: that if we or our allies should be attacked—by massive conventional means or otherwise—the United States has the will and the means to defend with the full range of American power. This by no means excludes the need to make improvements in our conventional forces in order to have increased confidence in our ability to defend effectively at the conventional level in many more situations, and thus to raise the nuclear threshold. Certainly mutual arms control agreements to reduce both sides' reliance on nuclear weapons should be pursued. But effective deterrence requires that early in any Soviet consideration of attack, or threat of attack, with conventional forces or chemical or biological weapons, Soviet leaders must understand that they risk an American nuclear response. . . .

In order to deter such Soviet threats we must be able to put at risk those types of Soviet targets—including hardened ones such as military command bunkers and facilities, missile silos, nuclear weapons and other storage, and the rest—which the Soviet leaders have given every indication by their actions they value most, and which constitute their tools of control and power. We cannot afford the delusion that Soviet leaders—human though they are and cautious though we hope they will be—are going to be deterred by exactly the same concerns that would dissuade us. Effective deterrence of the Soviet leaders requires them to be convinced in their own minds that there could be no case in which they could benefit by initiating war.

Effective deterrence of any Soviet temptation to threaten or launch a massive conventional or a limited nuclear war thus requires us to have a comparable ability to destroy Soviet military targets, hardened and otherwise. If there were ever a case to be made that the Soviets would unilaterally stop their strategic deployments at a level short of the ability seriously to threaten our forces, that argument vanished with the deployment of their SS-18 and SS-19 ICBMs. A one-sided strategic condition in which the Soviet Union could effectively destroy the whole range of strategic targets in the United States, but we could not effectively destroy a similar range of targets in the Soviet Union, would be extremely unstable over the long run. Such a situation could tempt the Soviets, in a crisis, to feel they could successfully threaten or even undertake conventional or limited nuclear aggression in the hope that the United States would lack a fully effective response. A one-sided condition of this sort would clearly not serve the cause of peace. In order, then, to pursue successfully a policy of deterrence and verifiable, stabilizing arms control we must have a strong and militarily effective nuclear deterrent. Consequently our strategic forces must be modernized, as necessary, to enhance to an adequate degree their overall survivability and to enable them to engage effectively the targets that Soviet leaders most value.

STRATEGIC FORCES AS A WHOLE

The development of the components of our strategic forces—the multiplicity of intercontinental ballistic missiles (ICBMs), submarine-launched ballistic missiles (SLBMs), and bombers—was in part the result of an historical evolution. This triad of forces, however, serves several important purposes.

First, the existence of several strategic forces requires the Soviets to solve a number of different problems in their efforts to plan how they might try to overcome them. Our objective, after all, is to make their planning of any such attack as difficult as we can. If it were possible for the Soviets to concentrate their research and development efforts on putting only one or two components of U.S. strategic forces at risk—e.g., by an intensive effort at anti-submarine warfare to attempt to threaten our ballistic missile submarines—both their incentive to do so and their potential gains would be sharply increased. Thus the existence of several components of our strategic forces permits each to function as a hedge against possible Soviet successes in endangering any of the others. For example, at earlier times uncertainties about the vulnerability of our bomber force were alleviated by our confidence in the survivability of our ICBMs. And although the survivability of our ICBMs is today a matter of concern (especially when that problem is viewed in isolation) it would be far more serious if we did not have a force of ballistic missile submarines at sea and a bomber force. By the same token, over the long run it would be unwise to rely so heavily on submarines as our only ballistic missile force that a Soviet breakthrough in anti-submarine warfare could not be offset by other strategic systems.

Second, the different components of our strategic forces would force the Soviets, if they were to contemplate an all-out attack, to make choices which would lead them to reduce significantly their effectiveness against one component in order to attack another. For example, if Soviet war planners should decide to attack our bomber and submarine bases and our ICBM silos with simultaneous detonations—by delaying missile launches from close-in submarines so that such missiles would *arrive* at our bomber bases at the same time the Soviet ICBM warheads (with their longer time of flight) would arrive at our ICBM silos—then a very high proportion of our alert bombers would have escaped before their bases were struck. This is because we would have been able to, and would have, ordered our bombers to take off from their bases within moments after the launch of the first Soviet ICBMs. If the Soviets, on the other hand, chose rather to *launch* their ICBM and SLBM attacks at the same moment (hoping to destroy a higher proportion of our bombers with SLBMs having a short time of flight), there would be a period of over a quarter of an hour after nuclear detonations had occurred on U.S. bomber bases but before our ICBMs had been struck. In such a case the Soviets should have no confidence that we would refrain from launching our ICBMs during that interval after we had been hit. It is important to appreciate that this would not be a "launch-on-warning," or even a "launch under attack," but rather a launch *after* attack—after massive nuclear detonations had already occurred on U.S. soil.

Thus our bombers and ICBMs are more survivable together against Soviet attack then either would be alone. This illustrates that the different components of our strategic forces should be assessed collectively and not in isolation. It also suggests that whereas it is highly desirable that a component of the strategic forces be survivable when it is viewed separately, it makes a major contribution to deterrence even if its survivability depends in substantial measure on the existence of one of the other components of the force.

The third purpose served by having multiple components in our strategic forces is that each component has unique properties not present in the others. Nuclear submarines have the advantage of being able to stay submerged and hidden for months at a time, and thus the missiles they carry may reasonably be held in reserve rather than being used early in the event of attack. Bombers may be launched from their bases on warning without irretrievably committing them to an attack; also, their weapons, though they arrive in hours, not minutes, have excellent accuracy against a range of possible targets. ICBMs have advantages in command and control, in the ability to be retargeted readily, and in accuracy. This means that ICBMs are especially effective in deterring Soviet threats of massive conventional or limited nuclear attacks, because they could most credibly respond promptly and controllably against specific military targets and thereby promptly disrupt an attack on us or our allies.

A more stable structure of ICBM deployments would exist if both sides moved toward more survivable methods of basing than is possible when there is primary dependence on large launchers and missiles. Thus from the point of view of enhancing such stability, the Commission believes that there is considerable merit in moving toward an ICBM force structure in which potential targets are of comparatively low value—missiles containing only one warhead. A single-warhead ICBM, suitably based, inherently denies an attacker the opportunity to destroy more than one warhead with one attacking warhead. The need to have basing flexibility, and particularly the need to keep open the option for different types of mobile basing, also suggests a missile of small size. If force survivability can be additionally increased by arms control agreements which lead both sides toward more survivable modes of basing than is possible with large launchers and missiles, the increase in stability would be further enhanced.

In the meantime, however, deployment of MX is essential in order to remove the Soviet advantage in ICBM capability and to help deter the threat of conventional or limited nuclear attacks on the alliance. Such deployment is also necessary to encourage the Soviets to move toward the more stable regime of deployments and arms control outlined above.

The Commission stresses that these two aspects of ICBM modernization and this approach toward arms control are integrally related. They point toward the same objective—permitting the U.S. and encouraging the Soviets to move toward more stable ICBM deployments over time in a way that is consistent with arms control agreements having the objective of reducing the risk of war. The Commission is unanimous that no one part of the proposed program can accomplish this alone.

ICBM Long-term Survivability: Toward the Small, Single-Warhead ICBM

The Commission believes that a single-warhead missile weighing about fifteen tons (rather than the nearly 100 tons of MX) may offer greater flexibility in the long-run effort to obtain an ICBM force that is highly survivable, even when viewed in isolation, and that can consequently serve as a hedge against potential threats to the submarine force.

The Commission thus recommends beginning engineering design of such an ICBM, leading to the initiation of full-scale development in 1987 and an initial operating capability in the early 1990s. The design of such a missile, hardened against nuclear effects, can be achieved with current technology. It should have sufficient accuracy and yield to put Soviet hardened military targets at risk. During that period an approach toward arms control, consistent with such deployments, should also seek to encourage the Soviets to move toward a more stable ICBM force structure at levels which would obviate the need to deploy very large numbers of such missiles. . . .

The key advantages of a small single-warhead missile are that it would reduce the value of each strategic target and that it is also compatible with either fixed or mobile deployments, or with combinations of the two.

Deployment of such small missiles would be compatible with arms control agreements reducing the number of warheads, in which case only a small number of such missiles would probably need to be deployed. If the Soviets proved unwilling to reach such agreements, however, the U.S. could deploy whatever number of small missiles were required—in whatever mix of basing modes—to maintain an adequate overall deterrent.

Immediate ICBM Modernization: Limited Deployment of the MX Missile— The MX in Minuteman Silos

There are important needs on several grounds for ICBM modernization that cannot be met by the small, single-warhead ICBM.

First, arms control negotiations—in particular the Soviets' willingness to enter agreements that will enhance stability—are heavily influenced by ongoing programs. The ABM Treaty of 1972, for example, came about only because the United States maintained an ongoing ABM program and indeed made a decision to make a limited deployment. It is illusory to believe that we could obtain a satisfactory agreement with the Soviets limiting ICBM deployments if we unilaterally terminated the only new U.S. ICBM program that could lead to deployment in this decade. Such a termination would effectively communicate to the Soviets that we were unable to neutralize their advantage in multiple-warhead ICBMs. Abandoning the MX at this time in search of a substitute would jeopardize, not enhance, the likelihood of reaching a stabilizing and equitable agreement. It would also undermine the incentives to the Soviets to change the nature of their own ICBM force and thus the environment most conducive to the deployment of a small missile.

Second, effective deterrence is in no small measure a question of the Soviets' perception of our national will and cohesion. Cancelling the MX, when it is ready for flight testing, when over $5 billion have already been spent on it, and when its importance has been stressed by the last four Presidents, does not communicate to the Soviets that we have the will essential to effective deterrence. Quite the opposite.

Third, the serious imbalance between the Soviets' massive ability to destroy hardened land-based military targets with their ballistic missile force and our lack of such a capability must be redressed promptly. Our ability to assure our allies that we have the capability and will to stand with them, with whatever forces are necessary, if the alliance is threatened by massive conventional, chemical or biological, or limited nuclear attack is in question as long as this imbalance exists. Even before the Soviet leaders, in a grave crisis, considered using the first tank regiment or the first SS-20 missile against NATO, they must be required to face what war would mean to them. In order to augment what we would hope would be an inherent sense of conservatism and caution on their part, we must have a credible capability for controlled, prompt, limited attack on hard targets ourselves. This capability casts a shadow over the calculus of Soviet risk-taking at any level of confrontation with the West. Consequently, in the interest of the alliance as a whole, we cannot safely permit a situation to continue wherein the Soviets have the capability promptly to destroy a range of hardened military targets and we do not.

Fourth, our current ICBM force is aging significantly. . . .

The existence of a production program for an ICBM of approximately 100 tons is important for two additional reasons. As Soviet ABM modern-ization and modern surface-to-air missile development and deployment proceed—even within the limitations of the ABM treaty—it is important to be able to match any possible Soviet breakout from that treaty with strategic forces that have the throw-weight to carry sufficient numbers of decoys and other penetration aids; these may be necessary in order to penetrate the Soviet defenses which such a breakout could provide before other compensating steps could be taken. Having in production a missile that could effectively counter such a Soviet step should help deter them from taking it. Moreover, in view of our coming sole reliance on space shuttle orbiters, it would be prudent to have in production a booster, such as MX, that is of sufficient size to place in orbit at least some of our most strategically important satellites.

These objectives can all be accomplished, at reasonable cost, by deploying MX missiles in current Minuteman silos. . . .

A program of deploying on the order of 100 MX missiles in existing Minuteman silos would, on the other hand, accomplish the objectives set forth in this section and it would do so without threatening stability. The throw-weight and megatonnage carried by the 100 MX missiles is about the same as that of the 54 large Titan missiles now being retired plus that of the 100 Minuteman III missiles that the MXs would replace. Such a deployment would thus represent a replacement and modernization of part of our ICBM force. It would provide a means of controlled limited attack on hardened targets but not a sufficient number of warheads to be able to attack all hardened Soviet ICBMs, much less all of the many command posts and other hardened military targets in the Soviet Union. Thus it would not match the overall capability of the recent Soviet deployment of over 600 modern ICBMs of MX size or larger. But a large deployment of several hundred MX missiles should be unnecessary for the limited but very important purposes set forth above. Should the Soviets refuse to engage in stabilizing arms control and engage instead in major new deployments, reconsideration of this and other conclusions would be necessary.

SUMMARY OF MODERNIZATION RECOMMENDATIONS

Strategic Forces Other than ICBMs

a. As first priority, vigorous programs should continue to improve the ability of the President to command, control, and communicate with the strategic forces under conditions of severe stress or actual attack.

b. The Trident submarine construction program and the Trident II (D-5) ballistic missile development program should continue with high priority; the work recommended on small submarines to avoid technological surprise in anti-submarine warfare should begin now.

c. No changes are recommended in the bomber and air-launched cruise missile programs.

d. Vigorous research and technology development on ABM should be pursued. The development of decoys and other penetration aids for our ballistic missiles is also recommended.

ICBM Programs

a. Engineering design should be initiated, now, of a single-warhead ICBM weighing about fifteen tons; this program should lead to the initiation of full-scale development in 1987 and an initial operating capability in the early 1990s. Deploying such a missile in more than one mode would serve stability. Hardened silos or shelters and hardened mobile launchers should be investigated now.

b. One hundred MX missiles should be deployed promptly in existing Minuteman silos as a replacement for those 100 Minutemen and the Titan II ICBMs now being decommissioned and as a modernization of the force.

c. A specific program to resolve the uncertainties regarding silo or shelter hardness should be undertaken, leading to later decisions about hardening MX in silos and deploying a small single-warhead ICBM in hardened silos or shelters. Vigorous investigation should proceed on different types of land-based vehicles and launchers, including particularly hardened vehicles. . . .

ARMS CONTROL

Over the long run, stability would be fostered by a dual approach toward arms control and ICBM deployments which moves toward encouraging small, single-warhead ICBMs. This requires that arms control limitations and reductions be couched, not in terms of launchers, but in terms of equal levels of warheads of roughly equivalent yield. Such an approach could permit relatively simple agreements, using appropriate counting rules, that exert pressure to reduce the overall number and destructive power of nuclear weapons and at the same time give each side an incentive to move toward more stable and less vulnerable deployments.

48. Address on Ballistic Missile Defense

Ronald Reagan

. . . My predecessors in the Oval Office have appeared before you on other occasions to describe the threat posed by Soviet power and have proposed steps to address that threat. But since the advent of nuclear weapons, those steps have been increasingly directed toward deterrence of aggression through the promise of retaliation.

This approach to stability through offensive threat has worked. We and our allies have succeeded in preventing nuclear war for more than three decades. In recent months, however, my advisers, including in particular the Joint Chiefs of Staff, have underscored the necessity to break out of a future that relies solely on offensive retaliation for our security.

Over the course of these discussions, I've become more and more deeply convinced that the human spirit must be capable of rising above dealing with other nations and human beings by threatening their existence. Feeling this way, I believe we must thoroughly examine every opportunity for reducing tensions and for introducing greater stability into the strategic calculus on both sides.

One of the most important contributions we can make is, of course, to lower the level of all arms, and particularly nuclear arms. We're engaged right now in several negotiations with the Soviet Union to bring about a mutual reduction of weapons. I will report to you a week from tomorrow my thoughts on that score. But let me just say, I'm totally committed to this course.

If the Soviet Union will join with us in our effort to achieve major arms reduction, we will have succeeded in stabilizing the nuclear balance. Nevertheless, it will still be necessary to rely on the specter of retaliation, on mutual threat. And that's a sad commentary on the human condition. Wouldn't it be better to save lives than to avenge them? Are we not capable of demonstrating our peaceful intentions by applying all our abilities and our ingenuity to achieving a truly lasting stability? I think we are. Indeed, we must.

After careful consultation with my advisers, including the Joint Chiefs of Staff, I believe there is a way. Let me share with you a vision of the future which offers hope. It is that we embark on a program to counter the awesome Soviet missile threat with measures that are defensive. Let us turn to the very strengths in technology that spawned our great industrial base and that have given us the quality of life we enjoy today.

What if free people could live secure in the knowledge that their security did not rest upon the threat of instant U.S. retaliation to deter a Soviet attack, that we could intercept and destroy strategic ballistic missiles before they reached our own soil or that of our allies?

I know this is a formidable, technical task, one that may not be accomplished before the end of this century. Yet, current technology has attained a level of sophistication where it's reasonable for us to begin this effort. It will take years, probably decades of effort on many fronts. There will be failures and setbacks, just as there will be successes and breakthroughs. And as we proceed, we must remain constant in preserving the nuclear deterrent and maintaining a solid capability for flexible response. But isn't it worth every investment necessary to free the world from the threat of nuclear war? We know it is.

In the meantime, we will continue to pursue real reductions in nuclear arms, negotiating from a position of strength that can be ensured only by modernizing our strategic forces. At the same time, we must take steps to reduce the risk of a conventional military conflict escalating to nuclear war by improving our non-nuclear capabilities.

America does possess—now—the technologies to attain very significant improvements in the effectiveness of our conventional, non-nuclear forces. Proceeding boldly with these new technologies, we can significantly reduce any incentive that the Soviet Union may have to threaten attack against the United States or its allies.

As we pursue our goal of defensive technologies, we recognize that our allies rely upon our strategic offensive power to deter attacks against them. Their vital interests and ours are inextricably linked. Their safety and ours are one. And no change in technology can or will alter that reality. We must and shall continue to honor our commitments.

I clearly recognize that defensive systems have limitations and raise certain problems and ambiguities. If paired with offensive systems, they can be viewed as fostering an aggressive policy, and no one wants that. But with these considerations firmly in mind, I call upon the scientific community in our country, those who gave us nuclear weapons, to turn their great talents now to the cause of mankind and world peace, to give us the means of rendering these nuclear weapons impotent and obsolete.

Tonight, consistent with our obligations of the ABM treaty and recognizing the need for closer consultation with our allies, I'm taking an important first step. I am directing a comprehensive and intensive effort to define a long-term research and development program to begin to achieve our ultimate goal of eliminating the threat posed by strategic nuclear missiles. This could pave the way for arms control measures to eliminate the weapons themselves. We seek neither military superiority nor political advantage. Our only purpose—one all people share—is to search for ways to reduce the danger of nuclear war.

49. The Case Against Strategic Defense: Technical and Strategic Realities

Sidney D. Drell and Wolfgang K. H. Panofsky

In his address to the nation on March 23, 1983, President Reagan advocated that the United States "embark on a program to counter the awesome Soviet missile threat with measures that are defensive." He called upon the American scientific community, "those who gave us nuclear weapons, to turn their great talents now to the cause of mankind and world peace, to give us the means of rendering these nuclear weapons impotent and obsolete." These words have rekindled the debate about defense against nuclear weapons—a subject that had been relatively dormant since the anti-ballistic missile (ABM) debates of 1969 and the ratification hearings for the ABM Treaty of 1972. . . .

In considering defenses against nuclear weapons, however, one must recognize that the enormous increase in the explosive power of nuclear bombs has wrought a fundamental discontinuity in the relative effectiveness of offensive and defensive measures. We must now recognize that a *single* relatively small nuclear bomb is a weapon of mass destruction. . . .

Given the staggering destructive potential of nuclear bombs, effective defenses must meet a much higher standard of performance today than at any other time in the history of warfare. During World War II, air defenses were considered effective if 10 percent of the airplanes were lost in each attack: after ten such sorties only one-third of the airplanes and their crews would survive. Indeed, approximately that level of defense effectiveness brought victory to the Royal Air Force in the Battle of Britain in 1942. In the nuclear age, however, if 90 percent of the attacking aircraft or missiles are destroyed, and thus only 10 percent reach their targets, their nuclear weapons would produce a catastrophe of unimaginable proportions to the country under seige. . . .

The strategic relationship between the United States and the Soviet Union currently rests upon the balance of offensive forces, a situation known as "offense dominance." If either superpower launches a nuclear attack, it faces the risk of a nuclear retaliatory strike that can endanger its very existence. While neither the United States nor the Soviet Union has made the destruction of enemy populations in response to enemy attack an explicit policy objective, both recognize that should a large fraction of the superpowers' arsenals be used—under any doctrine, any choice of pattern of attack, or for any purpose—then the threat to the survival of the two societies is very grave indeed. . . .

Many find this situation morally repugnant; moreover, it cannot be expected to prevent nuclear war for all time. Unhappily, however, we see no technical alternative to the mutual hostage relationship—the state of Mutual Assured Destruction (MAD)—as long as nuclear armaments remain anywhere near the levels now stockpiled or continue to grow as they are today.

We are emphasizing this point here because MAD is frequently represented as a doctrine promulgated by some misguided policymakers willing to gamble with the survival of their country. Nothing could be further from the truth. The history of the nuclear weapons competition is replete with attempts by policymakers in both the United States and the Soviet Union to avoid the mutual hostage situation. Each time, however, they concluded that technology simply does not permit such an escape unless the total level of nuclear arms in the world can be drastically lowered. . . .

In this situation of mutual vulnerability, maintaining the stability of the military balance is the principal tool for avoiding nuclear war. The United States and the Soviet Union should do their utmost to avoid a situation in which either would be tempted in time of crisis to use nuclear weapons first on the assumption that launching a preemptive strike would be preferable to facing a likely attack by the other. In other words, "crisis stability" calls for minimizing the perceived advantage to the party attacking first.

Much has already been done in the attempt to enhance crisis stability. Essential ingredients for ensuring stability are a reliable—and survivable—command and control system, the ability of each side's retaliatory forces to survive an attack, and the demonstrable ability of retaliatory forces to reach their targets. Efforts must also be made to minimize the likelihood that unauthorized, accidental, or third-party nuclear explosions might lead to all-out nuclear war.

In addition to the unilateral measures that each country can take to improve crisis stability, arms control measures can also be negotiated, based on the recognition that enhancement of stability is a matter of mutual interest to the Soviet Union and the United States. In the latter category, the ABM Treaty signed in 1972 and supplemented by a protocol in 1974 remains a key step, as we will discuss later in detail. In addition the agreement in 1963 to establish a "hot line" between Washington and Moscow, and the subsequent steps taken in 1971 and 1984 to improve that facility, is a noteworthy attempt to avoid nuclear war through miscalculation or misinformation.

While these measures can enhance the stability of the current offensive balance, they cannot guarantee it. In the long run, the primary thrust for relieving the situation must be political; the evolution of U.S.-China relations during the past 20 years shows what can be done. The current mutual hostage relationship between the Soviet Union and the United States cannot be expected to endure forever. Insofar as technical means can contribute to relieving this relationship, we conclude that the only tool available is a *drastic* reduction of the world's nuclear arsenals. Extending the hope that, at the current levels of these arsenals, the mutual hostage relationship can be avoided through changes in doctrine or by technical measures will detract from efforts to attack the problem at its source: the existence of vastly excessive nuclear stockpiles worldwide.

In his speech, President Reagan proposed nothing less than a defensive system sufficiently comprehensive and impenetrable to make the American people immune to nuclear attack. His contention was that once such a shield was erected, the nuclear weapons states could be persuaded that these weapons served no useful purpose and could therefore be abolished. . . .

MAKING NUCLEAR WEAPONS OBSOLETE

A defense meeting the president's vision would require near perfection in performance. If only 1 percent of the approximately 8,000 nuclear strategic warheads in the current Soviet force penetrated a defensive shield and landed on urban targets in the United States, it would cause one of the greatest disasters in recorded history.

To attain such a level of protection, *all* means of delivery of such weapons to the United States must be interdicted with near 100 percent efficiency. This would entail not only defenses against intercontinental and submarine-launched ballistic missiles, but also against strategic aircraft and cruise missiles from air, land, or sea. Even the introduction of "suitcase bombs" would have to be prevented. Yet the Strategic Defense Initiative specifically deals with only the ballistic missile threat. . . .

Even ignoring these other threats, we view the possibility of erecting an impenetrable shield against ballistic missiles as very remote indeed for many decades to come. It is difficult for a responsible scientist to say flatly that a task is impossible to achieve by technical means without being accused of being a "naysayer." Indeed, many instances can be cited in which prominent scientists have concluded that a task is impossible only to be proved wrong by future discoveries. One should recognize, however, that the deployment of an impenetrable defense over the nation is not a single technical achievement but the evolution of an extensive and exceedingly complex *system* that must work reliably in a hostile environment. Furthermore, one must have continued high confidence in the defense system, although it can never be tested under realistic conditions—especially against an offense that can adopt a broad repertoire of countermeasures against it. . . .

We strongly emphasize that the issue is not whether a specific technology for the interception of incoming ballistic missiles can be demonstrated. We fully agree that a single reentry vehicle from a ballistic missile can be destroyed by nuclear explosives lofted by interceptors, as was shown to be feasible in earlier developments, or by non-nuclear, high-velocity projectiles as was demonstrated by the U.S. Air Force in June 1984. We also believe that a demonstration can be staged in which an ascending ICBM boost vehicle can be damaged by airborne or spaceborne laser beams. However, such demonstrations of the interception of cooperating targets hardly have bearing on the feasibility of the overall system or the solution of operational problems. . . .

ENHANCED DETERRENCE

When submitting the five-year program for the Strategic Defense Initiative in March 1984, [Dr. Richard] DeLauer [Undersecretary of Defense for Research and Engineering] testified before the House Committee on Armed Services that Defense Department studies have "concluded that advanced defensive technologies could offer the potential to enhance deterrence and help prevent nuclear war by reducing significantly the military utility of Soviet preemptive attacks and by undermining an aggressor's confidence of a successful attack against the United States or our allies." Program advocates claim that this mission can be met by even a less-than-perfect area defense of population and industry.

It is indeed true that if a potential attacker faces the prospect of attrition of his forces by a defense, in addition to the expected retaliatory strike, his confidence in the success of his planned attack would decrease, and the complexity of planning such a move would increase. If one could correctly anticipate that this would be the *only* Soviet reaction to an expanded Strategic Defense Initiative, one might consider this to be sufficient reason to move ahead toward deployment of a nationwide defense against ballistic missiles. However, the much more likely Soviet response would be to initiate a variety of programs to counteract the effectiveness of such defenses to retain full confidence in their deterrent. More than likely they would also move ahead with intensified defense programs of their own. The net result of these moves and countermoves would be the addition of yet another component to the arms competition between the superpowers, in both offensive and defensive forces, and the end of the ABM Treaty of 1972, which explicitly prohibits the development and deployment of nationwide ballistic missile defenses. In consequence, the security of the United States would be diminished, not increased.

In contrast to "enhancing deterrence," the combination of our offensive strength and the damage-limiting potential of our defenses could be seen as giving us more flexibility to use nuclear weapons to punish, or threaten to punish, Soviet actions we consider unacceptable. In other words, the threshold at which we would consider the use of nuclear weapons could be lowered.

The Soviets may see our defensive programs, accompanied by our ongoing intensive effort to modernize and improve our offensive forces, as evidence of preparations for a first strike—a first strike that would leave them with a weakened retaliatory force against which our defenses, although imperfect, would be relatively more effective. Of course, the same would be true if the roles of the Soviet Union and the United States were interchanged. . .

Another distinct objective of ballistic missile defense, known as hardsite defense, is the protection of elements of the U.S. strategic retaliatory forces, in particular hardened missile silos and hardened command centers. Technically, this job is easier than the broader area defense of population and industry described above because only selected targets have to be protected. Moreover, hardened targets can be damaged only by a direct hit or a hit within the immediate vicinity. Consequently, there is a narrow "threat tube," or region through which the attacking missiles must pass, to destroy a hardened target. Furthermore, protection of the retaliatory forces does not require a high standard of effectiveness: if a reasonable fraction of the retaliatory forces survive, the attacker can expect assured retaliation with an unacceptable level of damage.

The strategic mission of protecting retaliatory forces with a hardsite defense is very different from that of complete area defense for population protection envisaged by the president. The president's goal was to shift the burden for maintaining a strategic balance from offensive to defensive weapons. By contrast, protection of the retaliatory forces maintains the dominant role of offensive weapons on which the current strategic military balance is based: it enhances offense dominance by preserving the survival of offensive missiles and ensuring their capability to retaliate against the opponent. . . . Furthermore, a technology development program for hardsite defense would have major differences from one directed toward realizing the president's vision of rendering "nuclear weapons impotent and obsolete."

. . . It has frequently been suggested that the United States should place greater emphasis on ballistic missile defense because the Soviets are undertaking an active program in this area. Yet if the United States concluded that the Soviet effort gave them a significant military edge or threatened strategic stability or U.S. security, the logical U.S. response should be to counter the Soviet threat, not to emulate it. Historically, the Soviet Union has dedicated a much larger fraction of its strategic military effort to defensive programs—motivated in part by the Russian tradition of "protecting the homeland." Yet there is little, if any, disagreement among military analysts that neither the Soviet's extensive air defense nor the Moscow missile defense offers effective protection against U.S. retaliatory forces. . . . In short, there is no need for the United States to match the Soviet ABM effort on the basis of its technical merit. Even if it were desirable for purely political reasons to keep pace with the Soviets, it would be difficult to justify an expansion of the current U.S. research and technology effort on that basis. The possibility that the Soviet program might result in new discoveries cannot be ignored, however. Therefore, a deliberate U.S. research and technology program can be justified to allow us to interpret Soviet progress and to prevent "technological surprise."

Ballistic missile defense systems can be deployed on the ground or in space, or both. . . . The Strategic Defense Initiative is considering a multitiered defense against ballistic missiles based in space and on Earth, with intercept layers proposed during the boost and post-boost (or bus) phase, during midcourse, and during warhead reentry. The purpose of such a multilayered defense is to achieve a very high level of overall system performance, beyond what is practical to achieve at each individual stage of intercept, and to attenuate the incoming force of warheads in successive steps.

Basing interceptors in space has advantages and disadvantages. A major advantage is that in space the atmosphere does not interfere with the propagation of the means of intercept—either such directed-energy weapons as light beams or particles, or more massive projectiles. A serious disadvantage is that space-based platforms are expensive and vulnerable—in fact, more vulnerable than ballistic missiles. Satellites in space move in precisely predictable orbits, which can be kept under observation for protracted periods. Thus an attack on space-based defenses can be executed much more deliberately than can attacks on ballistic missiles. This vulnerability is so fundamental that many have doubted the practicality of space-based battle stations for a viable ballistic missile defense.[1]

Inevitably, the basing of defensive systems in space would initiate a round of offense-defense competition in which the offense would seek to develop various options for attacking the space platforms (using a combination of antisatellite measures) while the defense would attempt to harden the space platforms and engage in evasive or otherwise protective tactics. Space-basing would not offer a simple way of deploying a defensive system. More likely it would stimulate further escalation of the arms race into space. Without more study and even in-depth experimentation one cannot draw firm conclusions on the outcome of such competition—if there were a clear outcome. One can conclude, however, that escalation of such competition into space would be expensive and also dangerous to those space systems now used for surveillance, early warning, communication, navigation, and commerce.

Based on those facts now known, our judgment is that placing defensive systems on space platforms would constitute a highly undesirable escalation of the arms competition. It also appears quite unpromising on technical grounds. Under no circumstances should the remote possibility of using space for ballistic missile defense purposes be accepted as a valid argument against proceeding in serious, good-faith negotiations with the Soviet Union toward banning weapons from space. . . .

An alternative would be to deploy a pop-up system based on ground. In principle, an x-ray laser pumped by a nuclear explosion is sufficiently small to be launched by a missile on detection of an enemy attack. Such a system would have to be based quite near Soviet territory, offshore from the United States, in order to make it geometrically possible to achieve an intercept. Moreover, like any other system for boost-phase intercept, a pop-up system would have to operate on an automated basis, given the minimal time available between detection of booster launch and commitment of the interceptors. This raises profound policy questions about whether such nuclear devices should be prepositioned close to Soviet shores and launched without human intervention.

To avoid the problems associated with the boost-phase intercept options discussed above, various hybrid systems have been proposed. The most prominent concept calls for large mirrors placed in space to focus laser beams onto enemy targets. The laser beams would be generated by a series of ground stations, widely dispersed to hedge against adverse weather conditions, or located on mountain tops above the weather. The beams from the ground would be focused to large relay mirrors in synchronous orbit, an altitude of 36,000 kilometers. To focus the beam would require active optics systems that could compensate for the atmospheric disturbances. The relay mirrors would also have to be aimed precisely and continuously at various mission mirrors in lower earth orbit. These, in turn, would focus the beam onto the boost vehicles. In principle, this system might indeed direct damaging levels of energy at the boost vehicle, although the technical requirements are quite severe. However, relay and mission mirrors are subject to all the vulnerabilities for space-borne stations discussed previously. If some of the mission mirrors were launched into lower earth orbit on detection of an attack, there would still be the problem of the short engagement time characteristic of all pop-up systems. The economics and practicality of this system cannot really be analyzed at this stage of development. . . .

DANGER OF CONFRONTATION WITH ARMS CONTROL TREATIES

The administration has been careful to describe its current program as research, and the activities it proposes for the next fiscal year appear to be in compliance with obligations under existing treaties. However, if the Strategic Defense Initiative were to expand in the future from its current research and technology focus through prototype and systems testing to actual deployment, several arms control treaties would be at risk. These are the 1972 ABM Treaty and its associated 1974 protocol, the 1963 Limited Test Ban Treaty, and the Outer Space Treaty of 1967. . . .

Neither renegotiation nor abrogation of the three treaties mentioned above can be taken lightly for a number of reasons. First, these treaties have served U.S. national security well. Second, should outright abrogation occur the concomitant political price would be heavy. Third, should renegotiation be initiated at the request of the United States, but with the Soviet Union disclaiming interest in such a move, the United States would have to pay a price in concessions to the Soviet Union to arrive at settlement. . . .

The Strategic Defense Initiative has not been greeted with expressions of support from the NATO alliance. This is not surprising, as the ABM Treaty is of great value to our nuclear-armed allies—especially the British and French—because it ensures the ability of their missiles to reach their targets. Were the treaty modified or abrogated, resulting in the expansion of Soviet missile defenses, the independent and much smaller deterrent forces of our allies would lose their effectiveness much sooner than would U.S. forces.

Clearly, careful consideration will have to be given to the future of the ABM Treaty should the activities of the Strategic Defense Initiative provide promising results.

THE RISK OF ESCALATION

Whatever the long-range objective of the Strategic Defense Initiative—be it President Reagan's vision of complete protection against nuclear weapons or the more recently stated goal of enhanced deterrence—during the actual deployment of a ballistic missile defense we will face a protracted time during which the offense dominance of the strategic balance is still a reality. Even optimistic projections of the development and deployment of only a partially effective missile defense, ignoring any treaty obstacles, span a decade or more. In contemplating the wisdom of deployment, we must factor in the likely Soviet responses and their impact on stability and escalation of the arms race.

The Soviets may respond by emulating the U.S. defense initiative or by taking specific offensive countermeasures against the U.S. moves, or both. Whichever choice they make, the Soviets will react during the development time to U.S. official statements as well as to emerging technologies and their potential military effectiveness to judge whether the United States remains deterred from initiating nuclear war under all conceivable conditions. To Soviet leaders a crucial question will be, "Does the U.S. leadership still recognize that it would be suicidal to start a nuclear exchange?"

. . . A further serious risk is that the effort to neutralize the effect of nuclear weapons by deploying nationwide missile defenses might make the use of such weapons appear to be more acceptable, thereby deflecting efforts to reduce through negotiations the dangers and burdens of arms competition.

On the positive side, *after* an effective arms control regime has been established through negotiated treaties and unilateral restraint, and once the level of nuclear weaponry has been reduced drastically from today's levels of more than 50,000 warheads, we can see a stabilizing role for ballistic missile defense. In particular, if the nuclear stockpiles were reduced to very low levels, a defense system would make the security of the United States and its allies less sensitive to the precise intelligence information on small numbers of weapons retained by an opponent or third parties.

In this context, the prospect of defense can add important support to negotiations leading to low levels of nuclear weaponry. However, missile defense deployments cannot *precede* such a reduction of offensive forces or be intertwined with it without incurring all the risks of instability and escalation discussed above. Therefore, we still conclude that a reversal from the current offense-dominated balance to defense-dominated stability between the United States and the Soviet Union cannot occur—"We can't get there from here"—until offensive nuclear weapons have been reduced to extremely low levels worldwide through negotiation and prudent restraint. . . .

NOTES

1. See Edward Teller, "Bringing Star Wars Down to Earth," in *Popular Mechanics*, July 1984, 84–122.

CHAPTER 10

Arms Control

Whether the pessimists were correct about the disappearance of comprehensive agreements depends too much on the evolution of Soviet-American relations to allow an early judgment. Their conclusions may prove to have been too gloomy, an outcome that would not, of course, disappoint them. However, the efforts of others to develop arms control methods that do not depend on comprehensive agreements deserve closer examination. Of the many suggested approaches, those of Alton Frye for *strategic build-down* and of four former high-ranking U.S. officials—McGeorge Bundy, George Kennan, Robert McNamara, and Gerard Smith—for *no first use* of nuclear weapons by the United States have attracted the most attention.

The decision of the four former U.S. policymakers to endorse no first use was the new element; the idea itself had been considered repeatedly over the years since Kennan's memorandum to Acheson in 1950. The strength of the proposal was that it touched U.S. and allied strategy directly, unlike the various proposals for nuclear freeze and nuclear free-zones. If the United States and its allies were to adopt no first use, they would immediately have to begin to make far-reaching changes in their approach to combat in various theaters, particularly in Western Europe, and to make major changes in their military forces, especially in the size and readiness of conventional troop formations and their support elements. The weakness of the idea lay in Western European opposition. There are three important reasons for this response by the allies. First, they fear the change would hand a decisive advantage to the Soviet Union, because of the large and potent conventional forces it has deployed in central Europe. Moreover, adoption of no first use would negate the value of the nuclear arsenals to which France and Britain have devoted so much of their resources. Finally, there is little sign of a willingness to pay the significantly higher costs of strengthening NATO conventional forces.

The stature of the U.S. advocates caused a stir in the United States and Europe. Partly in response to the criticism he and his fellow authors received, Robert McNamara revealed his belief that, when secretary of defense, he had persuaded Presidents Kennedy and Johnson that it would *never* be to the advantage of the Atlantic Alliance to initiate nuclear war.

That this revelation provoked none of the condemnation that had greeted the public proposal suggested that the commitment to first use derived from a determination to increase Soviet uncertainty rather than from an intention to resort to nuclear war.

Alton Frye's proposal for strategic build-down linked arms control to the modernization of strategic nuclear forces. For each new strategic delivery system deployed, the Soviet Union and the United States would each agree to make an appropriate reduction in its existing forces. The formula to be applied would favor a build-down toward smaller, un-MIRVed, and invulnerable weapons and toward fewer weapons. It is an attractive and ingenious approach. It even won inclusion in the formal public position of the Reagan administration on START. However, build-down and all the other plans await a significant improvement in Soviet-American relations.

50. Report to the Congress of the Communist Party of the Soviet Union

Leonid Brezhnev

Comrades! In the period under review, the USSR continued to actively pursue a Leninist policy of peaceful coexistence and mutually advantageous cooperation with the capitalist states, while dealing a firm rebuff to the aggressive intrigues of imperialism.

A further exacerbation of the general crisis of capitalism took place during these years. Capitalism's development has not frozen, of course. But it is going through its third economic slump in the past 10 years. . . .

Social contradictions have become noticeably more acute. In conditions of capitalist society, the employment in production of the latest scientific and technical achievements is turning against the working people and throwing millions of people out of the factories and plants. In the past decade, the army of unemployed in the developed capitalist countries has doubled. In 1980 it totaled 19 million.

Attempts to relieve the intensity of the class struggle through various social reforms are having no success either. The number of strikers grew by more than one-third over the decade, to a total, according to official figures, of 250 million.

Interimperialist contradictions are growing more acute and the struggle for markets and for raw-material and energy sources is intensifying. Japanese and West European monopolies are competing more and more successfully with American capital, including on the U.S. domestic market. In the 1970s, the U.S.'s share of the world exports declined by almost 20%.

The difficulties that capitalism is experiencing also have an effect on its policy, including foreign policy. The struggle over basic questions of the capitalist countries' foreign-policy course has become more acute. Opponents of détente, of limiting armaments and of improving relations with the Soviet Union and the other socialist countries have noticeably stepped up their activity of late. . . .

Unfortunately, since the change of leadership in the White House openly bellicose appeals and statements, specially designed, as it were, to poison the atmosphere of relations between our countries, have continued to emanate from Washington. However, we would like to hope that those who are setting American policy today will ultimately be able to see things more realistically. The military-strategic equilibrium that exists between the USSR and the U.S. and between the Warsaw Treaty and NATO objectively serves to preserve peace on our planet. We have not sought, and do not now seek, military superiority over the other side. This is not our policy. But neither will we allow such superiority to be created over us. Such attempts, as well as talking to us from positions of strength, are absolutely futile! (*Prolonged applause.*)

Not trying to upset the existing equilibrium, not imposing a new, still more expensive and dangerous round in the arms race—that would be a manifestation of truly statesmanlike wisdom. To this end, it is really high time to throw the decrepit scarecrow of the "Soviet threat" out the door of serious politics.

Let's look at the actual state of affairs.

Whether one takes strategic nuclear arms or medium-range nuclear weapons in Europe, in both cases approximate equality exists between the two sides. In some types of weapons, the West has a certain advantage. In others, we have the edge. This equality could be more enduring if relevant treaties and agreements were concluded.

There is also talk about tanks. It's true that the Soviet Union has more of them. However, the NATO countries also have a good many. Besides, they have considerably more antitank weapons.

The allegation of Soviet superiority in total number of armed forces doesn't correspond to reality either. The U.S. and the other NATO countries have slightly more troops than the USSR and the other Warsaw Treaty countries.

So, what kind of Soviet military superiority can one talk about?

The danger of war does indeed hang over the U.S., as it does over all other countries in the world. But its source is not the Soviet Union, not its mythical superiority, but the arms race itself and the continuing tension in the world. We are prepared to combat this genuine, not imaginary, danger—hand in hand with America, with the European states, with all countries on our planet. To try to prevail over the other side in the arms race or to count on victory in a nuclear war is dangerous madness. (*Applause.*)

It is generally recognized that the international situation depends in large part on the policies of the USSR and the U.S. In our opinion, the state of relations between the two at present and the crucial nature of the international problems that require solution dictate the need for an active dialogue on all levels. We are ready for such a dialogue.

Experience shows that the decisive factor here is summit meetings. This was true yesterday, and it remains true today. (*Applause.*)

The USSR wants normal relations with the U.S. From the standpoint of the interests of the peoples of both our countries and of mankind as a whole, there is simply no other sensible way. . . . (*Applause.*)

Probably no other state has in recent years put before mankind as broad a spectrum of concrete and realistic initiatives on highly important problems of international relations as the Soviet Union has.

I shall begin with the problem of the limitation of nuclear arms, the most dangerous to mankind. Throughout these years, the Soviet Union resolutely fought to put an end to the race in these arms and to rule out their proliferation on the planet. As you know, a vast amount of work was done to prepare the Treaty on the Limitation of Strategic Arms with the U.S. A great deal was accomplished during the talks with the U.S. and Britain on the complete prohibition of nuclear weapons tests. We took an important action when we stated and reaffirmed that we will not use nuclear weapons against nonnuclear countries that do not permit them to be deployed on their territory. But we have also proposed something more: that the production of nuclear weapons be stopped and the reduction of stockpiles of these weapons be started and continued to their complete elimination.

The Soviet Union has also actively worked for the prohibition of all other types of weapons of mass destruction. Some things along these lines were achieved during the period under review. A convention prohibiting the modification of the environment for military purposes went into effect. Preliminary agreement was reached on the basic provisions of a treaty banning radiological weapons. Talks on excluding chemical weapons from the arsenals of states are continuing, although at an impermissibly slow pace. Actions by peace-loving forces managed to halt the implementation of plans for the deployment of neutron weapons in Western Europe. This has made the peoples' indignation all the greater at the Pentagon's renewed attempts to hang over the European countries the sword of Damocles that these weapons represent. For our part, we reaffirm that we will not begin the production of neutron weapons if they do not appear in other states, and that we are prepared to conclude an agreement banning these weapons once and for all.

The Soviet Union and the other Warsaw Treaty countries have made a number of concrete proposals concerning military détente in Europe. In particular, we would like the participants in the all-European conference to pledge not to be the first to use either nuclear or conventional arms against any other participant; we do not want the existing military blocs in Europe and on other continents to be expanded or any blocs to be created. . . .

In Europe, for example, this purpose is to some extent served—and, on the whole, served fairly well—by the confidence-building measures in the military field that are being carried out by decision of the all-European conference. These measures include advance notification on military exercises by ground forces and the invitation to such exercises of observers from other countries. At present, these measures are carried out on the territory of the European states, including the western regions of the USSR. We have already said that we are prepared to go further—to also provide notification on naval and air exercises. We have proposed—and we propose again—that notification apply to large-scale troop movements as well.

Now we want to propose that the zone in which these measures apply be expanded substantially. *We are prepared to extend them to the entire European part of the USSR—provided that there is a corresponding expansion in the zone of confidence-building measures on the part of the Western states.*

The limitation and reduction of strategic arms is a problem of the highest importance. *For our part, we are prepared to continue appropriate talks with the U.S. without delay, preserving all the positive elements that have been achieved in this field so far.* Needless to say, the talks can be conducted only on the basis of equality and equal security. We will not enter into any agreement that gives a unilateral advantage to the U.S. There should be no illusions here. In our opinion, all the other nuclear powers should join these talks at the appropriate time. . . .

We propose that an agreement be reached to set a moratorium right now on the deployment in Europe of new medium-range nuclear missiles by the NATO countries and the USSR, i.e., that a quantitative and qualitative freeze be put on the existing level of these weapons—including, needless to say, the U.S.'s forward-based nuclear weapons in this region. This moratorium could go into effect at once, as soon as talks on this question begin, and last until a permanent treaty is concluded on the limitation or, better yet, the reduction of these nuclear weapons in Europe. In making this proposal, we proceed from the premise that both sides will stop all preparations for the deployment of additional weapons of the types covered, including American Pershing-2 missiles and ground-based strategic cruise missiles. . . .

51. Nuclear Weapons and the Atlantic Alliance

McGeorge Bundy, George F. Kennan, Robert S. McNamara, and Gerard Smith

For 33 years now, the Atlantic Alliance has relied on the asserted readiness of the United States to use nuclear weapons if necessary to repel aggression from the East. Initially, indeed, it was widely thought (notably by such great and different men as Winston Churchill and Niels Bohr) that the basic military balance in Europe was between American atomic bombs and the massive conventional forces of the Soviet Union. . . .

A major element in every doctrine has been that the United States has asserted its willingness to be the first—has indeed made plans to be the first if necessary—to use nuclear weapons to defend against aggression in Europe. It is this element that needs reexamination now. Both its cost to the coherence of the Alliance and its threat to the safety of the world are rising while its deterrent credibility declines.

This policy was first established when the American nuclear advantage was overwhelming, but that advantage has long since gone and cannot be recaptured. As early as the 1950s it was recognized by both Prime Minister Churchill and President Eisenhower that the nuclear strength of both sides was becoming so great that a nuclear war would be a ghastly catastrophe for all concerned. The following decades have only confirmed and intensified that reality. The time has come for careful study of the ways and means of moving to a new Alliance policy and doctrine: that nuclear weapons will not be used unless an aggressor should use them first. . . .

It is time to recognize that no one has ever succeeded in advancing any persuasive reason to believe that any use of nuclear weapons, even on the smallest scale, could reliably be expected to remain limited. Every serious analysis and every military exercise, for over 25 years, has demonstrated that even the most restrained battlefield use would be enormously destructive to civilian life and property. There is no way for anyone to have any confidence that such a nuclear action will not lead to further and more devastating exchanges. Any use of nuclear weapons in Europe, by the Alliance or against it, carries with it a high and inescapable risk of escalation into the general nuclear war which would bring ruin to all and victory to none.

The one clearly definable firebreak against the worldwide disaster of general nuclear war is the one that stands between all other kinds of conflict and any use whatsoever of nuclear weapons. To keep that firebreak wide and strong is in the deepest interest of all mankind. . . .

The largest question presented by any proposal for an Allied policy of no-first-use is that of its impact on the effectiveness of NATO's deterrent posture on the central front. In spite of the doubts that are created by any honest look at the probable consequences of resort to a first nuclear strike of any kind, it should be remembered that there were strong reasons for the creation of the American nuclear umbrella over NATO. The original American pledge, expressed in Article 5 of the Treaty, was understood to be a nuclear guarantee. It was extended at a time when only a conventional Soviet threat existed, so a readiness for first use was plainly implied from the beginning. To modify that guarantee now, even in the light of all that has happened since, would be a major change in the assumptions of the Alliance, and no such change should be made without the most careful exploration of its implications.

In such an exploration the role of the Federal Republic of Germany must be central. Americans too easily forget what the people of the Federal Republic never can: that their position is triply exposed in a fashion unique among the large industrial democracies. They do not have nuclear weapons; they share a long common boundary with the Soviet empire; in any conflict on the central front their land would be the first battleground. None of these conditions can be changed, and together they present a formidable challenge.

Having decisively rejected a policy of neutrality, the Federal Republic has necessarily relied on the nuclear protection of the United States, and we Americans should recognize that this relationship is not a favor we are doing our German friends, but the best available solution of a common problem. Both nations believe that the Federal Republic must be defended; both believe that the Federal Republic must not have nuclear weapons of its own; both believe that nuclear guarantees *of some sort* are essential; and both believe that only the United States can provide those guarantees in persuasively deterrent peacekeeping form. . . .

The quite special character of the nuclear relationship between the Federal Republic and the United States is a most powerful reason for defining that relationship with great care. It is rare for one major nation to depend entirely on another for a form of strength that is vital to its survival. It is unprecedented for any nation, however powerful, to pledge itself to a course of action, in defense of another, that might entail its own nuclear devastation. A policy of no-first-use would not and should not imply an abandonment of this extraordinary guarantee—only its redefinition. It would still be necessary to be ready to reply with American nuclear weapons to any nuclear attack on the Federal Republic, and this commitment would in itself be sufficiently demanding to constitute a powerful demonstration that a policy of no-first-use would represent no abandonment of our German ally. . . .

While we believe that careful study will lead to a firm conclusion that it is time to move decisively toward a policy of no-first-use, it is obvious that any such policy would require a strengthened confidence in the adequacy of the conventional forces of the Alliance, above all the forces in place on the central front and those available for prompt reinforcement. It seems clear that the nations of the Alliance together can provide whatever forces are needed, and within realistic budgetary constraints, but it is a quite different question whether they can summon the necessary political will. Evidence from the history of the Alliance is mixed. There has been great progress in the conventional defenses of NATO in the 30 years since the 1952 Lisbon communiqué, but there have also been failures to meet force goals all along the way. . . .

There should also be an examination of the ways in which the concept of early use of nuclear weapons may have been built into existing forces, tactics, and general military expectations. To the degree that this has happened, there could be a dangerous gap right now between real capabilities and those which political leaders might wish to have in a time of crisis. Conversely there should be careful study of what a policy of no-first-use would require in those same terms. It seems more than likely that once the military leaders of the Alliance have learned to think and act steadily on this "conventional" assumption, their forces will be better instruments for stability in crises and for general deterrence, as well as for the maintenance of the nuclear firebreak so vital to us all.

No one should underestimate either the difficulty or the importance of the shift in military attitudes implied by a no-first-use policy. . . .

An Allied posture of no-first-use would have one special effect that can be set forth in advance: it would draw new attention to the importance of maintaining and improving the specifically American conventional forces in Europe. The principal political difficulty in a policy of no-first-use is that it may be taken in Europe, and especially in the Federal Republic, as evidence of a reduced American interest in the Alliance and in effective overall deterrence. The argument here is exactly the opposite: that such a policy is the best one available for keeping the Alliance united and effective. Nonetheless the psychological realities of the relation between the Federal Republic and the United States are such that the only way to prevent corrosive German suspicion of American intentions, under a no-first-use regime, will be for Americans to accept for themselves an appropriate share in any new level of conventional effort that the policy may require.

Yet it would be wrong to make any hasty judgment that those new levels of effort must be excessively high. The subject is complex, and the more so because both technology and politics are changing. Precision-guided munitions, in technology, and the visible weakening of the military solidity of the Warsaw Pact, in politics, are only two examples of changes working to the advantage of the Alliance. Moreover there has been some tendency, over many years, to exaggerate the relative conventional strength of the U.S.S.R. and to underestimate Soviet awareness of the enormous costs and risks of any form of aggression against NATO. . . .

An especially arbitrary, if obviously convenient, measure of progress is that of spending levels. But it is political will, not budgetary pressure, that will be decisive. . . .

The direction of the Allied effort will be more important than its velocity. The final establishment of a firm policy of no-first-use, in any case, will obviously require time. What is important today is to begin to move in this direction.

The concept of renouncing any first use of nuclear weapons should also be tested by careful review of the value of existing NATO plans for selective and limited use of nuclear weapons. While many scenarios for nuclear war-fighting are nonsensical, it must be recognized that cautious and sober senior officers have found it prudent to ask themselves what alternatives to defeat they could propose to their civilian superiors if a massive conventional Soviet attack seemed about to make a decisive breakthrough. This question has generated contingency plans for battlefield uses of small numbers of

nuclear weapons which might prevent that particular disaster. It is hard to
see how any such action could be taken without the most enormous risk
of rapid and catastrophic escalation, but it is a fair challenge to a policy
of no-first-use that it should be accompanied by a level of conventional
strength that would make such plans unnecessary.

In the light of this difficulty it would be prudent to consider whether
there is any acceptable policy short of no-first-use. One possible example
is what might be called "no-*early*-first-use"; such a policy might leave open
the option of some limited nuclear action to fend off a final large-scale
conventional defeat, and by renunciation of any immediate first use and
increased emphasis on conventional capabilities it might be thought to help
somewhat in reducing current fears.

But the value of a clear and simple position would be great, especially
in its effect on ourselves and our Allies. One trouble with exceptions is
that they easily become rules. It seems much better that even the most
responsible choice of even the most limited nuclear actions to prevent even
the most imminent conventional disaster should be left out of authorized
policy. What the Alliance needs most today is not the refinement of its
nuclear options, but a clear-cut decision to avoid them as long as others
do. . . .

The first possible advantage of a policy of no-first-use is in the management
of the nuclear deterrent forces that would still be necessary. Once we escape
from the need to plan for a first use that is credible, we can escape also
from many of the complex arguments that have led to assertions that all
sorts of new nuclear capabilities are necessary to create or restore a capability
for something called "escalation dominance"—a capability to fight and "win"
a nuclear war at any level. What would be needed, under no-first-use, is
a set of capabilities for appropriate retaliation to any kind of Soviet nuclear
attack which would leave the Soviet Union in no doubt that it too should
adhere to a policy of no-first-use. The Soviet government is already aware
of the awful risk inherent in any use of these weapons, and there is no
current or prospective Soviet "superiority" that would tempt anyone in
Moscow toward nuclear adventurism. (All four of us are wholly unpersuaded
by the argument advanced in recent years that the Soviet Union could ever
rationally expect to gain from such a wild effort as a massive first strike
on land-based American strategic missiles.)

Once it is clear that the only nuclear need of the Alliance is for adequately
survivable and varied *second strike* forces, requirements for the modernization
of major nuclear systems will become more modest than has been assumed. . . .

A posture of no-first-use should also go far to meet the understandable
anxieties that underlie much of the new interest in nuclear disarmament,
both in Europe and in our own country. Some of the proposals generated
by this new interest may lack practicability for the present. For example,
proposals to make "all" of Europe—from Portugal to Poland—a nuclear-
free zone do not seem to take full account of the reality that thousands
of long-range weapons deep in the Soviet Union will still be able to target

Western Europe. But a policy of no-first-use, with its accompaniment of a reduced requirement for new Allied nuclear systems, should allow a considerable reduction in fears of all sorts. Certainly such a new policy would neutralize the highly disruptive argument currently put about in Europe: that plans for theater nuclear modernization reflect an American hope to fight a nuclear war limited to Europe. Such modernization might or might not be needed under a policy of no-first-use; that question, given the size and versatility of other existing and prospective American forces, would be a matter primarily for European decision (as it is today).

An effective policy of no-first-use will also reduce the risk of conventional aggression in Europe. That risk has never been as great as prophets of doom have claimed and has always lain primarily in the possibility that Soviet leaders might think they could achieve some quick and limited gain that would be accepted because no defense or reply could be concerted. That temptation has been much reduced by the Allied conventional deployments achieved in the last 20 years, and it would be reduced still further by the additional shift in the balance of Allied effort that a no-first-use policy would both permit and require. The risk that an adventurist Soviet leader might take the terrible gamble of conventional aggression was greater in the past than it is today, and is greater today than it would be under no-first-use, backed up by an effective conventional defense.

We have been discussing a problem of military policy, but our interest is also political. The principal immediate danger in the current military posture of the Alliance is not that it will lead to large-scale war, conventional or nuclear. The balance of terror and the caution of both sides appear strong enough today to prevent such a catastrophe, at least in the absence of some deeply destabilizing political change which might lead to panic or adventurism on either side. But the present unbalanced reliance on nuclear weapons, if long continued, might produce exactly such political change. The events of the last year have shown that differing perceptions of the role of nuclear weapons can lead to destructive recriminations, and when these differences are compounded by understandable disagreements on other matters such as Poland and the Middle East, the possibilities for trouble among Allies are evident.

The political coherence of the Alliance, especially in times of stress, is at least as important as the military strength required to maintain credible deterrence. . . .

There remains one underlying reality which could not be removed by even the most explicit declaratory policy of no-first-use. Even if the nuclear powers of the Alliance should join, with the support of other Allies, in a policy of no-first-use, and even if that decision should lead to a common declaration of such policy by these powers and the Soviet Union, no one on either side could guarantee beyond all possible doubt that if conventional warfare broke out on a large scale there would in fact be no use of nuclear weapons. We could not make that assumption about the Soviet Union, and we must recognize that Soviet leaders could not make it about us. As long as the weapons themselves exist, the possibility of their use will remain.

But this inescapable reality does not undercut the value of a no-first-use policy. That value is first of all for the internal health of the Western Alliance itself. A posture of effective conventional balance and survivable second-strike nuclear strength is vastly better for our own peoples and governments, in a deep sense more civilized, than one that forces the serious contemplation of "limited" nuclear scenarios that are at once terrifying and implausible.

There is strong reason to believe that no-first-use can also help in our relations with the Soviet Union. The Soviet government has repeatedly offered to join the West in declaring such a policy, and while such declarations may have only limited reliability, it would be wrong to disregard the real value to both sides of a jointly declared adherence to this policy. To renounce the first use of nuclear weapons is to accept an enormous burden of responsibility for any later violation. The existence of such a clearly declared common pledge would increase the cost and risk of any sudden use of nuclear weapons by either side and correspondingly reduce the political force of spoken or unspoken threats of such use.

A posture and policy of no-first-use also could help to open the path toward serious reduction of nuclear armaments on both sides. The nuclear decades have shown how hard it is to get agreements that really do constrain these weapons, and no one can say with assurance that any one step can make a decisive difference. . . .

Finally, and in sum, we think a policy of no-first-use, especially if shared with the Soviet Union, would bring new hope to everyone in every country whose life is shadowed by the hideous possibility of a third great twentieth-century conflict in Europe—conventional or nuclear. . . .

52. Nuclear Weapons and the Preservation of Peace

Karl Kaiser, Georg Leber, Alois Mertes, and Franz-Josef Schulze

A RESPONSE TO AN AMERICAN PROPOSAL
FOR RENOUNCING THE FIRST USE
OF NUCLEAR WEAPONS

When McGeorge Bundy, George F. Kennan, Robert S. McNamara and Gerard Smith submit a proposal to renounce the first use of nuclear weapons in Europe, the mere fact that it comes from respected American personalities with long years of experience in questions of security policy and the Alliance gives it particular weight. Their reflections must be taken particularly seriously in a country like the Federal Republic of Germany which has a special interest in preserving peace, because in case of war nuclear weapons could first be used on its territory. . . .

. . . What matters most is to concentrate not only on the prevention of nuclear war, but on how to prevent *any* war, conventional war as well. The decisive criterion in evaluating this proposal—like any new proposal—must be: Will it contribute to preserving, into the future, the peace and freedom of the last three decades?

Unfortunately, the current discussion on both sides of the Atlantic about the four authors' proposal has been rendered more difficult by a confusion between the option of the "first use" of nuclear weapons and the capability for a "first strike" with nuclear weapons. The authors themselves have unintentionally contributed to this confusion by using both terms. "First use" refers to the first use of a nuclear weapon regardless of its yield and place; even blowing up a bridge with a nuclear weapon in one's own territory would represent a first use. "First strike" refers to a preemptive disarming nuclear strike aimed at eliminating as completely as possible the entire strategic potential of the adversary. A first strike by the Alliance is not a relevant issue; such a strike must remain unthinkable in the future as it is now and has been in the past. The matter for debate should be exclusively the defensive first use of nuclear weapons by the Western Alliance.

The current NATO strategy of flexible response is intended to discourage an adversary from using or threatening the use of military force by confronting him with a full spectrum of deterrence and hence with an uncalculable risk. The strategy also aims at improving the tools of crisis management as a means of preventing conflict. The deterrent effect of the doctrine rests on three pillars:

- the political determination of all Alliance members to resist jointly any form of aggression or blackmail;
- the capability of the Alliance to react effectively at every level of aggression; and
- the flexibility to choose between different possible reactions—conventional or nuclear.

The primary goal of this strategy is the prevention of war. To this end it harnesses the revolutionary new and inescapable phenomenon of the nuclear age for its own purposes. Our era has brought humanity not only the curse of the unprecedented destructive power of nuclear weapons but also its twin, the dread of unleashing that power, grounded in the fear of self-destruction. Wherever nuclear weapons are present, war loses its earlier function as a continuation of politics by other means. Even more, the destructive power of these weapons has forced political leaders, especially those of nuclear weapons states, to weigh risks to a degree unknown in history.

The longest period of peace in European history is inconceivable without the war-preventing effect of nuclear weapons. . . .

The strategy of flexible response attempts to counter any attack by the adversary—no matter what the level—in such a way that the aggressor can have no hope of advantage or success by triggering a military conflict, be it conventional or nuclear. The tight and indissoluble coupling of conventional forces and nuclear weapons on the European continent with the strategic potential of the United States confronts the Soviet Union with the incalculable risk that any military conflict between the two Alliances could escalate to a nuclear war. The primary function of nuclear weapons is deterrence in order to prevent aggression and blackmail.

The coupling of conventional and nuclear weapons has rendered war between East and West unwageable and unwinnable up to now. It is the inescapable paradox of this strategy of war prevention that the will to conduct nuclear war must be demonstrated in order to prevent war at all. Yet the ensuing indispensable presence of nuclear weapons and the constantly recalled visions of their possible destructive effect, should they ever be used in a war, make many people anxious.

The case is similar with regard to the limitation of nuclear war: the strategy of massive retaliation was revised because, given the growing potential of destruction, the threat of responding even to low levels of aggression with a massive use of nuclear weapons became increasingly incredible. A threat once rendered incredible would no longer have been able to prevent war in Europe. Thus, in the mid-1960s the Europeans supported the introduction of flexible response, which made the restricted use of nuclear weapons—but also the limitation of any such use—an indispensable part of deterrence aimed at preventing even "small" wars in Europe.

A renunciation of the first use of nuclear weapons would certainly rob the present strategy of war prevention—which is supported by the government and the opposition in the Federal Republic of Germany, as well as by a great majority of the population—of a decisive characteristic. One cannot help concluding that the Soviet Union would thereby be put in a position where it could, once again, calculate its risk and thus be able to wage war in Europe. It would no longer have to fear that nuclear weapons would inflict unacceptable damage to its own territory. We therefore fear that a credible renunciation of the first use of nuclear weapons would, once again, make war more probable. . . .

It must be questioned, therefore, whether renunciation of first use represents a contribution to the "internal health of the Western alliance itself" or whether, instead, a no-first-use policy increases insecurity and fear of ever more probable war.

The conclusions that can be drawn from the four authors' recommendations with regard to the commitment of the United States to the defense of Europe are profoundly disturbing. To be sure, they assert that no-first-use does not represent an abandonment of the American protective guarantee for Western Europe, but "only its redefinition." Indeed, that would be the case, but in the form of a withdrawal from present commitments of the United States.

The opinion of the four American authors that "the one clearly definable firebreak against the worldwide disaster of general nuclear war is the one that stands between all other kinds of conflict and any use whatsoever of nuclear weapons" amounts to no less than limiting the existing nuclear guarantee of protection by the United States for their non-nuclear Alliance partners to the case of prior use of nuclear weapons by the Soviet Union. Even in the case of a large-scale conventional attack against the entire European NATO territory, the Soviet Union could be certain that its own land would remain a sanctuary as long as it did not itself resort to nuclear weapons. This would apply even more to surprise operations aimed at the quick occupation of parts of Western Europe which are hardly defensible by conventional means.

In such a case, those attacked would have to bear the destruction and devastation of war alone. It is only too understandable that for years the Soviet Union has, therefore, pressed for a joint American-Soviet renunciation of first use of nuclear weapons, on occasion in the guise of global proposals. If the ideas of the authors were to be followed, conventional conflicts in Europe would not longer involve any existential risk for the territory of the Soviet Union and—despite the increased American participation in the conventional defense of Europe suggested by the authors—would be without such risk for the territory of the United States as well.

The authors' suggestion that "even the most responsible choice of even the most limited nuclear actions to prevent even the most imminent conventional disaster should be left out of authorized policy" makes completely clear that a withdrawal of the United States from its previous guarantee is at stake. They thus advise Western Europe to capitulate should defeat threaten, for example if the Federal Republic were in danger of being overrun conventionally. The American nuclear guarantee would be withdrawn.

The authors assert that the implementation of their astonishing proposal would not be taken in Europe, and especially in the Federal Republic, "as evidence of a reduced American interest in the Alliance and in effective overall deterrence" but that, on the contrary, it would be the best means "for keeping the Alliance united and effective." On this point we beg to differ: the proposed no-first-use policy would destroy the confidence of Europeans and especially of Germans in the European-American Alliance as a community of risk, and would endanger the strategic unity of the Alliance and the security of Western Europe.

Given a renunciation of nuclear first use, the risks of a potential aggressor doubtlessly become more calculable. Moreover, the significance of Soviet conventional superiority would thereby increase dramatically. Conventional war in Europe would once again become possible. It could again become a continuation of politics by other means. Moreover, NATO would face a fundamentally different conventional threat. The elimination of the nuclear risk would free the Warsaw Pact from the necessity to disperse attack forces. As a result NATO would have to produce significantly higher numbers of combat forces than today.

The assertion of the four American authors that there is a tendency to overestimate the conventional strength of the Soviet Union does not correspond to the most recent East-West force comparison undertaken by NATO. They do admit, however, that a no-first-use policy requires stronger conventional forces; in their opinion the Alliance is capable of accomplishing such a buildup within realistic budgets. We believe the authors considerably underestimate the political and financial difficulties which stand in the way of establishing a conventional balance through increased armament by the West.

. . . The establishment of balance through the buildup of Western conventional forces would likewise be extremely difficult. The costs would be of a magnitude that would dramatically exceed the framework of present defense budgets. Suggestions by the authors about possible savings in the nuclear area in case of no-first-use are of little benefit for the non-nuclear weapons states of Europe. (Such savings, incidentally, imply a significant reduction of the Western nuclear arsenal.) In our judgment, the United States and Great Britain would have to introduce the draft, and the European countries would have to extend their period of military service. Because of the necessity for a significantly higher number of military forces, the Federal Republic of Germany would have to accept on its territory large contingents of additional troops, those of the allies and its own: the Federal Republic would be transformed into a large military camp for an indefinite period. Do the four American authors seriously believe that the preconditions for the buildup required by their proposal exist in Western Europe—and the United States?

Even if an approximate conventional balance could be achieved in Europe, two disadvantages to the detriment of Western Europe would remain: first, the Soviet Union has a geographic advantage, it can always quickly change the balance of forces from the relative proximity of its territory; second, there would always be the possibility, not even excluded by the American authors, that, despite no-first-use, conventional war could in an advanced phase degenerate into nuclear war. . . .

For Germans and other Europeans whose memory of the catastrophe of conventional war is still alive and on whose densely populated territory both pacts would confront each other with the destructive power of modern armies, the thought of an ever more probable conventional war is terrifying.

To Germans and other Europeans, an ever more probable conventional war is, therefore, no alternative to war prevention through the current strategy, including the option of a first use of nuclear weapons. While the four authors link their proposal with the laudable intention of reducing European anxieties about nuclear war, its implementation could result in anxieties about a more probable conventional war soon replacing anxieties about the much less probable nuclear war. . . .

The protection of a free society based on the rule of law is just as important a part of a policy of preserving peace as the prevention of war. War can always be avoided at the price of submission. It is naturally more obvious to Europeans, and in particular to Germans—in their precarious position within a divided country—than to the population of the American superpower that an actual military superiority of the Soviet Union, or a feeling of inferiority in Western Europe, can be exploited to put political pressure on Western Europe.

The feeling of vulnerability to political blackmail, as a result of the constant demonstration of superior military might, would be bound to grow considerably if the nuclear protector of the Atlantic Alliance were to declare—as suggested by the four authors—that it would not use nuclear weapons in case of a conventional attack against Europe. This applies in particular to those exposed areas which even with considerable improvements of conventional forces can only with great difficulty be conventionally defended, or not at all: these include, for example, North Norway, Thrace, and in particular, West Berlin. The protection of these areas lies solely in the incalculability of the American reaction.

The advice of the authors to renounce the use of nuclear weapons even in the face of pending conventional defeat of Western Europe is tantamount to suggesting that "rather Red than dead" would be the only remaining option for those Europeans then still alive. Were such advice to become policy, it would destroy the psychological basis necessary for the will to self-defense. Such counsel would strengthen tendencies in Europe to seek gradual voluntary and timely salvation in preventive "good conduct" and growing subservience vis-à-vis the Soviet Union for fear of war and Soviet superiority. The result would be to restrict the very freedom that the Alliance was founded to protect. . . .

Unlike the four American authors, we do not consider a renunciation of the option of a first use as the answer to the existing concerns and anxieties over nuclear weapons. Instead, we see the answer in a creative and realistic policy of arms control and disarmament. We consider the NATO double-track decision of December 1979, combining arms control negotiations and the announcement of armaments in case of failure, as an innovative step. . . .

We share many of the concerns about the risks of nuclear war. They lead us to conclude that an energetic attempt to reduce the *dependence on an early first use* of nuclear weapons must be undertaken. To be sure, the authors also mention a "no-early-first-use" policy as a possible alternative, but in the last analysis they discard it as a mere variation of nuclear options and therefore call for a clear decision in favor of a renunciation of "any first use of nuclear weapons. . . ."

In sum, we consider efforts to raise the nuclear threshold by a strengthening of conventional options to be urgently necessary. The reduction of the dependence on first use, in particular on early first use of nuclear weapons, should be a question of high political priority in our countries.

The Western Alliance has committed itself to a renunciation from the very beginning: the renunciation of the first use of *any* force. The entire military planning, structure and deployment of forces are geared exclusively toward defense. The presence of nuclear weapons has contributed essentially to the success of the Alliance in preventing war and preserving freedom for three decades. We are convinced that a reduction of the dependence on an early use of nuclear weapons would serve this purpose. Under the circumstances of the foreseeable future, however, a renunciation of the option of first use would be contrary to the security interests of Europe and the entire Alliance.

53. Address on Strategic Arms Reduction Talks (START)

Ronald Reagan

. . . During the 1970s, some of us forgot the warning of President Kennedy, who said that the Soviets "have offered to trade us an apple for an orchard. We don't do that in this country." But we came perilously close to doing just that.

If East-West relations in the détente era in Europe have yielded disappointment, détente outside of Europe has yielded a severe disillusionment for those who expected a moderation of Soviet behavior. The Soviet Union continues to support Vietnam in its occupation of Kampuchea and its massive military presence in Laos. It is engaged in a war of aggression against Afghanistan. Soviet proxy forces have brought instability and conflict to Africa and Central America.

We are now approaching an extremely important phase in East-West relations as the current Soviet leadership is succeeded by a new generation. Both the current and the new Soviet leadership should realize aggressive policies will meet a firm Western response. On the other hand, a Soviet leadership devoted to improving its people's lives, rather than expanding its armed conquests, will find a sympathetic partner in the West. The West will respond with expanded trade and other forms of cooperation. But all of this depends on Soviet actions. . . .

I believe such a policy consists of five points: military balance, economic security, regional stability, arms reductions, and dialog. Now, these are the means by which we can seek peace with the Soviet Union in the years ahead. Today, I want to set this five-point program to guide the future of our East-West relations, set it out for all to hear and see.

First, a sound East-West military balance is absolutely essential. . . .

The second point on which we must reach consensus with our allies deals with economic security. Consultations are under way among Western nations on the transfer of militarily significant technology and the extension of financial credits to the East, as well as on the question of energy dependence on the East, that energy dependence of Europe. We recognize that some of our allies' economic requirements are distinct from our own. But the Soviets must not have access to Western technology with military applications, and we must not subsidize the Soviet economy. The Soviet Union must make the difficult choices brought on by its military budgets and economic shortcomings.

The third element is regional stability with peaceful change. . . .

High on our agenda must be progress toward peace in Afghanistan. The United States is prepared to engage in a serious effort to negotiate an end to the conflict caused by the Soviet invasion of that country. We are ready to cooperate in an international effort to resolve this problem, to secure a full Soviet withdrawal from Afghanistan, and to ensure self-determination for the Afghan people.

In southern Africa, working closely with our Western allies and the African states, we've made real progress toward independence for Namibia. These negotiations, if successful, will result in peaceful and secure conditions throughout southern Africa. The simultaneous withdrawal of Cuban forces from Angola is essential to achieving Namibian independence, as well as creating long-range prospects for peace in the region.

Central America also has become a dangerous point of tension in East-West relations. The Soviet Union cannot escape responsibility for the violence and suffering in the region caused by accelerated transfer of advanced military equipment to Cuba. . . .

If martial law in Poland is lifted, if all the political prisoners are released, and if a dialog is restored with the Solidarity Union, the United States is prepared to join in a program of economic support. . . .

The fourth point is arms reduction. . . .

I wish more than anything there were a simple policy that would eliminate that nuclear danger. But there are only difficult policy choices through which we can achieve a stable nuclear balance at the lowest possible level. . . .

We must establish firm criteria for arms control in the 1980s if we're to secure genuine and lasting restraint on Soviet military programs throughout arms control. We must seek agreements which are verifiable, equitable, and militarily significant. Agreements that provide only the appearance of arms control breed dangerous illusions. . . .

The main threat to peace posed by nuclear weapons today is the growing instability of the nuclear balance. This is due to the increasingly destructive potential of the massive Soviet buildup in its ballistic missile force.

Therefore, our goal is to enhance deterrence and achieve stability through significant reductions in the most destabilizing nuclear systems, ballistic missiles, and especially the giant intercontinental ballistic missiles, while maintaining a nuclear capability sufficient to deter conflict, to underwrite our national security, and to meet our commitment to allies and friends.

For the immediate future, I'm asking my START—and START really means—we've given up on SALT—START means "Strategic Arms Reduction Talks," and that negotiating team to propose to their Soviet counterparts a practical, phased reduction plan. The focus of our efforts will be to reduce significantly the most destabilizing systems, the ballistic missiles, the number of warheads they carry, and their overall destructive potential.

At the first phase, or the end of the first phase of START, I expect ballistic missile warheads, the most serious threat we face, to be reduced to equal levels, equal ceilings, at least a third below the current levels. To enhance stability, I would ask that no more than half of those warheads be land-based. I hope that these warhead reductions, as well as significant reductions in missiles themselves, could be achieved as rapidly as possible.

In a second phase, we'll seek to achieve an equal ceiling on other elements of our strategic nuclear forces, including limits on the ballistic missile throw-weight at less than current American levels. In both phases, we shall insist on verification procedures to ensure compliance with the agreement. . . .

54. A Bilateral Nuclear-Weapon Freeze

Randall Forsberg

The proposal of a bilateral nuclear-weapon "freeze" by the U.S. and the U.S.S.R. has excited wide public discussion in this country during the past year. The idea is to stop the nuclear arms race quite literally, by stopping the development and production of all nuclear-weapon systems in the two countries. . . .

First put forward in the "Call to Halt the Nuclear Arms Race" drafted by me and published in April, 1980, by several public-interest groups, the freeze goes beyond other arms-control measures proposed in the past 25 years to put a stop to the production, testing and, implicitly, development of nuclear weapons as well as their deployment. By the same simplicity that has given it wide popular appeal the freeze proposal responds directly to an ominous turn in the arms race. The bilateral freeze would preclude the production of a new generation of "counterforce" weapons by the U.S. and the U.S.S.R. These are weapons designed to attack the opponent's nuclear forces. In the ultimate scenario they would disarm the other nation and hold its population hostage. The quest for improved counterforce capability has driven the arms race far past the point where each contender can destroy the other's society and much else besides. The production of counterforce weapons would increase the risk of nuclear war. Their deployment would put pressure on leaders to launch their weapons first in time of crisis, before they were attacked, and perhaps to place their nuclear forces in an automatic "launch on warning" status in peacetime. A freeze would prevent these dangerous developments. . . .

The time is propitious for a bilateral freeze. Today the U.S. and the U.S.S.R. are closer to parity in nuclear arms than they have been at any time since World War II. The U.S.S.R. has advantages in some elements of nuclear weaponry, the U.S. has advantages in others. The most frequently cited statistics compare the numbers of "strategic" ballistic missiles and bombers and the numbers of nuclear warheads and free-fall bombs they carry. The U.S.S.R. has more strategic missiles: 1,398 land-based intercontinental ballistic missiles (ICBMs) compared with 1,052 for the U.S. and 950 submarine-launched ballistic missiles (SLBMs) compared with 520 for the U.S. In addition a recent buildup has brought the strategic force of the U.S.S.R. abreast of that of the U.S. in the arming of these missiles with multiple independently targetable reentry vehicles (MIRVs). The land-based ICBMs of the U.S.S.R. carry more warheads and larger ones. On the other hand, the U.S. has more warheads in total, owing to the larger number of warheads on its SLBMs. The U.S. also has many more intercontinental bombers, with much larger payloads, and a five-to-ten-year lead in the new technology of small, long-range, low-flying cruise missiles.

More meaningful than comparisons of numbers of weapons is the fact that both countries have acquired enormous "overkill"; that is, each has many times the number of weapons necessary to annihilate the other's urban population. Thus even if the U.S.S.R. destroyed all U.S. ICBMs, all U.S. bombers and all U.S. submarines in port, the U.S. would still have about 2,400 nuclear warheads on submarines at sea, completely invulnerable to such preemptive attack. This is several times the number needed to destroy the 300 largest cities and towns in the U.S.S.R., which have one-third of the population and three-fourths of the industry. Conversely, a U.S. attack on ICBMs, bomber bases and submarines ports in the U.S.S.R. would leave an estimated 1,200 strategic warheads, more than enough to inflict equivalent damage on U.S. urban centers.

The bilateral freeze would preserve this parity. It would prevent the emergence of a new, destabilizing U.S. advantage in counterforce capability, projected in the buildup planned by the Reagan Administration.

. . . And it would forestall the inevitable effort by the U.S.S.R. to match U.S. developments.

As spelled out in the "Call to Halt the Nuclear Arms Race," a freeze on both sides would stop the following: the production of fissionable material (uranium 235 and plutonium) for nuclear weapons; the fabrication, assembly and testing of nuclear warheads; the testing, production and deployment of missiles designed to deliver nuclear warheads, and the testing of new types of aircraft and the production and deployment of any additional aircraft designed primarily to deliver nuclear weapons.

In order to achieve its promise as a new departure in arms control the freeze must be complete. One reason is to facilitate its verification. In all arms-control agreements to date the parties have relied primarily on "national" methods of verification, that is, methods that can be managed independently, without cooperation. This has meant principally surveillance by satellites. For this reason some have argued that the freeze should be limited to the deployment of large nuclear-weapon delivery systems, an activity that is confidently subject to satellite surveillance. Such a freeze would be better than none at all. There are, nonetheless, persuasive arguments in favor of a freeze that covers production as well as deployment and small systems as well as large.

If production is not banned, the military on both sides is likely to argue that missiles and bombers, warheads and bombs should continue to be produced, if only for storage in warehouses. Thus the 1963 Partial Test-Ban Treaty was followed by more vigorous testing underground than had ever been conducted aboveground, and the 1972 Anti-Ballistic-Missile Treaty has sheltered intensive developmental testing of those weapons. A "production but not deployment" race would continue the buildup on both sides, and along with it the confrontation, the cost and the growing destructive potential.

The production facilities for nuclear weapons are known at least as well as those for conventional weapons. Nuclear warheads are fabricated and assembled at just a handful of facilities in the U.S.S.R., which were identified in the 1950s, soon after they were established and years before the introduction of reconnaissance satellites. The closing down of these facilities could be monitored with high confidence from satellites; it would be no easier to replicate them secretly than it was to keep their construction secret the first time. . . .

A variety of nonintrusive on-site verification measures can supplement satellite observation. They can take such forms as occasional, unannounced inspections or continuously monitored, sensor-equipped, secure "black boxes" installed in shut-down or controlled factories. The U.S.S.R., reluctant to agree to on-site verification in the past, has recently shown signs of greater flexibility on this point in the negotiation of a comprehensive nuclear test ban. . . .

A strictly enforced freeze that includes production and testing as well as deployment could lead after some years to a decline in the reliability and readiness of existing nuclear armaments. . . .

Various factors in the aging of nuclear weapons must therefore be dealt with. For example, the tritium modules that initiate the fusion reaction in thermonuclear explosives must be replaced every few years. This implies that the freeze should allow the operation of tritium-component assembly facilities and the running of perhaps one military nuclear reactor to produce tritium. Special safeguards will be needed to ensure that the reactor does not produce plutonium for new warheads.

Among strategic delivery systems, submarines most clearly have a limited service life, generally estimated at 30 years. The "Call to Halt the Nuclear Arms Race" specifically excludes submarines from the freeze. It allows their replacement but requires the installation of existing missiles rather than new ones so that the quantitative and technical threat does not grow. . . .

Even missiles sitting in their silos deteriorate to some extent. The stored fuel can be corrosive and is subject to decomposition; the gyroscopes and electrical systems are in constant operation. By replacement of worn-out parts, however, missiles too can be maintained for long periods. . . .

Some have argued that if offensive nuclear-weapon systems are frozen, a country's existing weapons will gradually become vulnerable to the improved defenses and countermeasures of its opponent. The U.S.-U.S.S.R. strategic competition, however, pits the offense not so much against the defense as against the opposing offense, keeping up, as it were, with the Joneses. In ICBM technology no technical improvements are needed to ensure the penetration of defenses as long as the treaty prohibiting anti-ballistic-missile (ABM) systems is kept in force. (Reagan Administration officials are, to be sure, considering the abrogation of the treaty in order to provide ABM defense of MX-missile sites.)

. . . The advances in the technology of anti-ballistic-missile systems, anti-submarine warfare and air defense that can be foreseen for the remainder of this century will do little to decrease the capacity for devastation that exists in the offensive strategic nuclear forces of the two sides today. In fact, most of the technical advances planned for the offensive forces are not intended to offset improvements in defenses. They seek increases in offensive power only—in yield, accuracy and numbers. . . .

Even if U.S. land-based ICBMs are or become more vulnerable to preemptive attack than those of the U.S.S.R. (a disputable point), this would be offset by the invulnerability of U.S. bombers and submarines. On the U.S.S.R. side the high vulnerability of its strategic bomber and submarine forces is offset by the size of its ICBM force. Without a bilateral freeze this relatively stable balance will be eroded by the weapon programs planned for the next decade. . . .

A freeze would help to accomplish other desirable goals. The U.S. and the U.S.S.R., by fulfilling the pledge they made in the 1970 Nonproliferation Treaty to stop the arms race, would help to brake the spread of nuclear weapons to countries that do not already have them. A freeze would create an opportunity for the nations of the world to make further progress in arms control and other global issues. It would also save billions of dollars. . . .

The freeze would not eliminate the existing capacity of the U.S. and the U.S.S.R. to bring about a global nuclear holocaust. As few as 100 nuclear weapons on each side, half of 1 percent of the current arsenals, could devastate the U.S. and the U.S.S.R. beyond any previous historical experience and perhaps beyond recovery as industrial societies. To end the danger of nuclear war the nations must not merely freeze nuclear weapons but abolish them. The freeze represents a modest but significant step toward abolition. It would terminate the technological arms race and shut down entirely this wasteful and dangerous form of human competition.

55. Strategic Build-Down:
A Context for Restraint

Alton Frye

. . . In essence the build-down principle says that no new weapons should be deployed unless a larger number of existing weapons are destroyed. . . .

One needs to contrast the build-down proposal with the prior American position in START. Previously, the United States had concentrated on schemes to restructure the current Soviet forces, demanding not only that total missile warheads be cut to 5,000 on each side, but that the Soviets phase out large numbers of their biggest missiles. That would diminish, but not eliminate, the great disparity between the two nations in missile throw-weight, in which Moscow holds a considerable edge. Limits on heavy bombers, a category in which the U.S. lead is substantial and growing, were vague and inconsequential: the United States would have been free to go ahead with a planned growth in the number of air-launched cruise missiles (ALCMs) and other bomber weapons—one that, even with substantial cuts in missile warheads, would have given it *more* strategic nuclear weapons in the 1990s than it has now. However appealing from the standpoint of the United States and from the perspective of Western theories of strategic stability, those proposals were widely perceived as inequitable and non-negotiable.

Mr. Reagan's acceptance of the build-down concept modifies the U.S. position dramatically. Without demanding initial agreement on the ultimate composition of strategic missile forces—in effect leaving each side broad freedom to balance its forces as it deems best—the latest proposal tries to harness the momentum of force modernization to the declared goal of arms reductions. Specifically, Mr. Reagan now offers to build down ballistic missile warhead inventories from the current 8,000–9,000 range on each side to 5,000 by eliminating more than one warhead for each warhead newly deployed.[1]

1. Each warhead installed on a new land-based missile (ICBM) · with multiple independently targetable re-entry vehicles (MIRVs) would oblige a party to eliminate two existing warheads.

2. New warheads on submarine-launched ballistic missiles (SLBMs) or small, single-warhead ICBMs would force reductions at a lower ratio, perhaps three for two.

3. If a side were not modernizing and introducing new warheads—a highly unlikely contingency in the next few years—it would still have to make annual reductions at an agreed percentage rate, possibly five percent.

4. The President would also apply the build-down principle to deployment of new bombers (though not directly to individual weapons carried on bombers), reducing bomber forces to levels well below those permitted under the 1979 strategic arms limitation treaty (SALT II).

5. The United States would also accept limits on the number of air-launched cruise missiles each aircraft could carry and on the aggregate number of such missiles deployed. And finally,

6. The extent of reductions in missile throw-weight would be balanced against the reductions in bomber carrying capacity through a formula measuring potential destructive capacity. This feature of the proposed approach (to be explained below) is important. Since the two categories vary so basically from each other, and since the Soviet side would be more

affected by the missile warhead reductions and the U.S. side more affected by the reductions in bomber-carried weapons, a measure integrating these force components is necessary to permit precise trade-offs between them.

In short, the new approach seeks major reductions in both missile-carried and bomber-carried weapons—through the build-down route—and also explicitly links the two. In announcing this basic shift in the U.S. START position, President Reagan stated the central objective: "We seek limits on the destructive capability of missiles and recognize that the Soviet Union would seek limits on bombers in exchange. There will have to be trade-offs and the United States is prepared to make them, so long as they result in a more stable balance of forces."

. . . The build-down approach is not a magic panacea for all the problems that have piled up in the last 20 years. It accepts the continued existence of large numbers of nuclear weapons on both sides as a given, and assumes that the two countries will remain concerned, partly for political reasons, with both the substance and the appearance of parity. At root it addresses the goal endorsed by the Scowcroft Commission—moving toward reduced nuclear postures that are less vulnerable to a first strike. That goal requires a balanced melding of warhead reductions and selective modernization.

Build-down would retain the counting and verification provisions embodied in the SALT II Treaty while moving beyond it to bring about meaningful reductions in the arsenals on both sides. It would operate, not through rigid categories and subcategories (although previously agreed provisions would remain in effect), but through relatively free choice for the two military establishments to bring themselves within broadly defined warhead ceilings and other limits. . . .

The build-down approach is indeed couched in terms of *numbers of offensive* weapons. It neither includes nor precludes possible arms control measures expressly aimed to halt improvements in the quality of such weapons—such as a comprehensive nuclear test ban (CTB), limits on ballistic missile testing, or the suspension of fissile material production. All these are elements of proposals for a nuclear freeze, usually defined as a comprehensive curb on both the quantitative and qualitative escalation of nuclear armaments. The build-down is not a substitute for or an alternative to the full range of limitations envisioned by the nuclear freeze movement. But, if successful, it can help arrest the deterioration in Soviet-American relations and pave the way for more ambitious restraints. . . .

However one assesses the Soviet advantage in throw-weight, the crucial point in assessing the build-down approach is that *any agreement which cuts the number of warheads will itself tend to cut the throw-weight disparity.*

For instance, if the Soviets agreed to a 5,000-warhead limit for ballistic missiles, reaching that level would drop their missile throw-weight by more than 50 percent and lower the ratio of throw-weight between the two forces. That would be true even if Moscow elected to maximize its total throw-weight by holding onto its larger missiles, the SS-18s and SS-19s, in which case it would fill its permissible warhead quota with only about 650 missiles—

and it would have no other missiles at all on land or sea. Obviously, that posture is preposterous; the realistic range of Soviet forces would presumably spread the warheads over more launchers with relatively less gross throw-weight. In addition, technological trends on both sides point toward relatively compact and portable warheads, meaning that a reduction in the number of warheads will mean concomitant cuts in throw-weight and megatonnage. The two countries have such vast power at their disposal that these changes will still leave them with extraordinarily high levels of destructive capability, but there is little doubt that, even with modernization, reducing the number of warheads in the two forces will trim the currently deployed explosive yield quite significantly. . . .

There is a substantial consensus among the Western Allies that INF and START negotiations should sometime be merged, but much dispute about when and how.

Build-down could impose a common standard on both classes of systems, without requiring immediate merging of the talks. INF discussions could continue as a separate forum to determine what sub-limit would be desirable for intermediate forces within an overall warhead ceiling encompassing both START and INF forces. Under such an aggregate limit, build-down would automatically provide useful regulation of INF forces. While the United States would be able to allocate some of its warheads for the largely political purposes served by theater deployments, it would have strong reasons not to deploy very many weapons in Europe.

Likewise, the Soviet Union would need to reconsider how much of its declining warhead quota should go to systems capable of covering only part of its strategic target list. Gradually, operational considerations should lead planners to emphasize versatility over the limited functions of theater systems. Just as the logic of reductions favors dispersed systems with fewer warheads per launchers, it also argues for longer-range systems able to cover all targets. There would be a good case for preferring intercontinental missiles over intermediate-range launchers.

In short, and speaking personally, I believe the most promising route toward an early and worthwhile agreement would be to subject missiles and bombers to an aggregate warhead ceiling with differentiated counting rules for bombers. ALCM-carrying bombers would count as twenty warheads; non-ALCM bombers as one.

But there is a second difficult problem of measurement. The trade-off envisaged by President Reagan is only partly between numbers of warheads. It is also a trade-off between the throw-weight of missiles and the payload of bombers—the carrying capacity of these different vehicles which determines the overall destructive power they are able to deliver.

Wrestling with this problem over several months, Lt. Gen. Glenn Kent (USAF, Ret.) made an important analytical contribution by inventing a measure of "Potential Destructive Capacity" (PDC). Kent's analysis uses plausible factors to weight the potential number and size of weapons which missiles and bombers can carry. It then proposes the concept of "standard weapon stations" as a unit of account for reducing both types of systems. Kent's work is intellectually elegant and offers a novel approach to integrating missile and bomber weapons.[2]

The main merit in a consolidated measure of potential destructive capacity is not that it is needed to ensure reductions in missile throw-weight—the warhead cuts would accomplish that—but that it makes possible a fair trade-off between throw-weight and bomber payload.

General Kent's analysis illustrates one way to make equitable trade-offs between these asymmetrical forces, although the definition of Potential Destructive Capacity can and should be simplified.[3] The vital point is that the United States has expressed willingness to apply the build-down rule to bombers in a way that would assure Moscow that U.S. bomber payload is contracting commensurately with Soviet throw-weight. These would be substantial American concessions compared to the many thousand bomber weapons to which the United States is entitled under SALT II. They would warrant Soviet acceptance of the missile warhead and throw-weight reductions sought by the United States. . . .

The chances for strategic restraint depend, of course, on whether Moscow and Washington can frame compatible objectives. Do they wish to preserve the benefits of SALT II? If so, a build-down agreement would serve that purpose. Do they want to bolster stability by encouraging movement toward more dispersed and survivable systems? Build-down could promote that goal. Do they hope to rescue the arms control process from the gridlock caused by the counterproductive separation of the INF and START negotiations? Extension of build-down to cover all long-range nuclear forces under a common warhead aggregate would advance that objective. Would they prefer to confront each other in the next decade with thousands of additional warheads on their strategic launchers? Or would they find greater security in exploiting the momentum of modernization to impose reductions to a warhead level below that of 1983? Build-down provides a mechanism to make that choice real.

It is not a complete solution or the only solution to the dilemmas of nuclear competition. But those who find merit in reducing strategic arms and in restraining strategic modernization should welcome build-down as a useful approach. Above all, a mutual commitment to build-down would testify to acceptance of the first fact of our age: the American and Soviet peoples must find safety from the nuclear menace together—or they will not find it at all.

NOTES

1. In this discussion, the terms "warheads" and "weapons" are used interchangeably; unless otherwise noted, estimates of warhead inventories are based on the maximum number of warheads permitted on individual missiles under the SALT II Treaty.

2. See Glenn Kent and Edward L. Warner, "Key Aspects of Compulsory Double Build-Down Approach," unpublished, September 6, 1983.

3. On the assumption that bomber payload is worth about one-half an equivalent amount of missile throw-weight and that a bomber's payload is one-tenth the plane's gross takeoff weight, one could define Potential Destructive Capacity (PDC) as the sum of missile throw-weight plus one-half of bomber payload. Counting bombers identified in the SALT II data base (many of which are in fact moth-balled in the United States and obsolete in the Soviet Union), the potential destructive capacity totals are about 18,000,000 pounds for the United States and about 15,000,000 for the Soviet Union. In conjunction with reductions in missile warheads to 5,000, a parallel requirement to cut PDC to equality at 8,500,000 pounds should be workable and worthwhile. Under this limit, just as there would be constraints on the extent to which throw-weight was fractionated, there would presumably have to be limits on the divisibility of bomber payload, e.g., no more than 20 ALCMs per plane and no more than 2,500 ALCMs overall. These calculations are derived from the Kent-Warner data and appear in my unpublished technical analysis, "Constraining Potential Destructive Capability in Strategic Forces," September 1983.

56. Nordic Initiatives for a Nuclear Weapon-Free Zone in Europe

Sverre Lodgaard

INTRODUCTION

In 1961 Swedish Foreign Minister Undén suggested the creation of a "club" of states obligated not to acquire nuclear weapons and not to accept deployment of nuclear weapons on their territories. In 1963 President Kekkonen of Finland adapted and confined Undén's idea to the Nordic region, proposing a Nordic nuclear weapon-free zone (NWFZ).

The overriding concern behind the Kekkonen proposal was to keep the Nordic countries out of "the realm of speculation brought about by the development of nuclear strategy," and to maintain a state of low tension in the area. That same concern prompted a revised version of the proposal in 1978,[1] and has been an important impetus for the recent surge of interest in the zone issue in all the Nordic countries, precipitated by a programme declaration of the governing Labour Party in Norway.

OBJECTIVES

The overall objective of the Nordic NWFZ proposals is to strengthen the security of the countries in the region, and to stabilize relations between the big powers in this strategically important area.

The constellation of ground forces in northern Europe has remained stable for a number of years. Both Eastern and Western countries have shown restraint. However, military capabilities at sea and in the air are rapidly increasing in the region, threatening the security interests of all parties—Eastern, Western and neutral.

CHARACTERISTICS

There are three main characteristics of a NWFZ: non-possession, non-deployment and non-use of nuclear weapons. The non-possession requirement is already met by all the Nordic countries: they were among the first to ratify the Non-Proliferation Treaty (NPT). The non-deployment obligation, however, presents several difficulties for the two Nordic NATO members.[2]

Norway and Denmark do not allow the deployment of nuclear weapons on their territories in time of peace. This is a unilateral measure of restraint; therefore, they are free to change policy at will, and options for the use of nuclear weapons on or from Danish and Norwegian territory have existed for years. Unlike the NPT commitment, this is a policy that can be changed overnight. However, the broad consensus that has been formed around the non-deployment stand makes it hard for any government to back out of it under normal international conditions. Only a crisis could induce Norway and Denmark to ask for the transfer of nuclear weapons to their territories. Since the policy was instituted more than 20 years ago, technological developments have, moreover, rendered the exercise of the nuclear weapon option in time of crisis less important.

However, participation in a NWFZ would require an unqualified position against the deployment of nuclear weapons, applying in times of both war and peace, and embodied in an international legal instrument. While the policy of non-deployment in peace-time has never been challenged by other NATO members, non-deployment in wartime would impose a more substantial restraint on NATO nuclear planning for northern Europe. In important respects, Norway and Denmark would be decoupled from NATO's nuclear strategy, and their participation in NATO's military organization might have to be reconsidered also in other respects. . . .

A zone arrangement must be perfectly clear as to rights and obligations: lack of clarity may lead to misunderstandings and suspicion, and guarantor states can make use of ambiguous provisions to exert pressure on member states. Clarity would appear to be an overriding concern. However, it is difficult to find an extended definition of denuclearized status which discriminates as clearly between things permitted and things prohibited as the distinction between presence and non-presence of explosives. This difficulty therefore amounts to a strong argument for sticking to the established meaning of "nuclear weapon-free." Should a country like Norway ever want to go beyond this and eliminate U.S. or NATO-related facilities which may become nuclear targets in war, it could raise this question with other NATO members on a bilateral or alliance basis. In the NWFZ context, it would be another complication and, possibly, a major obstacle.

A zone arrangement implies, however, that all plans for the transfer of nuclear weapons to members of the zone must be scrapped. For instance, collocated operating bases (COBs) might be affected. The need for allied air support, essential for the defences of Norway and Denmark, must be made compatible with a credible non-nuclear status. This might be achieved either by changes in current agreements and practice, or by extended national verification rights, or by elements of both. At present, there are two Danish airfields in the COB programme and eight in Norway, in the total of some 70 for NATO Europe.

NATO members joining the zone may have to leave NATO's Nuclear Planning Group as well. Since they do not wish to be defended by nuclear weapons themselves, it might not be legitimate for them to participate in shaping the nuclear defences of other countries. On the other hand, in a nuclear war in Europe, the consequences would indeed be felt over the whole continent. Different countries would be differently affected, but there is no escape route for anyone. From that point of view, Nordic NATO members would still seem entitled to have a say in the formulation of nuclear strategies. The argument goes both ways. . . .

TRANSIT PROVISIONS

The Treaty of Tlatelolco does not contain any provision regarding the transit of nuclear weapons. The Preparatory Commission for the Denuclearization of Latin America (COPREDAL) argued that it should be the prerogative of the territorial state, in the exercise of its sovereignty, to grant or deny permission for transit. In signing Additional Protocol II to the Treaty, the U.S.A. and France emphasized that each party to a nuclear weapon-free zone should retain exclusive legal competence to grant or deny transit. (This was motivated mainly by the use of the Panama Canal by the U.S.A. and other major powers.) In ratifying the same Protocol, the Soviet Union stated its objection to any such permission for transit.

For the Nordic countries, the transit of nuclear weapons mainly entails sea transit, except for Iceland (Keflavik). Even thus confined, it is a complex issue: it could involve a nuclear-armed ship showing the flag in a Danish harbour, ships participating in joint exercises, or an attack submarine calling at a Norwegian port for supplies or repair. Since large parts of the great power navies are equipped with nuclear weapons, it might be difficult for NATO members to prohibit all kinds of transit. An absolute prohibition could hamper joint military exercises to such an extent that allied support for Norway and Denmark would be seriously weakened. Such a prohibition would, moreover, be a rather one-sided concession on the part of Denmark, Norway and other Western powers.

The Soviet Baltic Fleet, and the significance of Soviet shipyards there, practically excludes prohibition of transit through the straits. . . .

DEPLOYMENT LIMITATIONS IN AREAS ADJACENT TO THE ZONE

In its foreign policy declaration of 18 March 1981, the Swedish government reiterated its long-standing view that a NWFZ agreement must include nuclear weapons "which are intended for targets within the zone, are stationed near the zone and have ranges of a scale which makes them best suited for targets within the Nordic area."[3] Three months later, the Soviet Union stated its willingness to consider measures "applying to [Soviet] territory in the region adjoining the nuclear free zone in the North of Europe."[4] Today, the viability of the zone proposals hinges very much on the prospects for deployment limitations in areas adjacent to the zone.

There are two main perspectives on the issue of deployment limitations. First, such limitations may be seen as a consequence of the guarantees for the zone. To the extent that nuclear weapons are unambiguously directed at targets within the zone—because of their geographical position, range or other indicators—they have to be removed; otherwise, they would constitute proof that the guarantees are fictitious. For example, the dozen or so 800-km range SS-12 Scaleboard missiles in the Leningrad military district are, in all likelihood, intended for interdiction strikes against Nordic targets, because they do not reach continental Western Europe. The 350-km range Scud missiles in the same district are primarily intended for use against Nordic territory as well, in a tactical role. Scuds and Scaleboards belong to the standard Soviet weapon inventory at the Army and Front levels, respectively.

The elimination of weapons in this category is of special significance for the Nordic countries. The history of wars shows that belligerents usually do not surrender until all weapons have been used. Thus, weapons which can only be used against Nordic countries may, in an extreme situation, be used even against militarily insignificant targets on Nordic territory, as acts of terror. It is true that the elimination of these weapons would leave the nuclear powers with thousands of other weapons capable of striking targets in the Nordic area; but given that they can be used for a variety of important missions in other parts of the world, it is not certain that they would be used against a nuclear weapon-free zone. For the Nordic countries, the elimination of weapons without competing targets elsewhere is therefore more important than their relatively modest numbers would indicate.

However, few weapons can be used against Nordic countries only. Modern weapon systems are usable over varying distances and against different targets: they are becoming more mobile and more flexible, and can therefore meet a broad spectrum of military needs. It may therefore be more appropriate to seek deployment limitations as a matter of militarily significant confidence-building measures. . . .

VERIFICATION

The main problem is to verify that the deployment limitations are observed. While this is difficult to discuss until the limitations are determined, verifiability is an important parameter of the elaboration of restrictions.

One thing seems obvious: since the Nordic countries themselves do not possess adequate technical means of verification, co-operation with the guarantor states is important. Being parties to the same arrangement prescribing limitations and restraints on both sides, the great powers must be presumed to watch each other with the means they have at hand. By establishing a joint commission where all states involved may raise matters for clarification or submit charges of violations, the members of the NWFZ would be in a position to draw upon the verification capabilities of the guarantor states. New issues could be referred to the same commission for clarification, that is, to a multilateral setting, thereby avoiding bilateral exchanges between one or more Nordic countries on the one hand, and a nuclear weapon state on the other. . . .

THE EUROPEAN CONNECTION

A Nordic zone can be seen as a measure in its own right, although open-ended; as such, it may also be a first step towards a more comprehensive reduction of the numbers and roles of nuclear weapons in the European security system. Alternatively, it may be seen as an integral part of a broad European rearrangement, its fate being tied to developments on the larger European scene. . . .[5] The Nordic NWFZ idea is of political interest because it has received remarkable public support in all the Nordic countries.

Norway and Denmark would, as a matter of course, have to consult with their allies on the drawing up of any zone arrangement. Equally obviously, the Nordic countries themselves must kick the ball off by taking a joint decision to initiate a process aiming at the establishment of a nuclear weapon-free zone in northern Europe. The decision might be taken at a meeting of Nordic foreign ministers. Iceland, a Nordic country and a regular participant in Nordic ministerial meetings, naturally ought to take part. Should Icelandic membership in the zone be considered premature, the meeting might underline the desirability of including Iceland at a later stage. Accordingly, it might also wish to emphasize that actions drawing Iceland deeper into Western nuclear strategy, as compensation for the denuclearization of Norway and Denmark, should be avoided.

Alternatively, the process could be initiated by co-ordinated declarations of all the countries to be included in the zone. One way or another, the constitution of the zone must be a Nordic initiative, even if it were to be presented as a Nordic offer to the great powers and other European states in the pursuit of arms reductions in the wider European domain. Otherwise, it would not carry much weight on the diplomatic scene.

Should the zone be seen as a measure in its own right, or as a first step towards a more comprehensive rearrangement in Europe, the Nordic NATO members would become involved in a sensitive balancing act between membership in the zone, on the one hand, and continued NATO membership on the other. On the one hand, they would have to meet the non-deployment demand and discontinue all preparations for transfer of nuclear weapons to their territories in time of crisis or war. On the other hand, the initial, rudimentary design must be of such a character, and have enough built-in flexibility, that the United States and NATO can accommodate the new conditions. If the United States declines to give guarantees for the zone, and if NATO balks at the alliance obligation to render support if limited to conventional means only, the zone is unlikely to be established. . . .

NOTES

1. President Kekkonen, Address at the Swedish Institute of International Affairs, Stockholm, 8 May 1978.

2. J. Prawitz, speech before the Swedish Kungliga Krigsvetenskapsakademien [Royal Military Science Academy], 13 March 1979.

3. Protocol of the Parliamentary Foreign Affairs Committee, Stockholm, 18 March 1981.

4. *Suomen Sosialdemokraatti*, 26 June 1981.

5. J. J. Holst, "The challenge from nuclear weapons and nuclear-weapon-free zones," *Bulletin of Peace Proposals*, No. 3, 1981.

CHAPTER 11

Epilogue:
The Soviet Union
and the United States
Resume Arms Control
Negotiations

Two recent developments merit discussion. The first is the return of the Soviet Union and the United States to nuclear arms control negotiations. The second is the growing seriousness with which the scientific community has begun to regard the possibility of "nuclear winter."

Contrary to predictions, in the fall of 1984 the Soviet Union agreed to return to arms control negotiations with the United States. Secretary of State George Shultz and Foreign Minister Andrei Gromyko met in Geneva in January 1985 and agreed on a method of organizing the negotiations. There were to be three sets of negotiations—on strategic, theater, and space weapons—with a separate negotiator in charge of each. Formal talks began in March 1985. The seriousness with which both sides regarded the talks was apparent in the rigid secrecy they imposed on the proceedings.

At this writing, the two noteworthy features in the positions of both sides are the Soviet insistence on linking agreement on space weapons to progress in the other two areas and the U.S. insistence on preserving a major research and development program for space-based defensive systems. Neither side has said much in public about the substance of its goals in the negotiations. The Soviets have repeatedly emphasized the importance of linking the three sets of talks. The Americans have proclaimed a desire for deep reductions and the introduction of nonnuclear defense systems, which are seen as necessary to permit the reductions. Otherwise, as Paul Nitze (newly appointed special adviser to Secretary Shultz) has argued, neither country will be willing to make the cuts that should be the goal of the talks, because they will be too vulnerable to a surprise attack by their chief adversary or to nuclear blackmail by a small country with a primitive nuclear force led by an erratic leader such as Qaddafi of Libya.

The outlook for the new round of Soviet-American negotiations is more uncertain than it was about either SALT I or SALT II. The appeal of arms control as a solution to the arms race has been tarnished. From the Reagan administration come pessimistic warnings that no agreement will be possible and that this is not necessarily a bad thing. Unilateral actions in this view may produce a more stable nuclear relationship than prolonged and impossibly complicated big-package negotiations. Even so, one feels a surge of hope at the prospect of the two sides discussing arms control again. The greatest gain may come in the general area of Soviet-American relations rather than in breakthroughs in nuclear arms control. That is not without value in itself. It may eventually lead to helpful implicit understandings or even agreements that will contribute to the lowering of the risk of nuclear war.

The renewed concern over nuclear winter arises from the possibility that the detonation of a large number of nuclear weapons over cities in the Northern Hemisphere would cause such large fires that the smoke and particles released would blot out the sun. The temperature in the affected region would drop dangerously low, and because the affected land would include much of the world's most fertile agricultural areas, the prospect of worldwide famine must be added to the losses caused by blast, firestorm, radiation, breakdown in medical care, and social collapse.

Scientists do not agree about the meteorological effects of nuclear explosions over U.S., Soviet, and European cities. Moreover, the thousands of nuclear detonations postulated to trigger nuclear winter represent an extremely pessimistic view of Soviet-American conflict. Attacks on such a huge scale by Soviet and U.S. leaders would be ruinous, even without nuclear winter. For that reason, it is an unlikely future. However, what is new is that many scientists now believe the phenomenon should be carefully investigated, and policymakers have begun to argue that the possibility should be taken into account in setting strategy and targeting doctrine.[1] Plainly, the concept of nuclear winter has contributed an added urgency to the new round of Soviet-American negotiations in Geneva.

NOTES

1. See P. J. Crutzen and J. W. Birks, "The Atmosphere After a Nuclear War: Twilight at Noon," *Ambio*, no. 11 (June 1982):114–125; R. P. Turco, O. B. Toon, T. P. Ackerman, J. B. Pollock, and C. Sagan, "Nuclear Winter: Global Consequences of Multiple Nuclear Explosions," *Science* (December 23, 1983):1283–1292; and Paul R. Ehrlich, Carl Sagan, Donald Kennedy, and Walter Orr Roberts, *The Cold and the Dark: The World After Nuclear War* (New York: W. W. Norton, 1984).

57. The Soviet Position on Strategic Arms Control

Andrei Gromyko

If one strives to put an end to the arms race and to the removal of the threat of war, or, to put it briefly, if one strives for a lasting peace, it is necessary to ensure that the arms race does not begin in space and is stopped on Earth. That was the main task that faced the participants in the Geneva meeting.

The U.S. Administration pressed hard in a bid to prevent discussion on questions of space and to leave space open for the arms race. This position is absolutely unacceptable. We pointed this out on more than one occasion, in particular, before the Geneva meeting. Discussions to reach agreement on the terms of that meeting were not smooth either. It was finally agreed, however, that that question had to and should be discussed. The joint statement, as you know, also touches upon space, which means that the view has prevailed that the question of either strategic weapons or inter-mediate-range nuclear weapons cannot be examined without the question of space, or, to be more precise, without the question of preventing the arms race in space. . . .

We wanted to bring this home to the American Administration and those who supported and continue to support its viewpoint on questions of space. The conclusion we made—we also repeated it in Geneva—is: It is impossible to examine questions of strategic nuclear armaments and intermediate-range nuclear weapons productively without considering questions of space, outer space.

For a long time the proponents of the so-called large-scale antimissile defense of the USA were "drowning" individual components, individual phases of that problem, never differentiating between them. Later on they realized that, naturally, that position was weak and, facing criticism, they came to the conclusion that the deployment of relevant objects and their testing could be prohibited. They first stated this clearly in Geneva. . . .

As for research, they claimed that it should be conducted. Their motivation was that research could not be prohibited anyway since such a ban could not be verified. But there often is some proving ground next to a laboratory and it is used for relevant purposes.

Even assuming that verification is difficult, why should work be conducted, even though it is called research, when there is an accord that the goal of the ultimate and complete elimination of nuclear weapons should be pursued?

Why should research be conducted to develop a whole system of new types of weaponry for deployment in space? So this position of the U.S.

Administration, their position of conducting research, is vulnerable. Why should preparations be carried out, even at the initial stage, for the subsequent testing and deployment of new types of armaments? . . .

And who can guarantee that the line will be drawn after research has been completed? The line, political and moral, for conducting research with a view to developing a large-scale missile defense system stands no criticism. . . . It is vulnerable and must be rejected. . . .

No state has made a more radical or more far-reaching proposal on verification than the Soviet Union. Upon the emergence of nuclear weapons, when the question of verification arose, we made a proposal at the United Nations to ban those weapons forever.

How about verification, how about weapons of other types? Our answer was simple: Let us raise the question of weapons of other types and let us agree on verification. What sort of verification? The broadest and all-embracing verification, the most extensive verification possible, in short, universal verification. We made that proposal, we and no other state. Our friends supported it in full. It was our joint proposal. And we insisted for many years on the adoption of the proposal on universal disarmament and universal verification.

You would think the West, the United States of America would have agreed to it? No, they did not. They have not agreed either to universal disarmament and the prohibition of nuclear weapons or to universal verification. This proposal is still on the table of the U.S. Administration, the other NATO governments and the governments of the world as a whole.

Even today we are prepared to consider such a broad proposal, the proposal of universal and complete disarmament under universal and complete international control. I stress, international, which means that U.S. representatives can also participate in this control.

As for the hints that the Soviet Union fails to respect some commitments under the agreements concluded by it, it is invention. There are no direct statements with proof based on fact that the Soviet Union is really violating anything. We categorically reject it. No, the Soviet Union is not in the habit of violating its commitments under treaties and agreements signed by it and other states, be it a bilateral or a multilateral agreement. We take pride in this fact. . . .

The [U.S.] Administration itself realizes that the history of the establishment of strategic armaments in the USSR and the United States is such that these armaments were formed differently. The United States always had a powerful fleet of bombers, above all, aircraft as nuclear weapon-carriers. Another factor, submarines, was also in its favor. Eliminating the differences, including that in the structure of strategic nuclear forces, at one stroke is therefore not a serious matter. It seems that even official American circles at last grew to understand that their demand, which they stressed earlier in connection with our heavy missiles, is excessive and inappropriate. It is said more and more often that, indeed, it is necessary to understand the history of the establishment of the strategic nuclear forces of both sides.

If the United States continues to deploy its intermediate-range nuclear systems in Europe, the situation, one should put it straight, will become more complicated, much more complicated. . . . Our proposal for freezing all nuclear systems is well known. Moreover, at Geneva, we warned the United States in a rather lucid way that if it acted in this way, continued the stationing of the intermediate-range nuclear systems—and it stressed in all ways that it has such plans and intends to carry them through—it (the United States) would put in question the necessity of the talks that are to be started in accordance with agreement reached in Geneva. I repeat that we warned the United States about this. We hope that it will take this into consideration.

. . . It would be totally unjustified should the North Atlantic alliance receive a kind of allowance or bonus, if I may put it this way, in the form of the British and French nuclear systems. . . .

The Soviet Union talks to the Western countries and the USA on its own behalf and on behalf of its allies. We have one principled policy.

If the USA took the way of violating the agreement reached, the talks would be torpedoed. We made such a warning to the American representatives. There is no choice and no middle ground here. Either to deal seriously with space, to prevent an arms race in space and to keep it nonmilitarized, or there will be an arms race. This question cannot be solved by half-measures. . . . When people talk of so-called "Star Wars," a large-scale missile defense system, they talk of life and death. That is how the question stands. We consider it our duty to speak the truth not only to the United States of America but also to the whole world. This was said more than once by Konstantin Ustinovich Chernenko in his statements and addresses to the American President, when he called attention to the need to prevent a competition in deployment of arms in space. We shall fight resolutely on the question of space.

The American side characterizes its plan for space, the plan for a so-called large-scale missile defense, as defensive. We were often asked in Geneva as well, especially at the start of our conversations: "But why do you object, it is a defensive plan? We just want to develop systems which will shoot down missiles launched at the United States of America. It is indeed defense."

There is a rather devious and, generally speaking, perfidious stratagem here. Let us assume that they have developed a shield. They say that this shield is peaceable. It is intended to destroy missiles so that they should not reach their targets. As for the fact that missiles are hurled from behind that shield at the other side, at the Soviet Union—naturally, in an extreme situation—it does not mean anything, they believe; there is supposedly no danger here, and they even try to reassure us to this effect.

The fact that this plan has been called "defensive" by the U.S. Administration does not change anything. Some sections of the public are taken in—this should be admitted frankly. But, of course, there is nothing at all defensive about it. These are offensive weapons, and this whole plan, frankly speaking, is a plan of aggression, I repeat, aggression.

Even if there are people in the United States of America who believe that the United States will achieve dominance in one way or another, by carrying through its plan to develop the above system or without carrying it through, and that the Soviet Union will find itself in a subordinate position, they are mistaken. This will never happen. I want to stress anew, using the occasion, what Konstantin Chernenko pointed out in his statements and what he personally wrote to President Reagan: This is ruled out. We will never allow it. . . .

58. The U.S. Position on Strategic Arms Control

Kenneth W. Dam

KEEPING THE PEACE

Our fundamental objective is to avoid war while preserving our freedom and that of our allies. Keeping the peace requires a strategy with many facets.

First, we must have a strong defense. An equitable balance of forces (both nuclear and conventional) reduces the likelihood of conflict. The United States does not seek superiority over the Soviet Union, but we will do whatever is necessary to deny the Soviets superiority. We are modernizing both our nuclear and conventional forces to correct imbalances and give us the strength to deter aggression against the West.

Second, we are working closely with both our Atlantic and Pacific allies to strengthen our collective defenses. Last year NATO began deployment of U.S. intermediate-range nuclear missiles in response to the Soviet buildup of SS-20s. This year, special attention is being devoted to enhancing allied conventional defenses.

Third, we seek to negotiate equitable and verifiable agreements to reduce nuclear and conventional forces, to ban chemical weapons, and to enhance stability and reduce the risk of war.

These three efforts are mutually reinforcing. Our modernization program offers the Soviets an incentive to join us in negotiating significant reductions in nuclear arsenals. While all three must be pursued in a balanced way, today I shall focus on arms reduction negotiations. It is vital to the success of these negotiations that the Congress continue to support the modernization program, including the MX missile, and that the alliance remain strong and united.

U.S. APPROACH TO ARMS CONTROL

Four principles underlie our arms control proposals.

Substantial Reductions. That is, reductions to equal levels well below current levels. To repeat, the eventual goal is zero.

Equality. Agreements must result in overall equality between the United States and the Soviet Union in measures of military capability.

Stability. Special emphasis must be given to the most destabilizing systems, such as large, land-based MIRV [multiple independently-targetable reentry vehicles] missiles that threaten the forces of the other side.

Verification. Given the asymmetries between our two societies, agreements that cannot be verified can easily turn into unilateral constraints on ourselves. The United States would comply—with the Congress and the press monitoring closely—but we could not be certain about Soviet performance. Verification is intrinsically difficult, the more so as agreements become more ambitious and technology permits smaller, more mobile forces. A major obstacle is the Soviet record; they have not lived up to the terms of past agreements. There are serious problems with Soviet compliance with the ABM [Anti-Ballistic Missile] Treaty, with SALT II [strategic arms limitation talks], and with other agreements. We have discussed these problems at some length with the Soviets but have been unable to resolve them. This experience obliges us to redouble our efforts to make agreements that are verifiable. But verification can never be perfect, and we must not put ourselves in a situation in which the Soviets can gain a decisive advantage through violation or circumvention of agreements. . . .

SOVIET APPROACH

The Soviets have a different approach and different priorities. The Soviets acknowledge the need for reductions but give top priority to constraining weapons in space, which for the most part do not yet exist and will not for many years. The United States seeks to reduce offensive nuclear forces, which exist today in great numbers right here on Earth. . . .

STRATEGIC DEFENSE INITIATIVE

You have heard a great deal recently about that initiative. It is worthwhile to establish a few basic facts.

- The SDI is a research program. Its purpose is to explore new technologies that may lead to a reliable defense of the United States and our allies from ballistic missile attack. This program is completely consistent with the ABM Treaty and other agreements.
- The SDI puts primary emphasis on technologies that do not use nuclear weapons. This approach contrasts with the present Soviet ABM system, which relies on nuclear-armed interceptors.

- The SDI is a long-term program. No decisions on deployment of new defenses are expected for a number of years. If, in the future, we decide to pursue such defenses, this should be a matter for prior discussion with our allies and between the United States and the Soviet Union. If such defenses appear feasible and beneficial, we will want to discuss with them how these defenses can be fitted into a stable relationship between offense and defense.

- In the near term, the SDI program directly responds to the ongoing and extensive Soviet missile defense effort. The Soviets are expanding their ABM system to include all the actual deployments permitted under the ABM Treaty. They also have a major program to explore new defensive technologies, including technologies like those being pursued in our SDI program. The effort the Soviets devote to strategic defense is vastly greater than the relatively small U.S. program and even approximates the massive effort the Soviets devote to strategic offense. The SDI represents a prudent hedge against any Soviet decision to expand rapidly its ballistic missile defense capability beyond the bounds of the ABM Treaty.

- In the longer term, SDI offers the possibility of shifting away from total reliance on the threat of retaliation and toward greater reliance on defensive systems. As the President has said, "The human spirit must be capable of rising above dealing with other nations and human beings by threatening their existence. . . ."

GENEVA MEETING

So this is the situation Secretary Shultz and Foreign Minister Gromyko faced in Geneva:

The United States seeks negotiations where we can pursue equitable and verifiable agreements leading to deep reductions in offensive nuclear weapons, both strategic and intermediate range. We are prepared to discuss defenses, both space based and Earth based, but will protect our right to pursue research on defensive technologies that may provide a basis for a more stable relationship in the future. We do not believe there should be constraints on research that has such positive potential; and, in any event, constraints on research would not be verifiable. We also intend to protect our right to continue our modernization program to maintain deterrence, restore the balance of offensive forces, and provide incentives for the Soviets to agree to reductions. . . .

The agreement with the Soviets at Geneva to begin new negotiations is a useful first step. But to consider the outcome in Geneva as simply a resumption of formal dialogue misses the real significance of the process we are now embarking upon. We have established a forum where we can address the full spectrum of means for enhancing stability and reducing the risk of war. Our strategic concept can be summarized in the following single paragraph:

For the next 10 years, we should seek a radical reduction in the number and power of existing and planned offensive and defensive nuclear arms, whether land based, space based, or otherwise. We should even now be looking forward to a period of transition, beginning possibly 10 years from now, to effective non-nuclear defensive forces, including defenses against offensive nuclear arms. This period of transition should lead to the eventual elimination of nuclear arms, both offensive and defensive. A nuclear-free world is an ultimate objective to which we, the Soviet Union, and all other nations can agree.

The accord reached in Geneva is, of course, only a beginning. While Secretary Shultz and Foreign Minister Gromyko were successful in working out a basis for new negotiations, their discussions made clear that there are major differences of substance between us. There is a long road ahead. With patience, determination, and flexibility on both sides, the process set in motion last week in Geneva can successfully lead to a more stable peace.

———————— Bibliography: Part Six ————————

Betts, Richard K., ed. *Cruise Missiles: Technology, Strategy, Politics.* Washington, D.C.: Brookings, 1981.

Blechman, Barry. "Do Negotiated Arms Limitations Have a Future?" *Foreign Affairs* 59 (Fall 1980):102–125.

Brown, Harold, and Davis, Lynn E. "Nuclear Arms." *Foreign Affairs* 62 (Summer 1984):1145–1160.

Burrows, William E. "Ballistic Missile Defense." *Foreign Affairs* 62 (Spring 1984):843–856 (paired with Keith B. Payne and Colin S. Gray article).

Carter, Ashton B., and Schwartz, David N., eds. *Ballistic Missile Defense.* Washington, D.C.: Brookings, 1984.

Cochrane, Thomas B., et al. *The Nuclear Weapons Data Book.* vol. 1. Cambridge, Mass.: Ballinger, 1983.

Cole, Paul M., and Taylor, William J., eds. *The Nuclear Freeze Debate: Arms Control Issues for the 1980s.* Boulder, Colo.: Westview Press, 1983.

Durch, William J. *National Interests and the Military Use of Space.* Cambridge, Mass.: Ballinger, 1984.

Dyson, Freeman. *Weapons and Hope.* New York: Harper & Row, 1984.

Fletcher, James C. "The Technologies." *Issues in Science and Technology* 1 (Fall 1984):15–29.

Graham, Daniel. *High Frontier.* New York: Tom Doherty Associates, 1983.

Gray, Colin. *American Military Space Policy.* Cambridge, Mass.: Abt Associates, 1982.

Payne, Keith B., and Gray, Colin S. "Nuclear Policy and the Defensive Transition." *Foreign Affairs* 62 (Spring 1984):820–842 (paired with William E. Burrows article).

Steinbruner, John D., and Sigal, Leon V. *NATO and the No First Use Question.* Washington, D.C.: Brookings, 1983.

Talbott, Strobe. *Deadly Gambits.* New York: Knopf, 1984.

U.S. Arms Control and Disarmament Agency. *Arms Control and Disarmament Agreements: Texts and Histories of Negotiations.* Washington, D.C.: annual publication.

U.S. Congress. House. Subcommittee on International Security and Scientific Affairs of the Committee on Foreign Affairs. *Arms Control in Outer Space.* 98th Cong., 1983 and 1984.

U.S. Congress. Senate. Committee on Foreign Relations. *Controlling Space Weapons.* SH 98-141, 98th Cong., 1st sess., 1983, SR 43 and SJR 28.

Glossary

ABM. See *Anti-Ballistic Missile.*

ACDA. See *Arms Control and Disarmament Agency.*

Air-Launched Cruise Missile (ALCM). A cruise missile launched from aircraft. (See also *Cruise missile.*)

Airborne Launch Control System (ALCS). A secondary system backing up the primary control of the launch of missiles.

Airborne Warning and Control System (AWACS). A system carried in an airplane with the primary task of detecting and guiding the destruction of enemy aircraft. The sophisticated electronic equipment of AWACS also allows a variety of other surveillance and monitoring missions.

ALCM. See *Air-Launched Cruise Missile.*

ALCS. See *Airborne Launch Control System.*

Anti-Ballistic Missile (ABM). A missile weapon for destroying attacking ballistic missiles. ABMs can be designed to defend nuclear weapon silos, bases, or cities. The system includes radars for target acquisition, missile guidance, and weapons targetting.

Anti-Satellite (ASAT). A weapon for destroying satellites in orbit.

Anti-Submarine Warfare (ASW). The entire system of devices—radars, ships, planes, and weapons (nuclear and conventional)—for detecting and destroying submarines.

Arms Control and Disarmament Agency (ACDA). An executive agency of the U.S. government responsible for research, policy guidance, and negotiation on arms control and disarmament.

ASAT. See *Anti-Satellite.*

ASW. See *Anti-Submarine Warfare.*

AWACS. See *Airborne Warning and Control System.*

Ballistic Missile (BM). A rocket-launched missile that completes its trajectory in free fall; only gravitational and atmospheric forces affect the missile after the rocket ceases to fire.

Ballistic Missile Defense (BMD). A complex of radars, launchers, weapons, target acquisition, and missile guidance for the detection and destruction of ballistic missiles in flight. The many different proposed weapons systems include intercepting rockets, laser or particle beams, and debris thrown up by nuclear explosions.

Ballistic Missile Early Warning System (BMEWS). U.S. system of radars for the early detection of and warning about attacking ballistic missiles deployed in the path that enemy missiles would have to follow. Because of the need to provide the earliest possible warning, BMEWS was constructed, for example, in the arctic areas of northern Canada.

Ballistic Missile Nuclear Submarine (SSBN). A nuclear-powered submarine carrying ballistic missiles.

BM. See *Ballistic Missile.*

BMD. See *Ballistic Missile Defense.*

BMEWS. See *Ballistic Missile Early Warning System.*

CEP. See *Circular Error Probable.*

Circular Error Probable (CEP). The radius of the circle that includes 50 percent of the impact points of the incoming warheads. The use of estimated CEPs makes possible a simplified analysis of the survival probability (Ps) of a hardened silo:

$$Ps = 0.5^{[(16Y)^{2/3}/CEP^2 H^{2/3}]}$$

where Y = yield in megatons of attacking warhead, and H = hardness of silo in PSI (pounds per square inch). This formula gives a false impression of accuracy. First, although CEP assumes a circular error pattern, the scatter of impact points is in fact elliptical. Second, it assumes that the center of the elliptical scatter of errors falls on the target—in other words, that there is no systematic bias that significantly moves the scatter pattern of impact points off the target. Third, published estimates of CEP are significantly uncertain. Yet in spite of all these caveats, CEPs tend to provide reasonably accurate data on comparisons and trends. The formula given earlier can be generalized to yield the survival probability ($\bar{P}s$) for the case of n independently attacking warheads with reliability (r):

$$\bar{P}s = [1 - r + rPs]^n$$

Command, Control, Communications, and Intelligence (C³I). Refers to shelters, redundant capabilities, satellites, electronics, and other measures constructed to withstand the effects of nuclear attack, enhance effective control of nuclear weapons, and assess the damage inflicted by and against the enemy.

Counterforce. Refers to employment of weaponry against military targets, including nuclear weapons, troop formations, ports, railroad centers, airfields, and command centers.

Countervalue. Refers to an attack on cities, industry, and population.

Cruise missile. Self-propelled guided missile that flies in the atmosphere. Such missiles are capable of carrying either conventional or nuclear warheads, and they possess great accuracy because of terminal guidance.

C³I. See *Command, Control, Communications, and Intelligence.*

ECM. See *Electronic Counter-Measures.*

Electro-Magnetic Pulse (EMP). Sudden, strong surge of electromagnetic radiation generated by a nuclear explosion that threatens unprotected electronic equipment by inducing high voltages capable of burning out sensitive electronic components.

Electronic Counter-Measures (ECMs). The jamming, spoofing, and exploitation of the enemy's use of radios and radar detection systems.

EMP. See *Electro-Magnetic Pulse.*

Enhanced Radiation Weapon (ERW). Also known as the neutron bomb, the ERW is an atomic bomb designed to enhance the ratio of neutron output to blast. Primarily an antipersonnel weapon, it is regarded as potentially more suitable for use against an armored assault in a heavily urbanized area, such as West Germany.

ERW. See *Enhanced Radiation Weapon.*

FBS. See *Forward Based System.*

Forward Based System (FBS). Refers to systems based and operating close to enemy territory—for example, U.S. bombers, aircraft carriers, and missiles based in the European area.

GLCM. See *Ground-Launched Cruise Missile.*

Ground-Launched Cruise Missile (GLCM). A cruise missile launched from the ground. (See also *Cruise missile.*)

IAEA. See *International Atomic Energy Agency.*

ICBM. See *Inter-Continental Ballistic Missile.*

INF. See *Intermediate-Range Nuclear Forces.*

Inter-Continental Ballistic Missile (ICBM). A ballistic missile (BM) with inter-continental range.

Intermediate-Range Ballistic Missile (IRBM). A ballistic missile (BM) with a range of 1,500–3,000 nautical miles.

Intermediate-Range Nuclear Forces (INF). Nuclear weapons designed for use within a specific theater of operations, such as Europe, although their longer range permits their use as strategic weapons. (See also *Strategic Nuclear Delivery Vehicles.*)

International Atomic Energy Agency (IAEA). The United Nations Agency responsible for the monitoring and control of fissionable materials and nuclear reactors.

IRBM. See *Intermediate-Range Ballistic Missile.*

JCS. See *Joint Chiefs of Staff.*

Joint Chiefs of Staff (JCS). The agency in the U.S. Department of Defense responsible for all military strategy matters. The Joint Chiefs are the principal military advisers to the secretary of defense, to the president, and to the National Security Council. Each member, except the chairman, is the senior military officer in his respective service.

Kiloton (Kt). The equivalent in destructive force of the detonation of 1,000 tons of the conventional chemical explosive (TNT). The explosion of a 1 Kt atomic bomb would release the same amount of energy as the explosion of 1,000 tons of TNT.

Kt. See *Kiloton.*

Launch on Warning (LOW). Refers to a retaliatory strike launched on the detection of an enemy attack before the enemy weapons have arrived and exploded.

Limited and Selective Option, or Limited Strike Option (LSO). Refers to planning for nuclear attacks limited in number and types of targets.

LOW. See *Launch on Warning.*

LSO. See *Limited and Selective Option.*

MAD. See *Mutual Assured Destruction.*

Maneuverable Re-entry Vehicle (MARV). A system that is capable of maneuvering in the atmosphere and that may have terminal guidance to correct the flight path of warhead(s) in the final phase of the flight.

MAP. See *Multiple Aim Point.*

MARV. See *Maneuverable Re-entry Vehicle.*

Massive Retaliation. The nuclear strategy of the United States during the early 1950s, based on the threat to attack Soviet and Chinese cities and industry in the event of Communist aggression anywhere in the world.

MBFR. See *Mutual and Balanced Force Reductions.*

Medium-Range Ballistic Missile (MRBM). A ballistic missile with a range of 600–1,500 nautical miles.

Megaton (Mt). The equivalent in destructive force of the detonation of 1,000,000 tons of the conventional chemical explosive (TNT). The explosion of a 1 Mt atomic bomb would release the same amount of energy as the explosion of 1,000,000 tons of TNT.

MIRV. See *Multiple Independently-Targetable Re-entry Vehicle.*

Missile Experimental (MX). A U.S. MIRVed ICBM with ten highly accurate and powerful nuclear warheads. Construction of MX began in 1984, with deployment to begin in 1986.

MRBM. See *Medium-Range Ballistic Missile.*

MRV. See *Multiple Re-entry Vehicle.*

Mt. See *Megaton.*

Multiple Aim Point (MAP). Refers to a basing scheme once proposed for the MX in which each missile would be hidden and moved among a number of shelters.

Multiple Independently-Targetable Re-entry Vehicle (MIRV). A missile with two or more warheads, each of which can be separately targetted.

Multiple Re-entry Vehicle (MRV). A missile with two or more warheads that engage one target in shotgun fashion.

Mutual and Balanced Force Reductions (MBFR). Negotiations for the reduction of conventional arms in the forces of the Warsaw Pact and NATO.

Mutual Assured Destruction (MAD). Refers to the ability of the Soviet Union and the United States to destroy each other as modern twentieth-century societies even after they have endured a full-scale initial nuclear attack.

MX. See *Missile Experimental.*

National Command Authorities (NCA). Top national security decisionmakers.

National Security Council (NSC). The statutory function of the NSC is to advise the president with respect to the integration of domestic, foreign, and military policies relating to national security. The NSC was established in 1947 by the National Security Act and is chaired by the president. Statutory members in addition to the president are the vice-president, the secretaries of state and defense, the chairman of the Joint Chiefs of Staff, and the director of Central Intelligence.

National Technical Means of Verification (NTM). Unilateral methods of verifying treaty compliance such as satellite photo reconnaissance, radar, and intercept of telemetry.

NCA. See *National Command Authorities.*

NATO. See *North Atlantic Treaty Organization.*

Nautical Mile (NM). Equal to 1,852 meters or 1.15078 statute miles.

NM. See *Nautical Mile.*

North Atlantic Treaty Organization (NATO). An alliance for defense against the Soviet Union made up of Belgium, Canada, Denmark, France, the Federal Republic of Germany, Greece, Iceland, Italy, Luxemburg, the Netherlands, Norway, Portugal, Turkey, the United Kingdom, and the United States.

NSC. See *National Security Council.*

NTM. See *National Technical Means of Verification.*

PGM. See *Precision-Guided Munitions.*

Precision-Guided Munitions (PGM). Weapons of great accuracy that home in on the target with terminal guidance.

Re-entry Vehicle (RV). A device carrying a nuclear warhead through the atmosphere that provides protection against heat and buffeting. The RV may also provide additional guidance to target. (See also *Maneuverable Re-entry Vehicle*.)

RV. See *Re-entry Vehicle*.

SAC. See *Strategic Air Command*.

SACEUR. See *Supreme Allied Commander, Europe*.

SALT. See *Strategic Arms Limitation Talks*.

SAM. See *Surface-to-Air Missile*.

SCC. See *Standing Consultative Commission*.

Sea-Launched Cruise Missile (SLCM). A cruise missile launched from a naval vessel.

Short-Range Attack Missile (SRAM). An air-to-ground nuclear missile launched from a bomber, with a range on the order of 150 km.

Single Integrated Operational Plan (SIOP). The plan for the assignment of nuclear weapons to targets in a manner that corresponds to the strategic priority of the target and avoids duplication of attack.

SIOP. See *Single Integrated Operational Plan*.

SLBM. See *Submarine-Launched Ballistic Missile*.

SLCM. See *Sea-Launched Cruise Missile*.

SNDV. See *Strategic Nuclear Delivery Vehicle*.

SSBN. See *Ballistic Missile Nuclear Submarine*.

Standing Consultative Commission (SCC). A permanent Soviet-U.S. commission established to consider questions concerning compliance with the terms of the SALT I treaty and ABM agreements of 1972.

START. See *Strategic Arms Reduction Talks*.

Strategic Air Command (SAC). The U.S. Air Force organization whose mission is to train, equip, and prepare for combat strategic missiles, bombers, and strategic reconnaissance units, and to conduct strategic air operations.

Strategic Arms Limitation Talks (SALT). The name given by the U.S. government to the negotiations on strategic arms reduction initiated in 1970 between the United States and the USSR.

Strategic Arms Reduction Talks (START). The name given by the Reagan administration to negotiations on arms reductions initiated in 1982 between the United States and the USSR, broken off in 1983 by the Soviet Union and resumed in 1985.

Strategic Nuclear Delivery Vehicle (SNDV). Any bomber or missile for delivering a nuclear weapon to attack the enemy's population and its capacity to wage war.

Submarine-Launched Ballistic Missile (SLBM). An intercontinental ballistic missile capable of being launched from a submarine when submerged.

Supreme Allied Commander, Europe (SACEUR). Commander in Chief of NATO forces.

Surface-to-Air Missile (SAM). A missile launched from the ground and directed to a target in the air.

Tactical Nuclear Weapons. Nuclear-capable devices assigned to support the conduct of battles, with a shorter range than that of strategic weapons and deployed close to likely areas of military engagement.

Terminal Guidance. Correction of flight-path techniques during terminal approach to provide enhanced accuracy.

Theater Nuclear Forces (TNF). See Intermediate-Range Nuclear Forces.

TNF. See *Theater Nuclear Forces*.

About the Authors

Yuri Andropov (1914–1984) was general secretary of the Communist Party of the Soviet Union from 1982 to 1984; he served as ambassador to Hungary from 1954 to 1957 (during the suppression of the 1956 revolution) and headed the KGB from 1967 to 1982.

Desmond Ball is head of the Strategic and Defence Studies Centre at the Research School of Pacific Studies of the Australian National University.

Leonid Brezhnev (1906–1982) was general secretary of the Communist Party of the Soviet Union from 1964 to 1982.

Harold Brown was U.S. secretary of defense from 1977 to 1981.

McGeorge Bundy was special assistant to the president for National Security Affairs from 1961 to 1966 and president of the Ford Foundation from 1966 to 1979.

Kenneth Dam is U.S. deputy secretary of state in the Reagan administration.

Sidney D. Drell is deputy director of the Linear Accelerator Center and codirector of the Center for International Security and Arms Control, both at Stanford University.

John Foster Dulles (1888–1959) was U.S. secretary of state from 1953 to 1959.

Fritz W. Ermath has been an analyst of Soviet and U.S. strategic studies at the Central Intelligence Agency and the Rand Corporation and has been a member of the National Security Council.

Randall Forsberg is president and executive director of the Institute for Defense and Disarmament Studies (Brookline, Massachusetts).

French Catholic Bishops. The Conference of Catholic Bishops of France is an organization advisory to the French Catholic community.

Alton Frye is Washington director of the Council on Foreign Relations. He has written extensively on arms control and served as an assistant to Senator Edward Brooke of Massachusetts.

Raymond Garthoff is a senior fellow at the Brookings Institution. He is a retired foreign service officer and served as ambassador to Bulgaria from 1977 to 1979. He was a member of the U.S. SALT I delegation from 1969 to 1972.

Colin S. Gray is president of the National Institute for Public Policy in Fairfax, Virginia.

Andrei Gromyko served as foreign minister of the Soviet Union from 1957 to 1985, when he became president.

Lt. General H. R. Harmon headed a review committee of officers from all the services to examine the role of atomic weapons in U.S. foreign policy.

Major General S. K. Il'in is a Soviet officer and strategist.

Henry M. Jackson (1913–1984) was a senator from the state of Washington and a prominent member of the Senate Armed Services Committee. He was first elected to the House of Representatives in 1940 and to the Senate in 1952.

The Joint Chiefs of Staff (JCS) is the agency in the Department of Defense responsible for all military strategy matters. The JCS is the principal adviser to the secretary of defense, to the president, and to the National Security Council. Each member, except the chairman, is the senior military officer of his respective service.

General David C. Jones was chairman of the U.S. Joint Chiefs of Staff from 1978 to 1982.

Karl Kaiser is director of the Research Institute of the German Society of Foreign Affairs in Bonn.

George F. Kennan is professor emeritus at the Institute for Advanced Study at Princeton University. A career foreign service officer, he was director of the State Department Policy Planning Staff from 1947 to 1950. Other assignments include his service as minister-counselor in Moscow in 1944, ambassador to the Soviet Union in 1952, and ambassador to Yugoslavia from 1961 to 1963.

John F. Kennedy (1917–1963) was president of the United States from 1961 to 1963.

Henry A. Kissinger was national security adviser from 1969 to 1973 and secretary of state from 1973 to 1977.

Georg Leber is a Social Democratic member of the West German Bundestag and former defense minister of the Federal Republic of Germany.

Pierre Lellouche is head of the European Security Program at the Institut Français des Relations Internationales (IFRI) in Paris.

Sverre Lodgaard is on the staff of the Stockholm International Peace Research Institute (SIPRI) in Sweden.

Col. General N. A. Lomov is a Soviet officer and strategist.

Edward N. Luttwak is a senior fellow at the Center for Strategic and International Studies at Georgetown University and a consultant to the U.S. Defense Department.

Robert S. McNamara was secretary of defense from 1961 to 1968 and president of the World Bank from 1968 to 1981.

Alois Mertes is a Christian Democratic member of the West German Bundestag, a member of its foreign affairs committee, and foreign policy spokesman for the CDU/CSU parliamentary faction. He is a former member of the West German foreign service.

National Security Council (NSC). The statutory function of the NSC is to advise the president with respect to the integration of domestic, foreign, and military policies relating to national security. The NSC was established in 1947 by the National Security Act and is chaired by the president. Statutory members in addition to the president are the vice president, the secretaries of state and defense, the chairman of the Joint Chiefs of Staff, and the director of the Central Intelligence Agency.

John Newhouse has served on the staff of the Senate Committee on Foreign Relations. He has for many years published extensively on arms control and international politics from Washington and abroad.

Paul H. Nitze is special adviser to the president and the secretary of state for arms reduction negotiations. He was a member of the U.S. SALT delegation from 1969 to 1974; deputy secretary of defense from 1967 to 1969; secretary of the U.S. Navy from 1963 to 1967, and director of the State Department Policy Planning Staff from 1950 to 1953.

Marshal N. V. Ogarkov is marshal of the Soviet Union. He served as chief of staff of the Soviet armed forces from 1977 to 1984, when he was replaced.

Wolfgang K. H. Panofsky is director emeritus and professor of physics at the Stanford Linear Accelerator Center and has been a consultant to numerous government agencies since the 1950s.

Richard Perle was an assistant to Senator Henry Jackson. He has been a consultant to the Department of Defense and served as assistant secretary of defense for International Security Policy in both Reagan administrations.

Richard Pipes is Baird Professor of History at Harvard University. He was a member of the National Security Council from 1981 to 1983.

Ronald Reagan was elected to his second term as president of the United States in 1984.

Jeffrey Richelson is associate professor of government and public policy at the American University in Washington, D.C.

Andrei Sakharov is a leading Soviet dissident, former member of the Academy of Sciences of the USSR, and one of the developers of the Soviet nuclear arsenal. He received the Nobel Prize for Peace in 1975.

Helmut Schmidt was chancellor of the Federal Republic of Germany from 1974 to 1982. He served previously as defense minister and finance minister.

James R. Schlesinger was U.S. secretary of defense from 1973 to 1975.

General Franz-Josef Schulze was commander in chief of Allied Forces Central Europe from 1977 to 1979 and deputy chief of staff, Allied Command Europe, from 1973 to 1976.

Lt. General Brent Scowcroft is a retired Air Force general. He headed the President's Commission on Strategic Weapons and served as national security adviser in the Ford administration from 1975 to 1977.

Colonel A. A. Sidorenko is a Soviet officer and a leading Soviet strategist.

Gerard Smith was chief of the U.S. delegation to the Strategic Arms Limitation Talks (SALT) from 1969 to 1972. He has served as special assistant to the secretary of state for atomic energy affairs, as director of the Policy Planning Staff of the Department of State, as ambassador at large, and as special presidential representative.

Marshal V. D. Sokolovsky (1897–1968) was marshal of the Soviet Union and a Soviet strategist. He was chief of the General Staff of the Army and Navy from 1952 to 1960.

E. P. Thompson is professor of history at Oxford University and director of European Nuclear Disarmament (END).

U.S. Catholic Bishops. The National Conference of Catholic Bishops is an organization advisory to the U.S. Catholic community.

Albert Wohlstetter is director of Research at Pan Heuristics and president of the Euro-American Institute for Security Research. He has taught at the University of Chicago and Oxford and has had extensive government service over the past three decades.